"The study of far-right environmental politics presents unusual challenges. In this much-needed volume, a range of international scholars offer contemporary insights on the topic. In an era when the far right is once again on the rise, the critical perspectives gathered here could not be more timely."

Peter Staudenmaier
Department of History, Marquette University, USA

"Different far-right actors are in the political ascendancy; man-made environmental disasters are worsening. Understanding the links between the far-right and environmental themes could not be more topical or urgent."

James Painter
Department of Politics and International Relations, University of Oxford, UK

"This book makes an important contribution by bridging two of today's key concerns: the emergence and continuous growth of far right parties and movements, and the increasing awareness of climate change and environmental crises. Anyone who wants to understand the role of far-right parties in environmental politics, should read this informative and timely book."

Jens Rydgren
Professor of Sociology, Stockholm University, Sweden

"The Far Right is not conventionally considered an ecological movement – despite the (national) environment being one of their central ideological concerns. In this fascinating and timely book, Bernhard Forchtner has assembled an impressive range of contributors who amply dispel the myth that environmentalism is a new, or a fringe, feature of far-right discourse. Theoretically and analytically wide-ranging, and transnational in focus, this volume is highly recommended for scholars of discourse analysis, political science and environmental communications."

John E. Richardson
Language & Linguistics, University of the Sunshine Coast, Australia

"This welcome volume fills an important intellectual niche by examining the diverse ways in which far-right political parties and movements, especially those in Europe, deal with environmental and ecological issues. In some cases they promote meaningful environmental protection, but in many others environmental skepticism and anti-environmental policies. Tracking their impact will be vital, and this volume provides a baseline for future scholarship."

Riley E. Dunlap
*Regents Professor of Sociology and Dresser Professor Emeritus,
Oklahoma State University, USA*

THE FAR RIGHT AND THE ENVIRONMENT

At the beginning of the twenty-first century, both the crisis of liberal democracy, as visible in, for example, the rise of far-right actors in Europe and the United States, and environmental crises, from declining biodiversity to climate change, are increasingly in the public spotlight. Whilst both areas have been analysed extensively on their own, *The Far Right and the Environment: Politics, Discourse and Communication* provides much-needed insights into their intersection by illuminating the environmental communication of far-right party and non-party actors in Europe and the United States. Although commonly perceived as a 'left-wing' issue today, concerns over the natural environment by the far right have a long, ideology-driven history. Thus, it is not surprising that some members of the far right offer distinctive ecological visions of communal life, though, for example, climate-change scepticism is voiced too. Investigating this range of stances within their discourse about the natural environment provides a window into the wider politics of the far right and points to a close connection between the politics of identity and the imagination of nature. Connecting the fields of environmental communication and study of the far right, contributions to this edited volume therefore offer timely assessments of this often-overlooked dimension of far-right politics.

Bernhard Forchtner is associate professor at the School of Media, Communication and Sociology, University of Leicester, UK.

ROUTLEDGE STUDIES IN FASCISM AND THE FAR RIGHT

Series editors: Nigel Copsey, *Teesside University,* and **Graham Macklin**, *Center for Research on Extremism (C-REX), University of Oslo.*

This new book series focuses upon fascist, far right and right-wing politics primarily within a historical context but also drawing on insights from other disciplinary perspectives. Its scope also includes radical-right populism, cultural manifestations of the far right and points of convergence and exchange with the mainstream and traditional right.

Titles include:

The March on Rome
Violence and the Rise of Italian Fascism
Giulia Albanese

Aurel Kolnai's 'War Against the West' Reconsidered
Edited by Wolfgang Bialas

The Ku Klux Klan and Freemasonry in 1920s America
Fighting Fraternities
Miguel Hernandez

The Lives and Afterlives of Enoch Powell
The Undying Political Animal
Edited by Olivier Esteves and Stéphane Porion

Latin American Dictatorships in the Era of Fascism
The Corporatist Wave
António Costa Pinto

The Far Right and the Environment
Politics, Discourse and Communication
Edited by Bernhard Forchtner

Vigilantism against Migrant and Minorities
Edited by Tore Bjørgo and Miroslav Mareš

Trumping Democracy
From Ronald Reagan to Alt-Right
Edited by Chip Berlet

THE FAR RIGHT AND THE ENVIRONMENT

Politics, Discourse and Communication

Edited by Bernhard Forchtner

LONDON AND NEW YORK

First published 2020
by Routledge
2 Park Square, Milton Park, Abingdon, Oxon OX14 4RN

and by Routledge
52 Vanderbilt Avenue, New York, NY 10017

Routledge is an imprint of the Taylor & Francis Group, an Informa business

© 2020 selection and editorial matter, Bernhard Forchtner; individual chapters, the contributors

The right of Bernhard Forchtner to be identified as the author of the editorial material, and of the authors for their individual chapters, has been asserted in accordance with sections 77 and 78 of the Copyright, Designs and Patents Act 1988.

All rights reserved. No part of this book may be reprinted or reproduced or utilised in any form or by any electronic, mechanical, or other means, now known or hereafter invented, including photocopying and recording, or in any information storage or retrieval system, without permission in writing from the publishers.

Trademark notice: Product or corporate names may be trademarks or registered trademarks, and are used only for identification and explanation without intent to infringe.

British Library Cataloguing-in-Publication Data
A catalogue record for this book is available from the British Library

Library of Congress Cataloging-in-Publication Data
A catalog record has been requested for this book

ISBN: 978-1-138-47786-5 (hbk)
ISBN: 978-1-138-47789-6 (pbk)
ISBN: 978-1-351-10404-3 (ebk)

Typeset in Bembo
by Lumina Datamatics Limited

CONTENTS

List of figures   x
List of tables   xi
List of contributors   xii
Acknowledgements   xvii

1 Far-right articulations of the natural environment: An introduction   1
 Bernhard Forchtner

PART I
Two fields, many topics   **19**

2 The trajectory of far-right populism – a discourse-analytical perspective   21
 Ruth Wodak

3 Environmental communication research: Origins, development and new directions   38
 Anders Hansen

PART II
Western Europe   **55**

4 'Protecting our green and pleasant land': UKIP, the BNP and a history of green ideology on Britain's far right   57
 Emily Turner-Graham

5 From black to green: Analysing *Le Front National*'s
  'patriotic ecology' 72
  Salomi Boukala and Eirini Tountasaki

6 Environmental politics on the Italian far right: Not a party issue? 88
  Giorgia Bulli

PART III
Nordic countries 105

7 Wolves in sheep's clothing? The Danish far right and 'wild nature' 107
  Christoffer Kølvraa

8 The far right and climate change denial: Denouncing
  environmental challenges via anti-establishment rhetoric,
  marketing of doubts, industrial/breadwinner masculinities,
  enactments and ethno-nationalism 121
  Martin Hultman, Anna Björk and Tamya Viinikka

9 The allure of exploding bats: The Finns Party's populist
  environmental communication and the media 136
  Niko Hatakka and Matti Välimäki

PART IV
Central Europe 151

10 The ecological component of the ideology and legislative
   activity of the Freedom Party of Austria 153
   Kristian Voss

11 The environmental communication of Jobbik: Between
   strategy and ideology 184
   Anna Kyriazi

12 Is brown the new green? The environmental discourse
   of the Czech far right 201
   Zbyněk Tarant

13 Beyond the 'German forest': Environmental communication by the far right in Germany 216
Bernhard Forchtner and Özgür Özvatan

14 The environment as an emerging discourse in Polish far-right politics 237
Samuel Bennett and Cezary Kwiatkowski

PART V
Beyond Europe **255**

15 In the heartland of climate scepticism: A hyperlink network analysis of German climate sceptics and the US right wing 257
Jonas Kaiser

16 Alt-right ecology: Ecofascism and far-right environmentalism in the United States 275
Blair Taylor

17 The rhetorical landscapes of the 'alt right' and the patriot movements: Settler entitlement to native land 293
Kyle Boggs

18 Looking back, looking forward: Some preliminary conclusions on the far right and its natural environment(s) 310
Bernhard Forchtner

Index 321

LIST OF FIGURES

2.1	Results of recent elections of seven far-right populist parties	24
2.2	Conceptual map of the far-right mind-set	29
2.3	Right-wing populist perpetuum mobile	32
11.1	Major themes in Kepli Lajos' media appearances 2010–2018 (six-monthly)	194
11.2	'Jobbik Environmental Protection' posts per week (31 July 2017–5 April 2018)	196
11.3	Major themes in 'Jobbik Environmental Protection' Facebook page	197
14.1	'Nazi Roots'	245
15.1	Reduction of full network and comparison with seed sites (full network = 45,117 nodes, 77,674 edges, 106 seed sites; core network = 1,747 nodes, 13,155 edges, 61 seed sites; climate network = 906 nodes, 4906 edges, 43 seed sites	262
15.2	Labelled climate network by community	265
15.3	Selected sites for 'right-wing (extremist)', 'left-wing (extremist)', and '(alleged) Russian misinformation' sites	266

LIST OF TABLES

10.1	Salience of nature protection for activity in a legislature (motions) in Austria from 2006 to 2017	156
10.2	Nature protection in FPÖ party documents from 1975 to 2013 and legislative motions from 2006 to 2017	157
10.3	Vote results for individual opposition party sponsored nature protection motions from 2008 to 2017	179
11.1	Opinions related to the environment in Hungary (in percentage)	186
11.2	Jobbik electoral results, 2006–2019 (percentage and rank)	186
13.1	Overview of macro-topics in far-right environmental communication and their evaluation	223
15.1	Labelled communities per top five websites per indegree in each modularity class	263
15.2	Share of websites in each modularity class with IPs in Germany or the United States ($n = 906$)	268
15.3	Distribution of websites with IPs in Germany or the United States per modularity class ($n = 906$)	268
15.4	Links between different modularity classes in percent	269

LIST OF CONTRIBUTORS

Samuel Bennett is an assistant professor at Adam Mickiewicz University in Poznań, Poland. He is a linguist and social scientist who studies migrant integration, (non)belonging and exclusion, and populist politics. He is the author of *Constructions of Migrant Integration in British Public Discourse* (Bloomsbury 2018) and has been published widely in leading journals, including *Critical Discourse Studies* and the *Journal of Language & Politics*. He also serves on the board of a migrant integration non-governmental organisation (NGO).

Anna Björk holds a bachelor's degree in Environmental Science from Linköping University, Sweden. She currently works as an environmental consult in the field of contaminated soil and water.

Kyle Boggs is an assistant professor of rhetoric and composition at Boise State University, Idaho, in the United States. His research deploys a materialist rhetorical framework to make connections between environmental studies and settler colonialism in order to analyse everyday spaces and rhetorical practices as foregrounded by capitalism, heteropatriarchy and white supremacy. He is currently completing a manuscript for his first book, which uses this framework to interrogate outdoor recreational discourses, tentatively titled, *Recreational Colonialism: White Settlers in Spaces of Outdoor Recreation* (West Virginia University Press).

Salomi Boukala is assistant professor in the Department of Social Anthropology at Panteion University of Social and Political Sciences, Greece. She has been published widely in the field of critical discourse studies. She is the author of the book *European Identity and the Representation of Islam in the Mainstream Press: Argumentation and Media Discourse* (Palgrave, 2018). Her research interests include: argumentation, the discursive construction of political and (supra) national identities, political rhetoric, media discourses and ethnographic approaches.

Giorgia Bulli is a lecturer in 'Political Communication' and 'Discourse Analysis' at the Department of Political and Social Sciences, University of Florence. She is the author of publications in the field of right-wing parties and movements. Giorgia Bulli, Matteo Albanese, Pietro Castelli Gattinara, Caterina Froio authored *Fascisti di un altro Millennio? Crisi e partecipazione in Casa Pound Italia*. (Acireale-Roma, Bonanno Editor 2014); Giorgia Bulli has also authored 'Anti-Islamism and beyond: Pegida'. *Mondi migranti* (2017).

Bernhard Forchtner is associate professor at the School of Media, Communication and Sociology, University of Leicester, United Kingdom, and has previously worked as a Marie Curie Fellow at the Institute of Social Sciences, Humboldt University of Berlin, where he conducted a project on far-right discourses on the environment (project number 327595). Recent publications include 'Nation, nature, purity: extreme-right biodiversity in Germany' (*Patterns of Prejudice*, 2019), 'Being Skeptical? Exploring Far-Right Climate-Change Communication in Germany' (with A. Kroneder and D. Wetzel in *Environmental Communication*, 2018) and *The Routledge Handbook on Language and Politics* (with R. Wodak, 2017).

Anders Hansen is associate professor in the School of Media, Communication and Sociology, University of Leicester, United Kingdom. He is author of *Environment, Media and Communication* (2nd edition, Routledge, 2019); co-editor of *The Routledge Handbook of Environment and Communication* (2015); associate editor of *Environmental Communication*; founding member of the International Environmental Communication Association (IECA). He is co-editor of *Visual Environmental Communication* (Routledge, 2015), and co-editor of the *Palgrave Studies in Media and Environmental Communication* book series.

Niko Hatakka is a postdoctoral researcher at the Centre for Parliamentary Studies at the University of Turku, Finland. His research focuses on populist movements, online counter publics, and their interaction with institutional politics and political journalism.

Martin Hultman is associate professor at Chalmers University of Technology, Sweden, and has been published in journals such as *Environmental Humanities*; *NORMA: International Journal for Masculinity Studies*; *History & Technology* and *Hydrogen Energy*. He is the author of *Discourses of Global Climate Change* and *Ecological Masculinities*. Hultman leads the global *Centre for Studies of Climate Change Denial* and three research groups analysing masculinities and energy, ecopreneurship in circular economies and climate change denial. His latest book project titled *Rights of Nature* deals with nature.

Jonas Kaiser is an affiliate at the Berkman Klein Center for Internet and Society at Harvard University, Massachusetts, and associate researcher at the Humboldt Institute for Internet and Society, Berlin. He was a German Research Foundation (The Deutsche Forschungsgemeinschaft; DFG) Research Fellow and his work on the far right is funded by the DFG (KA 4618/1-1 & KA 4618/1-2). In his research, Kaiser analyses how the far right in Germany and the United States (ab)uses the internet and its affordances.

Christoffer Kølvraa is associate professor in the Section of European Studies, School of Culture and Society at Aarhus University, Denmark. His recent publications include: Psychoanalyzing Europe? Political Enjoyment and European Identity (*Political Psychology*, 2018), Limits of Attraction: The EU's Eastern Border and the European Neighbourhood Policy. (*East European Politics & Societies* 2017), Extreme right images of radical authenticity (with B. Forchtner in *European Journal of Cultural and Political Sociology* 2017).

Cezary Kwiatkowski is an independent researcher whose interest revolves around critical discourse analysis, critical metaphor theory and political discourse. A graduate of English Philology at Adam Mickiewicz University in Poznań, Poland, his most recent work includes a critical analysis of the discourse used by British nationalists and Euro-sceptics to create and reinforce British nationalist sentiment.

Anna Kyriazi is a Postdoctoral Research Fellow Juan de la Cierva-Formación at the Institut Barcelona d'Estudis Internacionals. She holds a PhD in Political and Social Sciences from the European University Institute. Her area of expertise is comparative ethnicity and nationalism, migration and political communication with particular emphasis on Eastern and Southern Europe. Her work has appeared in the *Journal of Ethnic and Migration Studies* and *Ethnicities*. Anna has received competitive scholarships from the Spanish Ministry of Science, Innovation and Universities, the Greek Scholarship Foundation and the Hungarian Ministry of Education.

Özgür Özvatan is a doctoral research fellow with the Department of Social Sciences and the Berlin Institute for Integration and Migration Research (BIM), Humboldt University of Berlin, Berlin, Germany. He is a PhD candidate in the International Doctoral Program of the Berlin Graduate School of Social Sciences (BGSS), Berlin, Germany, and a doctoral fellow with the Centre for Analysis of the Radical Right (CARR). Recent publications include 'Politics of Turkish European Belonging in the Era of National Rebirth' (*EuropeNow*, 2018), 'Games of Belonging: National Identity Contestation in Germany and Turkey' (with S. Metzger in *Nationalities Papers*, 2019) and 'The Far-Right

Alternative für Deutschland: Towards a 'Happy Ending'?' (with B. Forchtner in A. Waring *The New Authoritarianism* (Vol. 2), 2019).

Zbyněk Tarant works as an assistant professor at the Department of Middle-Eastern Studies, University of West Bohemia, Czech Republic. Along with his main subject in the field of Israeli studies, he became actively involved in the research of contemporary anti-semitism and political extremism in the Czech Republic since 2006. His specialty is monitoring cyber-hate and the analyses of emerging threats.

Blair Taylor is program director of the Institute for Social Ecology, a popular education centre for ecological scholarship and advocacy founded in 1974. He has lectured and published widely on political ecology, capitalism and social movements, especially contemporary far-right politics and the history of the left. His work has been featured in *Les Temps Modernes, American Studies, and City: Analysis of Urban Trends, Culture, Theory, Policy, Action*. He is co-editor of the Murray Bookchin anthology *The Next Revolution: Popular Assemblies and the Promise of Direct Democracy* (Verso, 2014).

Eirini Tountasaki is associate professor in the Department of Social Anthropology at Panteion University of Social and Political Sciences, Athens, Greece. Her research expertise is in anthropological theory of kinship and ethnographic approaches. She has researched and published extensively on Greek nationalism, the politics of culture and the construction of national identity through archaeological and folklore discourses. She has participated in several teaching and research networks; she is a member of the Scientific Committee of the *Greek Review of Social Research*.

Emily Turner-Graham is an independent academic researcher and writer on the far and extreme right. She is a former lecturer and researcher at Melbourne, Deakin and La Trobe universities. She has published on both the historical and contemporary extreme right in Britain, Germany, Austria and Australia. She is the author of *Never forget that you are a German: Die Brücke, Deutschtum and national socialism in interwar Australia* (Peter Lang Verlag, 2011).

Matti Välimäki is a postdoctoral researcher at the Centre for Parliamentary Studies at the University of Turku, Finland, and in the Migration Institute of Finland. His research focuses on the historical development of political parties' immigration policy outlooks, public discussion on immigration and immigrants and populist political communication.

Tamya Viinikka holds a bachelor's degree in Environmental Science from Linköping University, Sweden. She now studies gender and postcolonialism in a Latin American context while working with young people and education.

Kristian Voss obtained his PhD in political and social sciences from the European University Institute, Italy. His thesis, *Nature and Nation in Harmony: The Ecological Component of Far-Right Ideology*, focused on the historical and contemporary relationship between the protection of nature and the far right in Western Europe.

Ruth Wodak is Emerita Distinguished Professor of Discourse Studies at Lancaster University, United Kingdom, and affiliated with the University of Vienna, Austria. Besides other prizes, she was awarded the Austrian Wittgenstein Prize and an Honorary Doctorate from University of Örebro, Sweden. In 2017, she held the Willy Brandt Chair at Malmö University, Sweden and from 9/2018 until 6/2019, she was a senior visiting fellow at IWM, Vienna. She is co-editor of the journals *Discourse and Society*, *Critical Discourse Studies*, and *Language and Politics*. Her research interests focus on discourse studies, gender studies, language and/in politics, prejudice and discrimination and ethnographic methods of linguistic field work. Her recent publications include *Europe at the Crossroads* edited with Pieter Bevelander (Nordicum 2019), *The Routledge Handbook of Language and Politics* edited with Bernhard Forchtner (Routledge 2017) and *The Politics of Fear: What Right-wing Populist Discourses Mean* (SAGE 2015).

ACKNOWLEDGEMENTS

The far right, ranging from anti-liberal, (nominally) democratic radical-right actors to the anti-democratic extreme right, do think and talk about the natural environment. Given that both the far right and environmental issues are key concerns of our time, it might seem surprising that research on this thinking and talking is still rather limited in its extent. It is against this background that I would like to thank all the authors who have contributed to this edited volume and thus shed light on this area. Moreover, I want to thank the editors of the book series *Routledge Studies in Fascism and the Far Right* for their trust in this project, the (anonymous) reviewers who provided important input and the editorial team at Routledge for their guidance throughout the process. Additionally, I want to acknowledge, once again, the support I received from the People Programme (Marie Curie Action, FP7/2007–2013) [327595]), without which I would have been unable to dive into the far right's natural environment. I am also grateful for the financial support I received from the School of Media, Communication and Sociology at the University of Leicester, which covered both the creation of this volume's index and allowed the contributors in this volume to introduce and discuss their different cases at a workshop in July 2017 in Leicester.

1
FAR-RIGHT ARTICULATIONS OF THE NATURAL ENVIRONMENT

An introduction

Bernhard Forchtner

> Any national movement that is to be taken seriously can be seen as a national-ecological movement according to its own self-conception.
>
> *Michael Howanietz (2005: 25)*

Introducing *The Far Right and the Environment: Politics, Discourse and Communication*

The far right, ranging from the radical right which opposes some elements of *liberal* democracy to the anti-democratic extreme right, is commonly investigated for their stance on immigrants and refugees, as well as, for example, their memory politics and their views on gender. However, what is not usually looked at are their politics concerned with the natural environment, their environmental communication.[1] Indeed, given the common association of environmental issues with the (liberal) left today, environmental communication by the far right might come as a surprise. However, it should be anything but surprising – and it is this often overlooked dimension of far-right politics which this edited volume illuminates. After all, although it is true that both anti-liberal, but increasingly mainstream and (nominally) democratic, far-right parties, such as the Freedom Party of Austria and the French National Rally (formerly: National Front), as well as anti-democratic actors, including the National Democratic Party of Germany and various 'autonomous' extreme-right groupuscules, are not well-known for their environmental politics, the natural environment (and concerns over it) has – to varying extent – its place in these actors' ideology.

Indeed, meanings attributed to the natural environment are not inherent in nature itself, but, as Staudenmaier (2011[1995]: 41f) reminds us, are linked to ideology.[2] In fact, the natural environment has long played a role in nationalist and far-right political thinking. This has ranged from a full-blown ecological worldview which stresses the interconnectedness of flora, fauna, the nation and its homeland, including the naturalisation of social relations and the significance of 'the land' for the reproduction of 'the people', to the aesthetic idealisation of certain elements of the community's landscape. The significance of environmental, and sometimes even ecological, considerations is starkly visible in the 'epigraph by Michael Howanietz, a member of the Freedom Party of Austria, who has consistently thematised issues related to the natural environment.'[3]

It is against this background that contributions to this volume shed light on environmental communication by a diverse range of party and non-party actors populating the far-right spectrum. In so doing, the contributors to this book provide a reference volume concerning an area hardly investigated; an area, however, through which these actors have long reproduced their ideology. Indeed, it is because the 'politics of nature is at the same time a politics of identity' (Olsen 1999: 29) that far-right environmental communication has to be scrutinised too.

Inquiring into the far right and the environment is a timely endeavour as we are faced today with the intersection of (communication about) two crises: on the one hand, large sections of 'the West' are experiencing a crisis of liberal democracy. The rise of far-right parties (but also non-party actors) in Europe, parts of the Brexit debate in the United Kingdom, and the successful presidential campaigns of Donald Trump in the United States (US) and Jair Bolsonaro in Brazil illustrate this. On the other hand, anthropogenic environmental crises, although not just with us since industrialisation and urbanisation, are now a global phenomenon, with climate change, a paradigmatic unintended consequence of previous modernisation (Beck 2009), at the heart of public debates. Yet, this intersection of crises and the ways in which they are communicated has hardly been analysed, and it is here that this volume intervenes by providing a comprehensive account of far-right environmental communication in contemporary Europe and beyond. Importantly, however, we do not expect the far right to communicate in a uniform way. Thus, the goal is not to illuminate this intersection in general terms, but to highlight, first, differences and similarities within the far-right spectrum. That is, to illuminate differences and similarities between the anti-liberal radical right and the anti-democratic extreme right (Mudde 2007) in how these actors engage with the natural environment (if at all) due to ideology (more or less 'extreme'/organic ethnonationalist), diverging historical backgrounds, and diverse discursive and political contexts. Second, contributions to this volume furthermore point to equally relevant tensions within individual actors due to competing ideological elements.

This introduction constitutes a first step in this attempt and starts by offering a brief clarification of what is meant when speaking of the far right in the following. I will subsequently provide an overview of aspects related to the natural environment in the literature on nationalism and the far right before offering an overview of contributions to this volume.

The far right...

It has become common to mourn the lack of coherence concerning the name of the referent of those studying actors to the right of conservative parties. Besides, for example, 'right wing', 'radical right', 'extreme right' and 'right-wing extremist', and 'far right', we sometimes find the designator 'ultra-nationalist' and even '(neo-)fascist' and '(neo-)Nazi'. At times, this variety is simply due to linguistic and political preferences, but it is also rooted in actual differences. Anti-liberal, but (nominally) democratic parties such as the Danish People's Party and the aforementioned National Rally are increasingly 'mainstream' (a 'mainstream' which co-evolves with these parties, a 'mainstream' they co-shape). Other actors, however, are anti-democratic, for example Golden Dawn in Greece and so-called Autonomous Nationalists across Europe. Against this background, this volume, in line with the name of the series in which it is published, employs 'far right' as the overarching name of this continuum of actors.

At the core of far-right actors, we find ethnonationalism – linking membership in the nation to biological/racial and/or cultural traits – and authoritarianism (Bonikowski 2017; Rydgren 2018a). Besides these core ideological features, elements more or less dominantly present include, for example, ethnopluralism, anti-socialism, proclivity for scapegoating 'others', and an uncritical view on the community's historical past. The latter results in, for example, the denial of war guilt or, in Austria and Germany in particular, the Holocaust (see Holzer 1994; Salzborn 2014). In some cases, the far right is also 'populist'. Here, populism signifies a 'thin-centred ideology' which considers society to be divided in two camps, 'pure people' versus 'corrupt elite', and stresses the need to defend the general will of 'the people' (Mudde 2007: 23), though others conceptualise it as a political logic (Laclau 2005) or primarily a style (Moffitt 2016). Depending on how dominantly these core (and additional) criteria are present, far-right actors will populate different positions on a continuum. Accordingly, contributions to this volume deal with an array of actors, ranging from rather anti-liberal, today often mainstream radical right ones to anti-democratic, extreme-right actors.

As these contributions will indicate, a variety of elements appear to affect environmental communication by the far right, including the position the respective actor takes on the far-right continuum as well as the origins of these actors and their links to, for example, historical fascism and National Socialism; the respective party system, political opportunity structure and the media landscape; the intellectual abilities of these actors to engage with eco-theoretical questions (being able to go beyond 'green is left' knee-jerk reactions); the historical salience that environmental issues carry in particular contexts; and the configuration of discourses about the environment in these contexts. Dryzek (2012: 15–17), for example, identifies four basic environmental discourses (subsequently providing more specific types), understood by him as shared ways of apprehending the world, of providing legitimate knowledge and constructing meaning as well as relationships. These basic discourses are *problem-solving* (status quo needs adjustment, but no radical change is needed to cope with environmental problems), *limits and survival* (the limits of the Earth demand radical steps – though options to tackle these are set by industrialism), *sustainability* (economic growth and environmental protection can go hand in hand) and *green radicalism* (industrial society is rejected by Green romantics, deep ecologists and so forth).[4] While I cannot elaborate on this typology here, such discourses need to be acknowledged as the far right, even though it comes with its own background convictions, does operate in a wider, societal context. Each of the following chapters will illustrate this complexity – but it is a more general look at the relation between the ideology of ethnonationalist forces and the natural environment to which I turn next.

…on its natural environment

Communication concerned with the protection of the natural environment is often considered to be a relatively new phenomenon, a relatively new site through which society actively reproduces its symbolic boundaries. What Radkau (2014) has termed the *Age of Ecology* is furthermore primarily associated with political forces on the (liberal) left of the political spectrum. As such, the contemporary concern for the natural environment is commonly informed by a universalist perspective in the tradition of the Enlightenment, stressing one humanity and its responsibility for Earth. However, the history of environmental protection and campaigning does not begin in the 1960s and '70s. Instead, it can, at least, be traced back to the nineteenth century when the price of industrialisation and urbanisation was becoming apparent. Radkau (ibid.: 11–24), in fact, points to the period between Rousseau and the Romantic, arguing that it was in the 1790s that Europe saw a debate familiar to present-day controversies, back then concerning the shortage of wood and the potential risk of facing the destruction of forests. Indeed, it is since this period that universalist thought is criticised by various conservative and far-right actors, also in relation to nature and people. The Romantic response emphasised the particularity of both nature and peoples, spontaneity, originality and authenticity. According to this tradition, it is uniqueness and diversity which need to be celebrated, not 'cold' and 'abstract' universal reason.

Researchers of nationalism have long observed that such views of nature and the natural environment play a role in the nationalist imaginary.[5] It is in the sense of something not being corrupted by civilisation or, as far-right actors might say today, not being distorted by the *zeitgeist*, that nationalism's nature is one 'which rejects any suggestion of the contrived, of the consciously arranged' (Kedourie 1966: 57). This nature is, furthermore, territorially specific, it is the homeland, that is, the land where 'terrain and people have exerted mutual, and beneficial, influence over several generations' (Smith 1991: 9). These are 'poetic spaces' (ibid.: 78) or ''ethno-scape[s]' in which a people and its homeland become increasingly symbiotic' (Smith 2009: 50). As such, the nation's landscape is more than sheer matter, it is symbolically charged, creating a link between past, present and future (Palmer 1998: 191; Cosgrove 2004: 61). Memories are thus attached to sites, including natural ones (from forests to mountains and so on); and as communal being is projected onto sites, the latter are imputed with meaning and, as such, become a matter of identity. Schama, (1995: 10) too, emphasises the link between (national) identities and landscape, pointing out that the latter entered the English language via the Dutch *landschap* (the Netherlandish flood fields being a manifest site of human engineering with clear relevance for the identity of this particular community). In the Dutch and other cases of national identity, the latter would indeed 'lose much of its ferocious enchantment without the mystique of a particular landscape tradition: its topography mapped, elaborated, and enriched as a homeland' (ibid.: 15).

The significance of the local and particular is furthermore stressed by Barcena et al. (1997: 302), who argue that nationalism and ecologism share a fundamental philosophical stance in their 'rejection of the leveling perversion of the universal and defense of the particular'. Particular ecosystems appear as seemingly stable and orderly, with different species in different habitats, and the resulting proximity between the discourse about the nation and the discourse about the environment, between the protection of the homeland and the protection of the environment, easily leads to a rejection of so-called 'invasive species', both animals and plants. While also visible in wider public debates, a concern over such 'intruders' is arguably particularly prevalent in far-right discourses in which 'the supposed threat of foreign species, on the one hand, and, on the other, the perceived threat of foreign races and cultures to the native populations of their countries' (Olwig 2003: 61; for concerns over biodiversity and 'invasive species' in the far right's imaginary in particular, see Forchtner 2019a).

Today, such thinking draws regularly on, and is reinvigorated by, the *Nouvelle Droite* (Bar-On 2013). Like the 'old' far right, here too the social is naturalised and essentialised. Indeed, the *Nouvelle Droite* celebrates the 'right to difference' which leads to ethnopluralism, a concept forged by Henning Eichberg (Camus and Lebourg 2017: 130), that is, the protection of cultures and ethnicities by avoiding mixture. Thus, the *Nouvelle Droite* suggests concern for the preservation of cultures in general, but also in relation to the natural environment in particular. This derives from the *Nouvelle Droite's* view, as represented by its main intellectual Alain de Benoist, of liberalism as the main carrier of modernity, as the 'main enemy' (de Benoist and Champetier 1999). Modernity, it is said, denies human nature, while the *Nouvelle Droite* postulates the necessarily biological nature of our species – a nature which unites us (this is not the 'race' of Social Darwinist racism) – before stressing that '[m]an is rooted by nature in his culture'. Humans thus construct themselves 'historically and culturally' within the species' limitations; and diversity of cultures is consequently part of humanity's essence. In this neo- or cultural racism (Taguieff 1990; Balibar 1991), nature, landscape and soil are significant – though no longer simply in the sense of a biological connection between land and people ('blood and soil'), but in a symbolic way. Furthermore, de Benoist and

Champetier (1999) argue, that we must leave anthropocentrism behind and understand 'nature as a partner and not as an adversary or object'.

Here, one might be reminded of deep ecology (e.g. Naess 1973); and indeed, the social ecologist Murray Bookchin (1987) has forcefully addressed its mystifying tendencies, criticising the lack of focus on the genuinely social causes of environmental crises. In line with this argument, Peter Zegers (2002) subsequently focused on deep ecologist understandings of diversity, identifying a close resemblance between deep ecology thinking and the ethnopluralism advocated by the *Nouvelle Droite*.

As environmental as well as ecological thinking continue to exist on the far right, it would be a mistake to downplay these. Indeed, the far right does neither approach the natural environment purely strategically to attract a wider audience (though this, of course, happens too) nor as a simple object for exploitation (though some far-right actors certainly do so), but perceives land and landscape as being significant for and deeply linked to the nation. This results in a connection which can be, as Hurd and Werther (2016: 164) argue, 'perhaps more deeply ecocritical than the liberal tropes of "sustainable development"'. However, far-right concerns over the natural environment are not uniform. Indeed, there are several ways in which far-right thinking has drawn on and articulated the natural environment; and depending on, for example, national traditions and predecessors of the respective actor, attitudes will differ.

Looking back at such historical trajectories, the nineteenth and twentieth century saw organic ethno-nationalist, fascist and National Socialist concerns for the environment, which drew on particularistic interpretations of the relation between land and people already present in parts of the Romantic and nationalist thinking more generally. In Britain, for example, Jorian Jenks was not only an enthusiastic farmer, but also a stout fascist who became an important player in Oswald Mosley's British Union of Fascists during the 1930s. Even more significant, Jenks would become a pioneer of today's organic movement and a key player in the newly founded Soil Association after World War Two (Coupland 2017). This significance of the land is still present as recent studies on the British far right have illustrated (Forchtner 2016; Richardson 2017: 163–173).

Moving from fascists who failed to take power to those who succeeded in doing so, a special issue of *Modern Italy* (Armiero 2014) contains fascinating articles on environmental issues in fascist Italy. While illustrating that the policies of the time were not always of much environmental benefit, it also argues that Mussolini did view the land as key, an opportunity to regenerate the people. Similarly, Portuguese fascism did not simply propagate a romantic 'back to the land' agenda, but managed and recreated the nation's environment through increased wheat production, irrigation and afforestation (Saraiva 2016).

Yet, it was in Germany that the link between the far right and the natural environment became perhaps most clearly articulated – and certainly most intensively studied due to a long history of environmental concerns and ecology thinking being fused with ethnonationalist, Social Darwinist and racist ideas

(Geden 1996; Olsen 1999; see Riordan 1997 for a general overview of green thought in Germany). Following Romantic ideas by, for example, Ernst Moritz Arndt in the early nineteenth century, the late nineteenth century *Heimatschutz* movement (homeland protection) – resulting in the foundation of the *Bund Heimatschutz* (League for the Protection of the Homeland) in 1904 – appears to have, though in no way exclusively, been significant for the reproduction of *völkisch* ideas (Wolschke-Bulmahn 1996: 533 and Linse 2009: 158, but see Rollins 1997 on the *Heimatschutz's* aesthetic concerns as a possibly motivating force for progressive politics). The natural environment, being viewed as the fundament of the nation, must thus not be destroyed as this would, consequently, lead to the destruction of the *Volk*.[6] Many *Heimatschutz* activists continued under National Socialism, united in their dislike of 'cold materialism' while favouring the 'organic' and 'traditional'. Indeed, National Socialism integrally linked 'Volk, racism and conservation' (Brüggemeier et al. 2005: 8; see also Uekötter 2006). Early nature-protection initiatives resulted in the *Reichsnaturschutzgesetz* (Reich Nature Protection Law) of 1935, the 'most stringent and comprehensive environmental protection law in the world' (Lekan 2004: 168). Yet, the goals of autarky, economic revival and war preparations ultimately collided with environmental concerns and led to continuous subordination of the environment (Dominick 1992: 81–118; Lekan 2004: 204–251; Blackbourn 2006: 266–280; Uekötter 2006: 30–43).

In line with my argument concerning a diversity of attitudes vis-à-vis the natural environment, it is important to acknowledge Olsen's (1999: 78) point, that National Socialism's view of nature is only one, radical, nature-nationalism tradition, and that the echo of this particular tradition is only one among many ways in which the far right understands the natural environment today. Therefore, strands within the far right not, or to a lesser degree, influenced by organic ethnonationalism will not necessarily take the same stance as those which are committed to such ideas.[7]

Today, the non-uniformity of the far right's concerns over the natural environment is, for example, visible in what is the perhaps main environmental issue of our time: climate change. Although possibly the most severe threat to the homeland, many, though not all, of these actors appear to be, in one way or another, climate-change sceptics (for an overview, see Forchtner forthcoming).[8] For instance, Gemenis et al. (2012) analysis of environmental themes in far-right environmental communication includes 'Global warming is man-made'. The authors claim that findings concerning the National Democratic Party of Germany and the Sweden Democrats are inconclusive; while the Greek Popular Orthodox Rally agrees, the British National Party, the Danish Peoples Party, the Italian Northern League and the Belgian Flemish Interest disagree with the theory of anthropogenic climate change. Forchtner and Kølvraa (2015) confirm this regarding the British National Party and the Danish Peoples Party while Voss (2014: 163, 165) views, amongst others, the Danish Peoples Party, Northern League, Sweden Democrats (at the time), the Freedom Party of Austria and the National Democratic Party of Germany as agreeing with the thesis of anthropogenic climate change. Moreover, Forchtner et al. (2018) and Forchtner (2019b) report strong scepticism towards, but also some acceptance of, mainstream knowledge about climate change by far-right actors in Germany and Austria, respectively. Against this background, subsequent chapters will provide further insights into this complexity.

Having addressed the link between a range of far-right actors and environmental themes, I want to take a step back and close by briefly indicating ways of conceptualising the relation between the natural environment and the far right. First, and offering a broad way of conceptualising far-right concerns over the natural environment, Olsen (1999) identifies eco-naturalism (the natural world as a blueprint for the social order), eco-organicism (the *Volk*/people as an ecosystem) and eco-authoritarianism (the need for a strong state to deal with the environmental crises of our time) as key features of right-wing ecology. For example, far-right actors accepting mainstream views of anthropogenic climate change could call for harsh environmental laws and, at some point in the future, an eco-dictatorship so as to be able to deal with and respond to an accelerating climate-change crisis. It is, however, similarly interesting to consider why these features, which strongly affect far-right views on, for example, population growth, have not led to a uniform acceptance of the anthropogenic nature of climate change.

Concerning climate change in particular, Lockwood (2018) explores the relationship between far-right actors and climate-change scepticism by proposing two explanations. First, he considers a 'structuralist' approach to understanding the link between 'right-wing populism' and climate-change scepticism. This explanation addresses the appeal of right-wing populist parties to those 'left behind' by globalisation and technological modernisation. That is, it accounts for the 'marginalisation of specific groups in post-industrial societies through structural change in the global economy' (ibid.: 718). The second explanation is based on the ideological agenda of such actors, 'especially its antagonism between "the people" and a cosmopolitan elite, with climate change and policy occupying a symbolic place in this contrast' (ibid.: 712). Evaluating these two explanations, Lockwood ultimately argues that focusing on ideology is more compelling.

Third, and addressing both climate change and concerns for 'the land' and the countryside, Forchtner and Kølvraa (2015) speak of three dimensions through which the far right, like other ideological camps, make sense of the natural environment: the aesthetic, the symbolic and the material dimensions. Within nationalist ideology, these dimensions host specific contents; the *aesthetic dimension* foregrounds an idea of nature as being appreciable and enjoyable. Historically, the German *Heimatschutz* was concerned about (natural) monuments, ruins and billboards in the countryside while National Socialists, for example, claimed to 'sensitively' embed the *Autobahn* in the wider landscape. Today, this dimension is visible in, for example, protests against wind turbines ('a blight on the landscape'). The *symbolic dimension* is concerned with a community's claim to primacy and sovereignty in relation to a particular section of the Earth's surface, and thus the construction of its cultural difference from other communities. As such, in the symbolic dimension the historical primacy of the (pure) national community in this territory is asserted. For example, overpopulation has long been viewed in (neo-)Malthusian terms, in terms of being a major threat to the national ecosystem – in line with wider societal debates (see, e.g. Ehrlich 1968) – but appears to be less central at the moment (but see Forchtner and Kølvraa 2015 for the British context as well as Glättli and Niklaus 2014 for the Swiss one. Concerning the United States, this topic appears to be more central, see Bhatia 2004; Mix 2009; SPLC 2010; Hultgren 2015). This might, of course, change, if 'climate refugees' becomes a salient public issue. Indeed, it is in the discourse about climate change that sovereignty has played a central role, being often viewed as, for example, a hoax to install a 'one world government' or to increase the European Union's powers. Finally, the *material dimension* considers the land in terms of the resources it provides for its population and economy. As *we* are not supposed to be dependent on others, due to the ideal of self-sufficiency (both in relation to food and energy supplies), the nature-nation nexus is prominent. Indeed, the contemporary far right mobilises this dimension in various ways, for example when insisting on the significance of coal (or nuclear energy) for the national economy or when pointing to renewables to ensure autarky.

As such, the far right creates and relates in diverse ways to the natural environment, ways this volume will explore. Indeed, and against the aforementioned background, it is not surprising that contemporary far-right actors often, but not necessarily, show a concern for environmental protection. On the one hand, and as the particularities of the community need to be preserved, its natural environment needs to be protected. This results in a critique of what is viewed as being responsible for environmental degradation, of threats to the flora and fauna of ecosystems: 'materialism' and 'globalisation' (from the circulation of ideas to that of people and goods). As such, far-right criticism is based on the local, regional and/or national – and not the global, thus also reflecting the centrality of landscape to these identities. On the other hand, ideological, historical and contextual reasons mentioned above sometimes prevent actors from defending the natural environment. For example, worries about a loss of sovereignty (a key element linked to these actors' ideology) due to the alleged rise of a world government might trump environmental concerns. This environmental scepticism is not usually driven by neoliberal, free-market considerations, such as in contemporary conservative climate-change scepticism (see, e.g. Dunlap and McCright 2015; Krange et al. 2019), but by far-right ideological elements other than the environment. For example, the British National Party displays scepticism towards anthropogenic climate change, largely rooted in the above-mentioned fear that climate-change policies would undermine national sovereignty. In consequence, what could be perceived as endangering ecosystems around the world is instead viewed as a sinister plot to weaken the nation.

Different stances vis-à-vis the environment point to complexity characterising every ideology, and to tensions which arise as soon as ideologies are lived in a particular context. Illuminating these complexities, and tensions potentially resulting from them, requires detailed analyses.

Overview of contributions to the volume

Following this introduction, contributions to this volume focus primarily on environmental communication by far-right actors in Europe – as this is the place of origin of far-right ideology which has also experienced the forceful resurgence of such actors since the 1980s – but also include analyses of situations in the United States.[9]

Part I, Two Fields, Many Topics, provides two overview chapters on the two fields of research this volume connects: first, Ruth Wodak offers insights into how the far right has developed, more specifically, how its communication and contributions to discourses have evolved over the years. While acknowledging context-dependent, socio-political and historical differences, Wodak stresses four dimensions: nationalism/nativism/anti-pluralism, anti-elitism, authoritarianism and a charismatic leader, and conservatism/historical revisionism. More specifically, Wodak elaborates on mediatisation and communication strategies which have evolved over the past 30 years, pointing to, for example, discursive strategies of provocation, calculated ambivalence and scandalisation. This is followed by an introduction to environmental communication by Anders Hansen who traces the evolution of this subfield in communication studies. Emerging in the 1960s, the field has shed light on how, for example, communication strategies and public concerns have impacted on the construction and representation of the environment. While this is still relevant today, technological and other media developments, and this is of particular relevance in the case of media-savvy far-right activists, offer significant opportunities and challenges for communication about the environment.

After these two chapters, the volume presents four Parts which address a variety of far-right actors and their environmental communication. Each chapter deals with one state or site, but does not necessarily focus on only one actor within the respective context. These actors are not necessarily parties, though most chapters have the dominant far-right party of the respective country at their core. These chapters cover a variety of far-right actors with various stances towards different environmental issues, depending on what the authors consider to be most relevant. Parts are organised along geographical lines, starting with Part II, Western Europe.

Emily Turner-Graham looks at the United Kingdom, analysing material published by the British National Party, the United Kingdom Independence Party as well as more mainstream sources. She thus illustrates the workings and relevance of the former two's ideas – even though the parties themselves have become politically less relevant. Along the lines established above, she argues that the far right is not anti-environmental per se, but operates outside of the mainstream understanding of the environment, imagining a world of ordered (white) towns and a countryside which might, if at all, have existed in a distant past. This is followed by Salomi Boukala and Eirini Tountasaki's chapter on France, which focuses on what was back then the National Front, taking the party's adoption of environmental politics during the 2017 presidential elections as a case study.

Their analysis emphasises how the party's environmental vision is instrumental in the discursive revival of the French nation-state and national identity. The subsequent chapter by Giorgia Bulli analyses the Italian case, asserting that Italian parties do not much consider the natural environment, while non-party political movements have done so. She points to the influence of the *Nouvelle Droite* at the end of the 1970s, before turning to present-day *CasaPound Italia*, explaining how the promotion of ecological values, respect for the environment and actions to advertise their engagement play a role in the group's communication.

Part III is entitled Nordic Countries. Christoffer Kølvraa starts with a chapter on Denmark and the Danish People's Party in which he examines the party's environmental imaginary. He starts by exploring the party's idea of nature and its relation to human society by touching on environmental matters more generally, including their climate-change communication, before turning in much detail to the debate over the re-immigration of wild wolves to Denmark. Next, Martin Hultman, Anna Björk and Tamya Viinikka turn to Sweden, where they investigate the climate-change communication of the Sweden Democrats. The authors argue that the party's climate-change scepticism is based on anti-establishment rhetoric and the marketing of doubt, industrial/breadwinner masculinities and ethnonationalism. The final chapter in this Part is provided by Niko Hatakka and Matti Välimäki, who analyse the Finns Party's environmental performance and how, through populist representations, the party creates 'the people' it claims to represent. The authors do so by focusing on the party's campaign against wind power in 2016, which focuses on the alleged health and environmental hazards of wind turbines.

Part IV, Central Europe, brings together chapters on Austria, Hungary, the Czech Republic, Germany and Poland. Kristian Voss looks at the Freedom Party of Austria, analysing party documents and legislative motions in the Austrian parliament. He argues that the party is characterised by a comprehensive and fundamental ecological component and, furthermore, stresses the significance of ideology (the Freedom Party of Austria being ideologically more 'extreme' than its breakaway party, the Alliance for the Future of Austria). The chapter by Anna Kyriazi investigates the Hungarian party Jobbik, the Movement for a Better Hungary. Kyriazi argues, that through its profound environmental agenda the party confers its organic nationalism and criticises its political opponents. She furthermore points to Jobbik's long-standing aspiration to establish an 'eco-social economy', an economy dominated by ecological and societal considerations while also stressing strategic motivations. This chapter is followed by Zbyněk Tarant's analysis of the situation in the Czech Republic. Tarant introduces the reader to both far-right political parties and non-party actors, arguing that while environmental concerns exist, there is no established environmental agenda. Overall, nativist thinking prevents these actors from formulating proper responses to global environmental challenges. Bernhard Forchtner and Özgür Özvatan then look at Germany by analysing a range of far-right publications as well as the environmental communication of the newly formed radical-right Alternative for Germany and the older, extreme-right National Democratic Party of Germany. They discuss the historical dimension of environmental protection and offer a wide-ranging overview of the distribution of environmental topics and their evaluation in the present. The final chapter in this Part by Samuel Bennett and Cezary Kwiatkowski covers Poland and analyses manifestos as well as Facebook posts from relevant actors. Arguing that the environment is a relative new site of political and discursive contestation, they claim that there is little consistent green politics at the party level. Though little elite-level discourse about the environment thus exists, Bennett and Kwiatkowski claim that this has helped far-right actors to diffuse their discursive frames and strategies into mainstream Polish political discourse.

The final Part, Part V Beyond Europe, deals with environmental communication beyond the borders of individual European (nation-)states. First, Jonas Kaiser takes a fascinating look at how German climate-sceptic websites and blogs are linked to international sites, especially from the United States. Kaiser sheds light on the role the international right plays for German climate-change sceptics, contextualises the activities and positions of the latter, and stresses the internet's importance for transnational connections as well as the relevance for the connection of online counter-publics. Second, Blair Taylor examines the so called 'alt right' in the United States, pointing out how a variety of anti-egalitarian views, for example: anti-semitism, racism, gender traditionalism and homophobia have been harnessed to and justified by an ecological framework. Taylor argues that ecology represents one important political vector for the rejection of traditional pro-business conservative positions by the alt right. The final chapter in this Part by Kyle Boggs also looks at the situation in the United States, this time, however, specifically in terms of white settler colonialism. The author argues that the communicative practices of the United States far right can be traced back to the trope of 'white men and the frontier'; and Boggs carves out how the related issue of natural environment is present in these imaginaries.

In the final conclusion, Bernhard Forchtner reflects on the multifaceted phenomenon of far-right environmental communication and identifies areas of future research. As the contributions to this volume cover various (types of) actors and issues, *The Far Right and the Environment: Politics, Discourse and Communication* illustrates the varying importance attached to the natural environment. In so doing, the volume argues for more research on far-right environmental communication and invites readers to consider links between the far right and the natural environment, two areas of crucial importance for how people are going to live in the twenty-first century.

Notes

1 For seminal texts on the far right, see, for example, Rydgren (2018b), Mudde (2016) and (2007), Wodak (2015). However, these contributions do not consider environmental communication and environmental protection. Interestingly, Mudde (2000) lists possibly relevant ideological features, (environmentalism/ecologism), though they are only touched upon briefly in the context of the Dutch *Centrumpartij'86*. For examples of how nature protection has featured in some works on the far right, see Voss (2014: 7–10).

2 Ideology is here understood as 'a set of ideas by which men posit, explain and justify ends and means of organized social action', a belief system through which such action 'preserve[s], amend[s], uproot[s] or rebuild[s] a given social order' (Seliger 1976: 14).
3 For pragmatic reasons and due to public usage, I largely speak of 'environment'/'environmental', subsuming other notions such as 'nature' and 'ecology'/'ecological' under this umbrella. Thus, I speak of, for example, 'environmental communication' and 'environmental concerns' without implying a specific connotation. Whenever I use the term 'ecological', this, however, stresses an explicitly holistic perspective. Indeed, some foreground differences between these concepts. For example, 'environment' has only emerged in everyday language use since 1970 while 'nature' is older and signifies an understanding of flora and fauna as having intrinsic value (Radkau 2014). 'Ecology' goes back to Ernst Haeckel, a nineteenth century German scientist who referred to a holistic perspective according to which organisms should be studied in terms of their embeddedness in an interdependent system, an ecosystem. Here, (wo)man are *in* 'nature', not simply surrounded by it. Consequently, Dobson (1999: 235f) refers to 'ecologism' as addressing fundamental causes of environmental damage – while 'environmentalism' deals with symptoms and might thus take a managerial view.
4 For related concerns and typologies, see, for example, Anshelm and Hultman (2016) as well as Hajer (1995). Arguably, discourses are about representing the world, about being positioned as subjects (identities) and about being set in specific relations to other subjects (Fairclough and Wodak 1997). It should, however, be noted that 'discourse' is understood in many ways; for example, as natural spoken or written (or other modes) language in context or as more or less incommensurable sets of perspectives (with their particular knowledge and rules), while Reisigl and Wodak (2016) identify macro-topic-relatedness, pluri-perspective and argumentativity as constitutive elements of a discourse.
5 However, there are also other voices: for example, Fowler and Jones (2006: 315) analyse the relationship between nationalism and environmentalism in Wales during the early 1990s, claiming that this type of relationship will remain 'a situationally contingent phenomenon'. Focusing on Welsh (and Scottish) nationalism, Hamilton (2002) argues that nationalist movements mine environmental discourse to legitimise their claims. Still, he hopes for a green and civic nationalism to enhance nature protection while nevertheless claiming that (ibid.: 38) classic nationalism (authoritarian, militaristic and self-worshiping) is 'necessarily anthropocentric and utilitarian as regards the natural world'. De-Shalit and Talias (1994: 290), in their analysis of environmental controversies in Israel, argue that the latter are characterised by anthropocentric modes of reasoning due to the Zionist 'ethos of development'.
6 The fact that such an understanding of 'the people' and their natural environment still persists is visible in the manifesto apparently produced by the suspect of a terrorist attack on Christchurch mosques in March 2019 who killed 51 people. The manifesto rages against what the author calls 'white genocide', but also speaks of 'Green nationalism' and states on page 38: '…there is no nationalism without environmentalism, the natural environment of our lands shaped us just as we shaped it. We were born from our lands and our own culture was molded by these same lands. The protection and preservation of these lands is of the same importance as the protection and preservation of our own ideals and beliefs. (…)Each nation and each ethnicity was melded by their own environment and if they are to be protected so must their own environments.'. Indeed, the author referred to himself as 'eco-fascist' (see Staudenmaier 2011[1995]). However, the latter concept lies not at the heart of this volume as it is only one amongst many positions, far-right actors might take towards the natural environment.
7 See, for example, Schaller and Carius' (2019: 39f) claim concerning 'green patriotism' of some far-right parties.
8 Research on climate-change scepticism more generally has largely focused on conservative actors (see, for example, the review in Krange et al. 2019).

9 This limitation is primarily due to pragmatic reasons and I do not suggest that environmental communication by the far right does not also happen in other countries/regions of the world than those addressed in this volume (or is less relevant there). For example, Galbreath and Auers (2009), Schwartz (2005) and Malloy (2009) discuss the Latvian case where the struggle for independence had an environmental dimension. Concerning Russia, Davidov (2015) has recently pointed to the eco-nationalist 'Anastasia' or 'Ringing Cedars' movement. Russia is also one of those cases analysed by Dawson (1996), furthermore including Lithuania and Ukraine, who discusses the convergence of environmentalism and nationalism in anti-nuclear activism under the label of 'eco-nationalism'. Similarly, Cederlöf and Sivaramakrishnan (2014) investigate how struggles over nature and conservation are linked to citizenship subjectivity and nationalism, how identity is linked to territory and the assertion of territory in South Asia. Sharma (2012) and Mawdsley (2006) trace the connection between Hindu nationalism and environmentalism in India.

References

Anshelm, Jonas and Hultman, Martin (2016): *Discourses of Global Climate Change: Apocalyptic Framing and Political Antagonisms*. Abingdon: Routledge.

Armiero, Marco (2014): Fascism and nature. *Special Issue of Modern Italy*. 19(3).

Balibar, Étienne (1991): Is there a 'neo-racism'? In: Etienne Balibar and Immanuel Wallerstein (Eds). *Race, Nation, Class: Ambiguous Identities*. London: Verso, pp. 17–28.

Barcena, Inaki, Ibarra, Pedro and Zubiaga, Mario (1997): The evolution of the relationship between ecologism and nationalism. In: Michael R. Redclift and Graham Woodgate (Eds). *The International Handbook of Environmental Sociology*. Cheltenham: Edward Elgar, pp. 300–315.

Bar-On, Tamir (2013): *Rethinking the French New Right. Alternatives to Modernity*. Abingdon: Routledge.

Beck, Ulrich (2009): Global public sphere and global subpolitics or: how real is catastrophic climate change? In: Ulrich Beck (Ed). *World at Risk*. Cambridge: Polity Press, pp. 81–108.

Bhatia, Rajani (2004): Green or brown: white nativists environmental movements. In: Abby Ferber (Ed). *Home-grown Hate: Gender and Organized Racism*. Abingdon: Routledge, pp. 194–214.

Blackbourn, David (2006): *The Conquest of Nature. Water, Landscape and the Making of Modern Germany*. London: Vintage.

Bonikowski, Bart (2017): Ethno-nationalist populism and the mobilization of collective resentment, *The British Journal of Sociology*, 68: 181–213.

Bookchin, Murray (1987): Social ecology versus deep ecology: A challenge for the ecology movement, *Green Perspectives: Newsletter of the Green Program Project*, 4–5. http://dwardmac.pitzer.edu/Anarchist_Archives/bookchin/socecovdeepeco.html [5 September 2018].

Brüggemeier, Franz-Josef, Cioc, Mark and Zeller, Thomas (2005): Introduction. In: Franz-Josef Brüggemeier, Mark Cioc and Thomas Zeller (Eds). *How Green Were the Nazis? Nature, Environment, and Nation in the Third Reich*. Athens, OH: Ohio University Press, pp. 1–17.

Camus, Jean-Yves and Lebourg, Nicolas (2017): *Far-Right Politics in Europe*. Cambridge: Belknap Press.

Cederlöf, Gunnel and Sivaramakrishnan, Kalyanakrishnan (Eds) (2014): *Ecological Nationalisms: Nature, Livelihoods, and Identities in South Asia*. Washington, DC: University of Washington Press.

Cosgrove, Denis (2004): Landscape and Landschaft, *GHI Bulletin*, 35: 57–71.

Coupland, Philip M. (2017): *Farming, Fascism and Ecology: A Life of Jorian Jenks*. Abingdon: Routledge.
Davidov, Veronica (2015): Beyond formal environmentalism: Eco-nationalism and the "Ringing Cedars" of Russia, *Culture, Agriculture, Food and Environment*, 37(1): 2–13.
Dawson, Jane I. (1996): *Eco-Nationalism: Anti-Nuclear Activism and National Identity in Russia, Lithuania, and Ukraine*. Durham: Duke University Press.
de Benoist, Alain and Champetier, Charles (1999): The French New Right in the year 2000. http://home.alphalink.com.au/~radnat/debenoist/alain9.html [25 June 2017].
De-Shalit, Avner and Talias, Moti (1994): Green or blue and white? Environmental controversies in Israel, *Environmental Politics*, 3(2): 273–294.
Dobson, Andrew (1999): Ecologism. In: Roger Eatwell and Anthony Wright (Eds). *Contemporary Political Ideologies*. London and New York: Pinter.
Dominick, Raymond H. (1992): *The Environmental Movement in Germany. Prophets and Pioneers, 1871–1971*. Bloomington: Indiana University Press.
Dryzek, John S. (2012): *The Politics of the Earth: Environmental Discourses*. Oxford: Oxford University Press.
Dunlap, Riley E. and McCright, Aaron M. (2015): Challenging climate change: The denial countermovement. In: Riley E. Dunlap and Robert J. Brulle (Eds). *Climate Change and Society: Sociological Perspectives*. Oxford: Oxford University Press, pp. 300–332.
Ehrlich, Paul R. (1968): *The Population Bomb*. New York: Ballantine Books.
Fairclough, Norman and Wodak, Ruth (1997): Critical discourse analysis. In: Teun A. van Dijk (Ed). *Discourse as Social Interaction*. London: Sage, pp. 258–284.
Forchtner, Bernhard (2016): Longing for communal purity: Countryside, (far-right) nationalism and the (im)possibility of progressive politics of nostalgia. In: Christian Karner and Bernhard Weicht (Eds). *The Commonalities of Global Crises: Markets, Communities and Nostalgia*. Basingstoke: Palgrave, pp. 271–294.
Forchtner, Bernhard (2019a): Nation, nature, purity: Extreme-right biodiversity, *Cultural Imaginaries of the Extreme Right. Special Issue of Patterns of Prejudice*, 53(3): 285–301.
Forchtner, Bernhard (2019b): Articulations of climate change by the Austrian far right: A discourse-historical perspective on what is 'allegedly manmade'. In: R. Wodak and P. Bevelander (Eds.), *Europe at the Crossroad: Confronting Populist, Nationalist and Global Challenges*. Lund: Nordic Academic Press, pp. 159–179.
Forchtner, Bernhard (forthcoming): Climate change and the far right, *Wires Climate Change*.
Forchtner, Bernhard and Kølvraa, Christoffer (2015): The nature of nationalism: Populist radical right parties on countryside and climate, *Nature and Culture*, 10(2): 199–224.
Forchtner, Bernhard, Kroneder, Andreas and Wetzel, David (2018): Being skeptical? Exploring far-right climate change communication in Germany, *Environmental Communication*, 12(5): 589–604.
Fowler, Carwyn and Jones, Rhys (2006): Can environmentalism and nationalism be reconciled? The Plaid Cymru/Green Party alliance, 1991–95, *Regional & Federal Studies*, 16(3): 315–331.
Galbreath, David J. and Auers, Daunis (2009): Green, black and brown: Uncovering Latvia's environmental politics, *Journal of Baltic Studies*, 40(3): 333–348.
Geden, Oliver (1996): *Rechte Ökologie. Umweltschutz zwischen Emanzipation und Faschismus*. Berlin, Germany: Elefanten.
Gemenis, Kostas, Katsanidou, Alexia and Vasilopoulou, Sofia (2012): The politics of anti-environmentalism: Positional issue framing by the European radical right. *Paper prepared for the MPSA Annual Conference*, 12–15 April 2012, Chicago.
Glättli, Balthasar and Niklaus, Pierre-Alain (2014): *Die unheimlichen Ökologen: Sind zu viele Menschen das Problem?* Zürich: Rotpunkt Verlag.

Hajer, Maarten A. (1995): *The Politics of Environmental Discourse: Ecological Modernization and the Policy Process*. Oxford: Oxford University Press.
Hamilton, Paul (2002): The greening of nationalism: Nationalising nature in Europe, *Environmental Politics*, 11:2, 27–48.
Holzer, Willibald I. (1994): Rechtsextremismus – Konturen, Definitionsmerkmale und Erklärungsansätze. In: Dokumentationsarchiv des österreichischen Widerstands (Ed). *Handbuch des österreichischen Rechtsextremismus. Aktualisierte und erweiterte Neuausgabe*. Wien: Deuticke, pp. 12–96.
Howanietz, Michael (2005): Ideologischer Biosprit. Natur – Umwelt – Heimat sind keine Versatzstücke einer Blut- und Bodenromantik, *Zur Zeit*, 24: 25.
Hultgren, John (2015): Border Walls Gone Green. Nature and Anti-Immigration Politics in America. University of Minnesota Press..
Hurd, Madeleine and Werther, Steffen (2016): The militant media of neo-Nazi environmentalism. In: Heike Graf (Ed). *The Environment in the Age of the Internet Activists, Communication and the Digital Landscape*. Cambridge: Open Book Publishers, pp. 137–170.
Kedourie, Elie (1966): *Nationalism*. London: Hutchinson.
Krange, Olve, Kaltenborn, Bjørn P. and Hultman, Martin (2019): Cool dudes in Norway: Climate change denial among conservative Norwegian men, *Environmental Sociology*, 5:1, 1-11.
Laclau, Ernesto (2005): *On Populist Reason*. London: Verso.
Lekan, Thomas M. (2004): *Imagining the Nation in Nature. Landscape Preservation and German Identity 1885–1945*. Cambridge: Harvard University Press.
Linse, Ulrich (2009): "Fundamentalistischer" Heimatschutz. Die "Naturphilosophie" Reinhard Falters. In: Uwe Puschner and Ulrich Großmann (Eds). *Völkisch und National*. Darmstadt: WBG, pp. 156–178.
Lockwood, Matthew (2018): Right-wing populism and the climate change agenda: Exploring the linkages, *Environmental Politics*, 27(4): 712–732.
Malloy, Tove H. (2009): Minority environmentalism and eco-nationalism in the Baltics: Green citizenship in the making? *Journal of Baltic Studies*, 40(3): 375–395.
Mawdsley, Emma (2006): Hindu nationalism, neo-traditionalism and environmental discourse in India, *Geoforum*, 37(3): 380–390.
Mix, Tamara L. (2009): The greening of white separatism: use of environmental themes to elaborate and legitimize extremist discourse, *Nature and Culture*, 4(2): 138–166.
Moffitt, Benjamin (2016): *The Global Rise of Populism. Performance, Political Style, and Representation*. Redwood City: Stanford University Press.
Mudde, Cas (2000): *The Ideology of the Extreme Right*. Manchester: Manchester University Press.
Mudde, Cas (2007): *Populist Radical Right Parties in Europe*. Cambridge: Cambridge University Press.
Mudde, Cas (2016): *The Populist Radical Right: A Reader*. Abingdon: Routledge.
Naess, Arne (1973): The shallow and the deep, long-range ecological movement. A summary, *Inquiry*, 16: 95–99.
Olsen, Jonathan (1999): *Nature and Nationalism: Right-wing Ecology and the Politics of Identity in Contemporary Germany*. Basingstoke: Palgrave.
Olwig, Kenneth (2003): Natives and aliens in the national landscape, *Landscape Research*, 28(1): 61–74.
Palmer, Catherine (1998): From theory to practice: experiencing the nation in everyday life, *Journal of Material Culture*, 3(2): 175–199.
Radkau, Joachim (2014): *The Age of Ecology*. Cambridge: Polity Press.

Reisigl, Martin and Wodak, Ruth (2016): The discourse-historical approach (DHA). In: Ruth Wodak and Michael Meyer (Eds). *Methods of Critical Discourse Studies*. London: Sage, pp. 23–61.

Richardson, John (2017): *British Fascism. A Discourse-Historical Analysis*. Stuttgart: ibidem Press.

Riordan, Colin (Ed) (1997): *Green Thought in German Culture. Historical and Contemporary Perspectives*. Cardiff: University of Wales Press.

Rollins, William H. (1997): *A Greener Vision of Home: Cultural Politics and Environmental Reform in the German Heimatschutz Movement, 1904–1918*. Ann Arbor: University of Michigan Press.

Rydgren, Jens (2018a): The radical right: An introduction. In: Jens Rydgren (Ed). *The Oxford Handbook of the Radical Right*. New York: Oxford University Press, pp. 1–16.

Rydgren, Jens (Ed) (2018b): *The Oxford Handbook of the Radical Right*. New York: Oxford University Press.

Salzborn, Samuel (2014): *Rechtsextremismus*. Baden-Baden: Nomos.

Saraiva, Tiago (2016): Fascist modernist landscapes: Wheat, dams, forests, and the making of the Portuguese New State, *Environmental History*, 21(1): 54–75.

Schaller, Stella and Carius, Alexander (2019): *Convenient Truths: Mapping Climate Agendas of Right-Wing Populist Parties in Europe*. Berlin: Adelphi.

Schama, Simon (1995): *Landscape and Memory*. New York: Vintage Books.

Schwartz, Katrina (2005): *Nature and National Identity after Communism: Globalizing the Ethnoscape*. Pittsburgh: University of Pittsburgh Press.

Seliger, Martin (1976): *Ideology and Politics*. London: Allen & Unwin.

Sharma, Mukul (2012): *Green and Saffron. Hindu Nationalism and Indian Environmental Politics*. Ranikhet: Permanent Black.

Smith, Anthony D. (1991): *National Identity*. Reno: University of Nevada Press.

Smith, Anthony D. (2009): *Ethno-Symbolism and Nationalism*. Abingdon: Routledge.

SPLC (2010): *Greenwash: Nativism, Environmentalism and the Hypocrisy of Hate*. www.splcenter.org/20100630/greenwash-nativists-environmentalism-and-hypocrisy-hate [3 December 2017].

Staudenmaier, Peter (2011[1995]): Fascist ecology: The 'green wing' of the Nazi party and its historical antecedents. In: Janet Biehl and Peter Staudenmaier (Eds). *Ecofascism Revisited: Lessons from the German Experience*. Porsgrunn, Norway: New Compass Press, pp. 13–42.

Taguieff, Pierre-André (1990): The neo cultural racism in France, *Telos*, 83: 29–39.

Uekötter, Frank (2006): *The Green and the Brown: A History of Conservation in Nazi Germany*. Cambridge: Cambridge University Press.

Voss, Kristian (2014): *Nature and Nation in Harmony: The Ecological Component of Far Right Ideology*. Unpublished PhD thesis. Florence, Italy.

Wodak, Ruth (2015): *The Politics of Fear: What Right-Wing Populist Discourses Mean*. London: Sage.

Wolschke-Bulmahn, Joachim (1996): Heimatschutz. In: Uwe Puschner, Walter Schmitz and Justus H. Ulbricht (Eds). *Handbuch zur "Völkischen Bewegung" 1871–1918*. Oldenbourg: De Gruyter, pp. 533–545.

Zegers, Peter (2002): The dark side of political ecology, *COMMUNALISM: International Journal for a Rational Society*, 3. https://archive.org/details/TheDarkSideOfPolitical EcologyPeterZegers [1 June 2019].

PART I
Two fields, many topics

2
THE TRAJECTORY OF FAR-RIGHT POPULISM – A DISCOURSE-ANALYTICAL PERSPECTIVE

Ruth Wodak

Introduction

At the beginning of the twenty-first century, the European Union (EU) is confronted with the rise of far-right – sometimes also populist – parties (also labelled as 'right-wing populist', 'radical-right', 'alt-right' or 'anti-liberal democratic' parties; see below for a terminological debate). To avoid misunderstandings, I start off with a working definition, which will be discussed later. Generally, the far-right populism I will be dealing with in this chapter attempts to reduce social and economic structures in their complexity and proposes simple explanations for complex and often global developments (Wodak 2015a; Muller 2016; Pelinka 2018). In doing so, far-right populist discourses oppose 'the true people' to an allegedly corrupt 'élite' and regularly draw on well-known and established stereotypes of 'the Other' and 'the Stranger', whose discursive and socio-political exclusion is supposed to create a sense of community and belonging within the allegedly homogenous 'people' or '*Volk*'. The fact that these 'strangers' may, indeed, be right in the middle of the respective society marks far-right populism as a pseudo-democratic battleground for internal conflicts of interest within that society. The real political and economic contradictions, however, are not addressed directly, since far-right populism as an ideological strategy does not seek to situate social conflict where it originates but to obscure or externalize it.

Far-right populist actors, parties and movements, across Europe and beyond, draw on and combine different *political imaginaries* and different traditions, evoke (and construct) different nationalist pasts in the form of *identity narratives* and emphasize a range of different issues in everyday politics: Some parties gain support via flaunting an ambivalent relationship with *fascist* and *Nazi* pasts (e.g. in Austria, Hungary, Italy, Romania and France); some parties, in contrast, focus primarily on one or two issues, such as a *perceived threat from Islam* (e.g. in the

Netherlands, Denmark, Austria, Germany, Poland, Sweden and Switzerland); some parties primarily stress a *perceived danger to their national identities* from ethnic minorities (e.g. in Hungary, Greece, Italy, and the UK); and some parties primarily endorse a *traditional Christian (fundamentalist) conservative-reactionary agenda* (e.g. in the US, Poland and Russia). In their free-for-all rush for votes, most far-right parties evidently pursue several such strategies at once, depending on the specific audience and context; thus, the aforementioned distinctions are primarily of an analytic nature. Left-wing populist parties such as *Syriza* in Greece and *Podemos* in Spain have also succeeded in winning in national elections due to the severe economic recession since 2008, the negative impact of austerity politics in Greece and Spain, and the Euro-zone crisis. While both far-right and left-wing populist parties regard the EU as part of the élite, the latter do so with social rather than nationalist-nativist arguments (Stavrakakis and Katsambekis 2014; Zakaria 2016: 9).

A vast amount of research and publications[1] from many disciplines (sociology, political science, communication studies, discourse studies, media studies, anthropology and so forth) are investigating this phenomenon, some of which unfortunately neglect specific historical contexts. Indeed, as Finchelstein (2014: 474) claims, 'framing populism historically helps us understand why its return to Europe actualized this continent's past and anti-democratic characteristics'. Therefore – in the framework of the discourse-historical approach (DHA; Wodak 2015a) – I first sketch out some relevant developments in the histories of far-right populism in the twentieth and twenty-first centuries (Histories and foundations of populisms) before turning to an overview of definitions and characteristics of far-right populist parties in Europe and beyond (Defining 'far-right populism'). Subsequently, I elaborate the *micro-politics* of such parties (Micro-politics of the far right) while presenting some of the salient elements of their discourse and performance. Finally, in the conclusion I trace important stages and discursive shifts in the communicative performances of far-right parties since 1945, related to socio-political tipping points in European as well as global developments.

Histories and foundations of populisms

It is impossible to present a comprehensive history of 'populisms', which would start with actors in ancient Greece, continue with Marat at the time of the French Revolution and proceed to the American and Russian populists of the nineteenth century (e.g. Rosanvallon 2007: 260–1; Deiwiks 2009; Pelinka 2018). Instead, I focus on some aspects of the recent history of far-right populist parties, in Europe and beyond, following Finchelstein's proposal (2014: 468) to view populism 'as an outcome of a modern historical process'. Moreover, it is important to emphasize that the re-emergence of the far right and its trajectory from the margins into the political mainstream, the so-called *normalization* of previously tabooed ideologies and political programmes (e.g. Link 2013; Wodak 2015b, 2018a), is not solely a European-based development. It is also not a US-American invention – as some scholars and journalists focusing on Trumpism tend to assume (Wodak 2017; Fuchs 2018).

Quite the contrary: As many political scientists have argued, the history of post-World War Two populism (left- and right-wing) started in South America (de la Torre 2014). The so-called first wave of Latin-American right-wing populism is related to Juan Domingo Perón; a second wave followed in the 1990s, including Alberto Fujimori in Peru; the third wave (left-wing populism) is represented by Hugo Chávez in Venezuela, followed by Bolivian Evo Morales and the neo-populism of the Kirchner administrations in Argentina. Perónism is perceived as a response to the new Western post-World War Two liberal democratic consensus and had its roots in the expansion of the industrial economy after the recession in the 1930s, from which the working class, however, did not benefit (James 1988). In contrast to Perónism's 'third way' (Finchelstein 2014: 471ff.) – neither liberal West nor communist East – the more recent populism à la Chávez described itself as anti-imperialist and anti-neoliberal, using the high oil revenues of the 2000s to implement a range of social programmes. Both Perón and Chávez constructed themselves as incarnations of the 'people', thus conflating state and movement (see below).

There have also been several versions of populist parties in the United States, some inspired by the *Populist Party* of the 1890s which pushed for an anti-trust agenda (Pelinka 2013: 3, 15). Other early populist political parties in the United States included the *Greenback Party*, the *Progressive Party* of 1912 led by Theodore Roosevelt, the *Progressive Party* of 1924 led by Robert M. La Follette, Sr. and the *Share Our Wealth* movement (1933–35). Populism and far-right ideologies continue to be an important force in modern U.S. politics, especially in the 1992 and 1996 presidential campaigns of billionaire Ross Perot and in the so-called 'Tea-Parties' since 2008 (Wodak 2015a). Donald Trump's election campaign and his success in 2016 mark a new stage in U.S. far-right politics. Trump's promises of easy solutions to complex problems, without any need for compromise or negotiation, appeal to a highly disaffected section of the American public, as do his constant challenges to supposedly hegemonic 'political correctness'. In fact, he operates a movement outside of political institutions, and he detests (and evades) independent media (Greven 2016; Fuchs 2018).

European far-right populism also emerged as a post-World War Two phenomenon, rooted in Italian, Hungarian, Rumanian and Spanish fascisms, as well as German National Socialism, as illustrated in the two oldest far-right populist parties, Austria's *VDU* (*Verband der Unabhängigen* founded 1949 as a melting pot for former Austrian Nationalsozialistische Deutsche Arbeiter Partei [NSDAP] members and redefined as the *Austrian Freedom Party* [FPÖ] in 1955/56) and France's *Front National* (founded 1972, drawing on *Poujadism*) (Pelinka 2018). Other parties have since established themselves successfully (although at times only temporarily) in Europe, such as Berlusconi's *Forza Italia*, the *United Kingdom Independence Party* (UKIP), Germany's *Alternative für Deutschland* (AfD) and a range of far-right/national-conservative populist parties in post-communist countries (Wodak et al. 2013; see Figure 2.1, below). *Fidesz* (founded 1988) began moving towards an '*illiberal democracy*'[2] in 2010 (Grabbe and Lehner 2017). The governing coalition under Hungarian Prime Minister Viktor Orbán adopted a new constitution and restricted the freedom of the media. Nowadays, Fidesz can be considered a national-conservative populist party, favouring interventionist economic

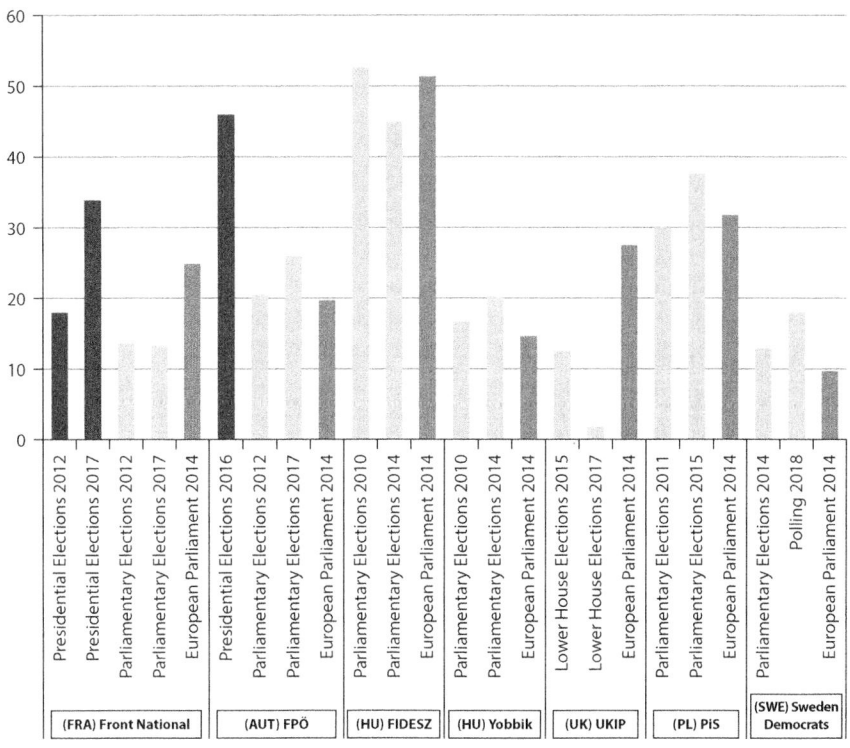

FIGURE 2.1 Results of recent elections of seven far-right populist parties.

policies. Fidesz has partly adopted policies from the extreme-right party *Jobbik*, these target the Roma and Jewish minorities. In the context of the 2015 influx of refugees to the EU, the Orbán government has increased its anti-EU and anti-immigrant rhetoric, reversing a slip in the polls (Greven 2016). Other far-right populist parties have followed Orbán's model and achieved significant electoral wins since 2015.

However, as I have argued elsewhere, there is no one-size-fits-all genealogy or simple explanation for the re-emergence of the far-right in different European countries (Wodak 2015a, 2017). More specifically, different context-dependent responses to the many recent European crises, such as the Financial Crisis of 2008 (and subsequent austerity politics) and the so-called refugee crisis of 2015, should be accounted for as well as the non-simultaneity of developments in former Eastern Europe (Central and Eastern European [CEE] countries) and in the West. Figure 2.1 illustrates the results of recent elections for seven far-right populist parties; it becomes obvious that huge differences exist, for example, FPÖ, Front National (FN), Fidesz and Prawo i Sprawiedliwość; i.e. Law and Justice (PiS) have continued to gain votes during recent years whereas United Kingdom Independence Party (UKIP) has drastically lost votes. Many scholars explain this phenomenon by pointing to UKIP having already achieved their main goal (i.e., the outcome of the Brexit referendum, June 2016) and being colonized by the governing Tory Party.[3]

Defining 'far-right populism'

The question as to far-right populism's exact nature cannot be answered in a simple and straightforward manner: Questions remain as to whether far-right populism even exists and how to define it, especially as distinct from other ideologies and social movements, such as right-wing extremism/the radical right, 'alt-right', 'right-wing/far-right', populism, fascism and left-wing populism. Even more fundamentally, there is no consensus on whether far-right populism is an ideology (*thin* or *thick*; Kriesi and Pappas 2015: 5), a philosophy (Priester 2007: 9), a specific media phenomenon (Pajnik and Sauer 2017) or a political style (Moffitt 2017) that manifests itself mainly in performance and communication. Indeed, Deiwiks (2009: 1) maintains that

> ...strikingly, even some of the works on populism regarded as groundbreaking and substantial like Ionescu and Gellner (1969) fail to state explicitly what they mean by the term. Likewise, Margaret Canovan's *Populism* (1981) comes up with a typology of populism which basically consists of two categories, namely *agrarian populism* and *political populism*. They are further subdivided into a total of seven different kinds of populism – yet, what they have in common is left to the reader to ponder.

In the preface to the new *Handbook of the Radical Right* (2018: 1–2), Rydgren claims that the term 'right-wing populism' is obsolete; rather, he continues, we are dealing with 'ethno-nationalist' parties which also always contain a populist element. In contrast to such ethno-nationalist parties, he claims, radical-right parties are characterized by their rejection of the democratic system and its institutions. However, Rydgren concedes, the boundaries between ethno-nationalist and radical-right parties sometimes become blurred. A similar argument is advanced by Benjamin van de Cleen (2017: 8).

In stark contrast to this approach is the position of Brubaker (2017: 3), who perceives populism as 'a discursive and stylistic repertoire'. He focuses on the discursive, rhetorical and stylistic commonalities, which in his view characterize all populist party movements as part of a broad, comprehensive 'discursive and stylistic turn' (ibid.). From a discourse-oriented perspective, however, and unlike Brubaker, I want to stress that far-right populism should be seen *not only* as a rhetorical style or a *purely* media performance phenomenon (although the significance of [media] staging should never be underestimated; see below), but the ideological content communicated in specific instances is crucial. Indeed, following Pels (2012: 32), it would be wrong to think that there is no substance behind this populist style; and it is precisely the *dynamic mixing of content and form* that has led to the success of far-right populist politics with voters in today's media democracies.

To return to a frequently cited definition of populism: According to Mudde and Kaltwasser (2017: 9–12), it constitutes a (thin) ideology, realized in various discursive and material practices. They emphasize three parameters: first, the opposition between 'the people' and 'the corrupt élite'; second, a grounding in the *volonté générale* of the people; third, a *thin* ideology because it does not constitute a coherent structure of beliefs but assembles contradictory ideologemes in an eclectic fashion. As they do not restrict their definition to *right-wing* populism, the notion of 'the people' refers to both the people as sovereign (*demos*) and to the common people. Moreover, it can refer to the people as *ethnos*. Furthermore, the notion of 'the élite' is differentiated into élites with (cultural, economic or social) power and élites defined on ethnic grounds. And, finally, the *volonté générale* is equated with the general will of the people in the sense of Jean-Jacques Rousseau.

This rather general definition must be differentiated and specified with respect to several relevant issues. Four dimensions seem crucial to me (Wodak 2015a: 20–22, 25–33):

- *Nationalism/Nativism/Anti-pluralism*: Far-right populist parties stipulate a seemingly *homogenous ethnos*, a *populum* (community, *Volk*), which can be arbitrarily defined – often in nativist (blood-related) terms. Such parties value the *homeland* or *Heimat* (or *heartland*, if an internal distinction within the nation is sought), which seems to require protection from dangerous invaders. In this way, *threat scenarios* are constructed – the homeland or 'we' is/are threatened by 'others': Strangers within society *and/or* from outside, i.e. migrants, refugees, Turks, Jews, Roma, bankers, Muslims etc.
- *Anti-élitism*: Such parties share an anti-élitist and anti-intellectual attitude ('*arrogance of ignorance*'; Wodak 2015a), related to strong EU scepticism. Moreover, they prefer plebiscites and strive for a 'true democracy', to which the so-called 'formalistic democracy' is opposed as an antonym. According to these parties, democracy should essentially be reduced to the majoritarian principle, i.e. the rule of the (arbitrarily defined) 'people'.
- *Authoritarianism*: A *saviour, a charismatic leader* is worshipped, alternating between the roles of Robin Hood (protecting the welfare state, supporting the 'simple folk') and the 'strict father' (Lakoff 2004). Such charismatic leaders require a hierarchically structured party and *authoritarian structures* to guarantee 'law and order' and 'security'.
- *Conservativism/Historical revisionism*: Far-right populist parties represent *traditional, conservative values* (traditional gender roles and family values) and insist on preserving the status quo or a return to former, supposedly better times. The aim of protecting the homeland also builds on the *shared narrative of the past*, in which '*we*' are either heroes or victims of evil (a conspiracy of enemies of the *fatherland* etc.). This transforms past sufferings or defeats into stories of successes of the *people* or into stories of betrayal and treachery by others. Social welfare, in the resulting *welfare chauvinism*, should be given only to the so-called 'true' members of the *ethnos*.

Although not all far-right populist parties endorse all the above contents, they – realized in specific combination – can be generalized as typical elements of the far-right's ideology. In all cases, such parties will advocate *change*, and moving away from an allegedly highly dangerous path that would lead straight to an apocalyptic end (Özvatan and Forchtner 2019).

Fear-mongering is therefore a constant political, persuasive strategy in far-right populism and an overarching pattern of argumentation in their campaigns. They wilfully choose someone to be responsible for misery or threats, identified as specific *scapegoats*. Thus, it is sometimes Jews, sometimes Muslims, Roma or other minorities who serve as scapegoats, sometimes capitalists, socialists, career women, non-governmental organizations, the EU, the United Nations, the United States or communists, the governing parties, the élite, the media and so forth. In a third step, the *saviour* appears: the respective party leader, ready to 'solve' the problems identified in a simple manner, for instance by closing borders, deporting so-called 'illegal migrants' etc. A new, positive narrative is created, which should raise the people's hopes in contrast to the dreaded apocalypse. However, this new vision, mostly advertised as an unspecified *change*, is backward-looking, founded on an anachronistic longing for an ethnically homogenous patriarchal society.

Furthermore, extreme-right content, which was taboo for Western-European publics or even illegal since 1945 (as, for example, in Austria since 1947[4]) is voiced as provocation. The deliberate *shamelessness* that accompanies the *normalization* of previously taboo content and behaviour patterns creates – as I claim by drawing on Scheff's theory (Scheff 2000) on the relevance of shame as a basic emotion of group identity – a new *group cohesion*, thus strengthening the voters of far-right populist parties vis-à-vis the (often moralizing) élites. Indeed, some voters for far-right populist parties have repeatedly emphasized how important it is to them that such politicians are '*authentic*', finally voicing what they themselves have always been thinking, and thus they feel taken seriously (Wodak 2015a: 141).

The micro-politics of the far right

As aforementioned, far-right politicians work to create an image of themselves as the 'true representatives of the people' in contrast to 'the untrustworthy political classes', perceived by them as having failed. In these parties' efforts to substantiate their claims, their discourse becomes

> ...magically non-falsifiable, as only factual statements could be verified or falsified. Right-wing populist communication style creates its own 'genre' as a mix of scandal, provocation, transgression, and passion.
> *(Sauer et al. 2017: 28)*

In other words, they strategically create their own visions, beliefs, threat scenarios and nationalistic/nativist identities. In this section, I am concerned with the *micro-politics* of far-right political parties – *how they produce and reproduce their ideologies and exclusionary politics in everyday politics, in the media, when campaigning, on posters, in slogans and in speeches* while employing the framework of the

DHA (see Wodak 2015a). To date, many discourse analysts and applied linguists have studied far-right rhetoric by drawing on a range of qualitative and quantitative data, using various methods, and have analysed relevant pragmatic, semantic, lexical and rhetorical features, as well as argumentation schemes, some of which I have already outlined above. Below, I list a few widely used discursive strategies and performative elements of the far right, which are characteristic for genres such as party programmes, political speeches, campaign rallies and events, posters and slogans, websites, social-media posts and tweets, television and radio interviews as well as TV debates.

The 'people', 'the élites' and 'the others'

Far-right political rhetoric relies on the construction of a distinct dichotomy which aims to divide the people living in a country into two quasi-homogenous blocs: '*the people*' are juxtaposed with '*the élites*' within a specific narrative of threat and betrayal, accusing the so-called 'establishment' of having intentionally or subconsciously neglected the so-called 'people', having instead pursued only their own interests, thus failing to protect the people and to voice their interests, and having ignored the obvious anxieties of the people. Indeed, this narrative arbitrarily constructs two groups via texts and images in manifold ways. Such a Manichean opposition portrays these two groups as vehemently opposed to each other, two epistemic communities, one defined as powerless, the other as powerful; the former described as good, innocent and hard-working, the latter as bad, corrupt, criminal, lazy and unjustly privileged, and so forth. For far-right populists, immigration constitutes a threat to the presumed (constructed) identity of the Volk and their traditional values. The definition of 'the other' varies pursuant to nationally specific conditions. In Hungary, the targets include Roma and Jewish minorities, while the Tea Parties and US President Trump focus on Mexicans and other immigrants from Latin America.

Accordingly, the mechanism of 'scapegoating' constitutes an important feature of such parties' discourse. Sometimes the scapegoats are Jews, sometimes Muslims, sometimes Roma or other minorities, sometimes capitalists, socialists, career women, non-governmental organizations (NGOs), the European Union, the United Nations, the United States or Communists, the governing parties, the élites, the media and so forth. 'They' are foreigners, defined by 'race', religion or language. 'They' are élites, not only within the respective country but also on the European stage ('Brussels') or at the global level ('Financial Capital'). Important fissures and divides within a society, such as class, caste, religion, gender and so forth, may be neglected in focusing on such internal or external 'others', when expedient, and are interpreted as the result of 'élitist *conspiracies*'. Following an aggressive campaign mode implies the use of *ad hominem arguments* as well as other fallacies such as the *straw-man fallacy* or the *hasty generalization fallacy* (an intentionally deceptive argument). Politicians tend to deny and justify even obvious failures (euphemistically labelled as 'mistakes') and quickly find somebody else to blame; under much pressure, *ambiguous, evasive* and *insincere*

DIVIDING THE WORLD

decent, honest, good, industrious, dutiful, charismatic, honourable, noble, brave, trustworthy, incorruptible

amoral, deceitful, lazy, without conscience, evil, bad, cowardly, criminal

The good
The true
The upright
The victims

WE

The 'good' fight

THE OTHERS

The bad
The fake
The liars
The perps

'The true people', represented by the populists

'The others' are a threat to us!
We must fear 'the others'!
We have the right to defend ourselves against 'the others'!

Those up there
élites, politicians, upper classes, 'East Coast', fake media
Those out there
Asylum seekers, economic refugees, welfare tourists
Those down there
'Spongers', 'parasites', the work-shy

FIGURE 2.2 Conceptual map of the far-right mind-set.

apologies may be made; or no apologies given at all (Wodak 2017, 2018b) (see Figure 2.2, below, which maps 'us' and 'them' in the populist mindset).

'Bad manners' (Moffitt 2016: 61–63; Montgomery 2017: 632; Wodak 2017: 559–560) also play a much bigger role, as do deliberate impoliteness, lies, insults, destructive (*eristic*) argumentation and intentional breaches of taboos. Norms of *political correctness* are not merely violated without apology, but explicitly challenged as restricting free speech; this offers identification with anti-élitist behaviour (Scheff 2000). Breaching norms facilitates conveying messages by opening up space for '*calculated ambivalence*' (Engel and Wodak 2013). The latter is defined as a phenomenon whereby one utterance carries at least two more-or-less contradictory meanings, oriented towards at least two different audiences. This not only increases the overall audience, but also enables the speaker/writer to deny any responsibility: after all, 'it wasn't meant that way'. Finally, the power of discourse creates regimes of 'quasi-normality'.

Some far-right populist parties have become more and more explicitly racist (anti-Muslim, antisemitic and anti-Ziganist). These parties tend to emphasize the violence of immigrants while also vindicating violence against immigrants, for example the Aktion für Deutschland (AfD). They deny the discrimination to which immigrants are subjected, but maintain that the native population is discriminated against (Fennema 2004). This 'blaming the victim' may end up in denying the social and historical reality altogether, as frequent Holocaust denials show (Wodak 2015a). Moreover, Fennema (2004: 9) argues that if a party defines the criminality of immigrants or the number of asylum-seekers as the one and only agenda, then this party may be labelled racist, even if their public statements regarded in isolation are not explicitly racist. In distinguishing far-right populist parties from the extreme right and even neo-Nazis (such as *Golden Dawn, Nationaldemokratische Partei Deutschlands, British National Party* or *Yobbik*), Fennema (2004: 15) concludes that, in the former, 'front-stage' activities are conscientiously screened, while 'back-stage', explicitly racist activities are usually hidden from public view (e.g. Rheindorf and Wodak 2019).

Conspiracy theories

Constructing conspiracies necessitates *unreal scenarios* where some perpetrators (lobbies, parties, bankers and the 'Other') are allegedly pulling the strings; these are frequently dramatized and exaggerated. Lies and rumours are spread which denounce, trivialize and demonize 'Others', following the slogan of 'Anything goes'. Rheindorf (2019, in press) maintains that while 'Donald Trump may be credited with popularizing the term "fake news" for this goal [i.e. disseminating conspiracy theories], he was certainly not the first populist to pursue it.' Of course, online media and social media – due to their globalized outreach – have been instrumental in constructing conspiracy theories, because such media construct the kind of immediacy between populist actors and 'the people' that enables strong identification (Moffitt 2016: 88–94; Fuchs 2018). Moreover, as Krzyżanowski and Ledin (2017) argue, the 'antagonistic sphere' and obvious lack of accountability of the internet support an 'uncivil society' and lend themselves to populism's dichotomization of politics and society.

Conspiracy theories draw on the traditional antisemitic world-conspiracy stereotype which also characterized Nazi and fascist ideologies. For example, Hungarian Prime Minister Victor Orbán published a list of 200 so-called 'Soros mercenaries'[5] (including scholars, journalists, intellectuals and NGOs that allegedly supported the Hungarian-American philanthropist, who is Jewish) who are trying to help refugees in Hungary. Indeed, Soros has been demonized via traditional antisemitic conspiracy stereotypes, and subsequently by all Visegrad countries in Europe (i.e. the Czech Republic, Slovakia, Hungary and Poland), and even further afield (Wodak 2018c). Moreover, recent research observes that far-right populists endorse a sense of '*looming crises*' which are threatening 'the people' (Mazzoleni 2008; Triandafyllidou et al. 2009; Rheindorf and Wodak 2019). Conspiracy theories lend themselves to supporting such apocalyptic scenarios, i.e. a continuous state of siege of the 'homeland' (Muller 2016: 43). Ultimately, such constructions of crisis advertise strong leadership as a method for overcoming the crisis while creating hope and promising change (Wodak 2015a), and thus they recontextualize the political agenda into a simplistic common-sense choice of either-or (Müller, ibid.).

Charisma, leadership and mediatization

The *form of the performance* is only one – though important – part of the specific far-right populist habitus. In such rhetoric, several elements are thus combined:

specific topics which are addressed; *specific ideologies* which feed into and constitute utterances and performances; *strategies of calculated ambivalence* and *provocation* which are used to create and de-escalate intentionally provoked scandals; and a *continuous campaigning style*, an overall antagonistic habitus that does not comply with hitherto conventional rules of negotiation and compromise (Kienpointner 2009; Forchtner et al. 2013).

Moreover, party leaders perfectly *instrumentalize the media* attempting to come across as *authentic*. *Authenticity* implies and presupposes that they represent, know and understand how 'normal and true' Austrian, British or Hungarian citizens feel and live. They are part of the in-group, not strangers, not élitist or intellectual, but firmly rooted in common-sense opinions and beliefs. They visit the same pubs as everybody else; they travel to similar places, drive similar cars, have similar problems in their family lives and speak the same language, i.e. the mother tongue. Simultaneously, they are also constructed as being representatives of the common/ordinary people, having the necessary courage to say what the woman/man in the street only thinks; they dare to oppose the powerful and to be direct and explicit, not minding the rules of political correctness and politeness (Wodak 2015a: 132).

The political stage has obviously moved to television, social media, YouTube and so forth, a phenomenon termed the *'mediatization of politics'* or even *Berlusconization* (Forchtner et al. 2013). This development may be one of the reasons why far-right populism has entered mainstream politics. It might also relate to how the media represent scandals; negative and sensationalist angles are likely to receive more coverage, which normally plays into the hands of populists (Deiwiks 2009; Greven 2016). Charisma in politics should be linked to the audience's recognition of the 'right' set of social and cultural capital (habitus), within the 'right' context. Charisma is thus socially constructed and publicly recognized (Wodak 2015a: 131–4). Eatwell (2007: 6–11) conceptualizes charisma via four leadership traits, all of which should be fulfilled by the politician in question:

- Charismatic leaders have a mission, as saviours of the people.
- Charismatic leaders portray themselves as ordinary men, as merely obeying the wishes of the people, and thus also as having a symbiotic relationship with the people whom they represent.
- Charismatic leaders target and indeed demonize enemies.
- Charismatic personalities have great personal presence, which is frequently described as 'magnetism'.

Individual studies have thus far described the persona and rhetoric of specific leaders as the 'Le Pen effect' (Christofferson 2003), the 'Haider phenomenon' (Wodak and Pelinka 2002) or 'Trumpism' (Moffit 2016: 6; Fuchs 2018), the latter being linked to the celebrity image that some far-right populists have built on in becoming political actors.

> - The scandal is first denied.
> - Once evidence is produced, the scandal is redefined and equated with different phenomena.
> - The provocateurs then claim the right to freedom of speech for themselves ('Why can one not utter critique?', 'One must be permitted to criticize Turks, Roma, Muslims, Jews...!' or '*We* dare to say what everybody thinks').
> - Such utterances trigger another debate – not related to the original scandal – about *freedom of speech* and *political correctness.*
> - Simultaneously, victimhood is claimed by the original provocateur, the event is dramatized and exaggerated.
> - This leads to the construction of a conspiracy: somebody must be 'pulling the strings' against the original producer of the scandal: scapegoats (e.g. Muslims, Jews, Turks, Roma or foreigners) are quickly discovered.
> - Once the accused member of the respective minority finally receives a chance to present substantial counter-evidence, a new scandal is launched.
> - Possibly, a 'quasi-apology' might follow if a 'misunderstanding' has occurred; and the entire process starts all over again.

FIGURE 2.3 Right-wing populist perpetuum mobile.

By means of constant *provocation,* attention is drawn to the respective leader and their political agenda – what I label the '*right-wing populist perpetuum mobile*' (Wodak 2013a, 2013b; Figure 2.3).

This dynamic implies that right-wing populist parties strategically manage to frame media debates; other parties and politicians are thus forced to continuously react and respond to ever more newly staged scandals. Few opportunities remain to present other agendas, frames, values and counter-arguments. In this way, far-right populist parties succeed in dominating the media and public debates; moreover, the dissemination of discriminating rhetoric and 'fake news' persists and is continuously (re)produced.

Conclusion

Despite significant contextual differences, it is possible to indicate some post-World War Two tipping points – besides the rise of new communication technologies – in transnational and global politics which have supported both the rise of the far-right as well as salient policy and rhetorical changes.

After 1945, as aforementioned, fascist and Nazi ideologies, symbols and rhetoric were taboo – at least officially. This led to a coding of such content illustrated in the early FPÖ's and FN's agendas; antisemitism continued to be widespread, but could only be uttered explicitly at the *Stammtisch* (dinner table). Of course, scandals with respect to, for example, Holocaust denial and the euphemization of war crimes occurred from time to time (Wodak et al. 1990); however, these far-right parties remained relatively small until the mid-1980s.

In 1989, i.e. with the fall of the Iron Curtain, this salient date not only marked the end of the Cold War, the unification of Europe implied migration to the West from former Eastern European countries (Matouschek et al. 1995). Thus, after 1989, the FPÖ, the *Schweizerische Volkspartei* (SVP), the *Dansk Folkeparti* (DF) and other far-right populist parties started to gain more votes with the help of xenophobic attitudes and rhetoric. This stage could be labelled *Haiderization* (Wodak and Pelinka 2002), widely supported by scandalous incidents and clever mediatization by media-savvy and charismatic leaders. The opposition from mainstream parties and civil society was huge, a *cordon sanitaire* seemed to be the best counter-policy. Nevertheless, mainstream parties tended to appropriate some of the xenophobic agenda because they hoped to keep their voters on board – a strategy doomed to fail as voters usually sanction preference changes.

Then, 9/11 (2001) indicates the next stage: the beginning of the US-American led 'war on terror' fuelled by anti-Muslim sentiments. Ever since, anti-Muslim rhetoric has gained traction, sometimes substituting antisemitism, sometimes simultaneously with anti-Jewish prejudices. Single-issue parties (primarily anti-immigration in regards to Muslims and Eastern Europeans) also attracted more voters (e.g. UKIP, PVV, Sweden Democrats, Norwegian Progress Party). Moreover, if in government, far-right populist parties typically seemed to fail, for example in Austria and the Netherlands, due to mismanagement, major corruption cases and inexperience (Reisigl 2012). During this stage, the agenda and related rhetoric focused mainly on identity issues, on the protection of the Occident and its Christian heritage (Wodak and Köhler 2010; Wodak et al. 2013). The threat of immigration was visualized, for example, as boots trampling on the nation's flag or conflating Islam with Islamist terrorism by picturing minarets as missiles piercing the national body (Kallis 2013; Wodak 2015a). Nativist slogans and apocalyptic scenarios were packaged in comic books, rap songs, national symbols and so forth (Wodak and Forchtner 2014), as symbols that intimate unity, homogeneity and cohesion among 'the true people'. A strong multimodal component thus becomes part and parcel of far-right populist performances (Machin and Richardson 2012; Richardson and Colombo 2013).

The financial crisis of 2008 further enhanced the growth of far-right and left-wing populist parties: on the one hand, people observed how banks were rescued but much less money was spent on helping 'the people', thus fuelling inequality and massive anger with the 'élites' (Sayer 2014). On the other hand, fear of losing out and of the destruction of the welfare state attracted people to the far right (in the rich North) or the far left (in the poor South). Anti-globalization and EU-scepticism lent themselves as agendas for anti-elitist populist movements (and parties such as *Pegida, Syriza, Podemos* and the *AfD*). Interestingly, in recent surveys about 'Fears and values in the EU' (Bertelsmannstiftung 2016), the main fears of EU citizens relate to globalization, operationalized as immigration; and not to financial crises, climate change or war. This indicates a clear tendency to favour *identity politics* instead of economic and class politics. During this stage, the *mediatization and commodification* of politics went centre-stage, also influenced by the impact of social media.

The next stage was triggered by the so-called refugee crisis of 2015 (Krastev 2017; Rheindorf and Wodak 2018). Borders, fences and walls are erected, often supported by mainstream parties. The othering now targets refugees, frequently fallaciously labelled 'illegal migrants' (see Micro-politics of the far right). Far-right parties enter government coalitions (Austria) or even gain majorities (Italy, Poland, Hungary, the US, Turkey). Authoritarianism and 'illiberal democracies' are becoming more and more acceptable (Snyder 2018). Democratic institutions are being threatened and undermined, for example the media, justice and education in Hungary, Poland, Turkey and the United States. Although more 'direct democracy' is called for by far-right populist parties, new legislation is quickly implemented without accounting for transparency, expert opinions, minority opposition and so forth.

Notes

1 See, for example, Rydgren (2017), Wodak and Krzyżanowski (2017), Wodak and Pelinka (2002), Rheindorf (2019), Mudde and Kaltwasser (2017), Salzborn (2014), Kriesi and Pappas (2015). Moffitt (2016), Muller (2016), Wodak (2015a).
2 See Zakaria (1997) for a definition of 'illiberal democracy'.
3 For more information on the election results of all far-right European parties since 1989 and the fall of the Iron Curtain, see Davis and Deole (2017) (www.cesifo-group.de/DocDL/dice-report-2017-4-davis-deole-december.pdf).
4 As per the National Socialism Prohibition Act 1947 www.ris.bka.gv.at/GeltendeFassung.wxe?Abfrage=Bundesnormen&Gesetzesnummer=10000207 (accessed 4 March 2017).
5 www.dw.com/en/hungarys-viktor-orban-targets-critics-with-soros-mercenaries-blacklist/a-43381963 (accessed 20 June 2018).

References

Brubaker, Rogers (2017): Why populism? *Theory and Society*, 46(5): 357–385.
Canovan, Margaret (1981): *Populism*. New York: Harcourt.
Christofferson, Thomas R. (2003): The French elections of 2002: The issue of insecurity and the Le Pen effect, *Acta Politica*, 38(2): 109–123.
de Cleen, Benjamin (2017): Populism and nationalism. In: Cristóbal Rovira Kaltwasser, Paul Taggart, Paulina Ochoa Espejo and Pierre Ostiguy (Eds). *Handbook of Populism*. Oxford: Oxford University Press, pp. 342–362.
Davis, Lewis and Deole, Sumit S. (2017): Immigration and the rise of far-right parties in Europe. www.cesifo-group.de/DocDL/dice-report-2017-4-davis-deole-december.pdf [6 August 2018].
De la Torre, Carlos (2014): Populism in Latin American politics. In: Dwayne Woods and Barbara Wejnert (Eds). *The Many Faces of Populism: Current Perspectives*. Bingley: Emerald. pp. 79–100.
Deiwiks, Christa (2009): Populism, *Living Reviews in Democracy*, 1–9.
Eatwell, Roger (2007): The concept and theory of charismatic leadership. In: Antonio Costa Pinto, Roger Eatwell and Ugelvik Larsen Stein (Eds). *Charisma and Fascism in Interwar Europe*. London: Routledge, pp. 3–18.
Engel, Jakob and Wodak, Ruth (2013): 'Calculated Ambivalence' and Holocaust denial in Austria. In: Ruth Wodak and John E. Richardson (Eds). *Analysing Fascist Discourse: European Fascism in Talk and Text*. London: Routledge, pp. 73–96.

Fennema, Meindert (2004): Populist parties of the right. *Working Paper* No.4/1. Amsterdam, the Netherlands: Amsterdamse school sociaal wetensch.

Finchelstein, Federico (2014): Returning populism to history, *Constellations*, 21(4): 467–482.

Forchtner, Bernhard, Krzyżanowski, Michal and Wodak, Ruth (2013): Mediatisation, right-wing populism and political campaigning: The case of the Austrian Freedom Party (FPÖ). In: Mats Ekström and Andrew Tolson (Eds). *Media Talk and Political Elections in Europe and America*. Basingstoke: Palgrave, pp. 205–228.

Fuchs, Christian (2018): *Digital Demagogue. Authoritarian Capitalism in the Age of Trump and Twitter*. London: Pluto Press.

Grabbe, Heather and Lehner, Stefan (2017): Defending EU values in Poland and Hungary. *Open Society European Policy Institute* (Brussels).

Greven, Thomas (2016): *The Rise of Right-wing Populism in Europe and the United States*. Washington, DC: Friedrich-Ebert-Stiftung.

Ionescu, Ghița and Gellner, Ernest (Eds) (1969): *Populism: Its Meaning and National Characteristics*. London: Weidenfeld & Nicolson.

James, Daniel (1988): *Resistance and Integration: Peronism and the Argentine Working Class, 1946–1976*. Cambridge: CUP.

Kallis, Aristotle (2013): Breaking taboos and 'mainstreaming the extreme': The debates on restricting Islamic symbols in contemporary Europe. In: Ruth Wodak, Majid KhosraviNik and Brigitte Mral (Eds). *Right-wing Populism in Europe: Politics and Discourse*. London: Bloomsbury, pp. 55–70.

Kienpointner. Manfred (2009): Plausible and fallacious strategies to silence one's opponent. In: Frans van Eemeren (Ed). *Examining Argumentation in Context: Fifteen Studies on Strategic Manoeuvring*. Amsterdam: Benjamins, pp. 61–75.

Krastev, Ivan (2017): *Europadämmerung*. Frankfurt/Main: Suhrkamp.

Kriesi, Hanspeter and Pappas, Takis S. (Eds) (2015): *European Populism in the Shadow of the Great Recession*. Colchester: ECPR Press.

Krzyżanowski, Michał and Ledin, Per (2017): Uncivility on the web. Populism in/and the borderline discourses of exclusion. In Ruth Wodak and Michał Krzyżanowski (Eds). *Right-Wing Populism in Europe & USA: Contesting Politics & Discourse beyond 'Orbanism' and 'Trumpism'*. Special issue of *Journal of Language and Politics*, 16(4): 566–581.

Lakoff, George (2004): *Don't Think of an Elephant: Know Your Values and Frame the Debate*. White River Junction: Chelsea Green.

Link, Jürgen (2013): *Versuch über den Normalismus: Wie Normalität produziert wird*. Göttingen: Vandenhoeck & Ruprecht.

Machin, David and Richardson, John E. (2012): Discourses of unity and purpose in the sounds of fascist music: A multimodal approach, *Critical Discourse Studies*, 9(4): 329–345.

Matouschek, Bernd, Wodak, Ruth and Januschek, Franz (1995): *Notwendige Maßnahmen gegen Fremde?* Vienna: Passagen Verlag.

Mazzoleni Gianpietro (2008): Populism and the media. In: Daniele Albertazzi and Duncan McDonnell (Eds). *Twenty-First Century Populism*. London: Palgrave, pp. 49–64.

Moffitt, Benjamin (2016): *The Global Rise of Populism. Performance, Political Style, and Representation*. Palo Alto: Stanford University Press.

Montgomery, Martin (2017): Post-truth politics? Authenticity, populism and the electoral discourses of Donald Trump, *Journal of Language and Politics*, 16(4): 619–639.

Mudde, Cas and Kaltwasser, Cristobal R. (2017): *Populism*. Oxford: Oxford University Press.

Müller, Jan-Werner (2016): *What Is Populism?* Philadelphia: University of Pennsylvania Press.

Ötsch, Walter and Horaczek, Nina (2017): *Populismus für Anfänger: Anleitung zur Volksverführung*. Frankfurt/Main: Westend.

Özvatan, Özgür and Forchtner, Bernhard (2019): The Far-Right *Alternative für Deutschland* in Germany: Towards a 'Happy Ending'? In: Alan Waring (Ed). *The New Authoritarianism: Vol 2. A Risk Analysis of the European Alt-Right Phenomenon*. Stuttgart: ibidem Press.

Pajnik, Mojca and Sauer, Birgit (Eds) (2017): *Populism and the Web: Communicative Practices of Parties and Movements in Europe*. London: Ashgate.

Pelinka, Anton (2013): Right-wing populism: Concept and typology. In: Ruth Wodak, Majid KhosraviNik and Brigitte Mral (Eds). *Right-Wing Populism in Europe. Politics and Discourse.* London: Bloomsbury, pp. 3–22.

Pelinka, Anton (2018): Identity politics, populism and the far right. In: Ruth Wodak and Bernhard Forchtner (Eds). *The Routledge Handbook of Language and Politics*. London: Routledge, pp. 618–629.

Pels, Dick (2012): The new national individualism–populism is here to stay. In: Eric Meijers (Ed). *Populism in Europe*. Linz: Planet, pp. 25–46.

Priester, Karin (2007): *Populismus. Historische und aktuelle Erscheinungsformen*. Frankfurt/Main: Campus Verlag.

Reisigl, Martin (2012): Rechtspopulistische und faschistische Rhetorik: Ein Vergleich, *Totalitarismus und Demokratie*. 9: 303–323.

Rheindorf, Markus (2019): Populism, rhetoric, discourse. In: Anna da Fina (Ed). *Handbook of Discourse Analysis*. Oxford: Oxford University Press (in press).

Rheindorf, Markus and Wodak, Ruth (2018): Borders, fences and limits: Protecting Austria from refugees. Metadiscursive negotiation of meaning in the current refugee crisis, *Journal Immigrant & Refugee Studies*, 16(1–2): 15–38.

Rheindorf, Markus and Wodak, Ruth (2019): 'Austria First' revisited: A diachronic cross-sectional analysis of the gender and body politics of the extreme right. In: Christoffer Kølvraa and Bernhard Forchtner (Eds). *Cultural Imaginaries of the Extreme Right*. Special issue of *Patterns of Prejudice*, 53(2). https://doi.org/10.1080/0031322X.2019.1595392.

Richardson, John E. and Colombo, Monica (2013): Continuity and change in anti-immigrant discourse in Italy: An analysis of the visual propaganda of the Lega Nord, *Journal of Language and Politics*, 12(2): 180–202.

Rosanvallon, Pierre (2007): *La Contrademocracia. La politica en la era de la desconfianza*. Buenos Aires: Manantial.

Rydgren, Jens (2018): The radical right. An introduction. In: Jens Rydgren (Ed). *The Oxford Handbook of the Radical Right*. Oxford: Oxford University Press, pp. 1–15.

Salzborn, Samuel (2014): *Rechtsextremismus: Erscheinungsformen und Erklärungsansätze*. Baden-Baden: Nomos.

Sauer, Birgit, Krasteva, Anna and Saarinen, Aino (2017): Post-democracy, party politics and right-wing populist communication. In: Monica Pajnik and Birgit Sauer (Eds). *Populism and the Web*. London: Routledge, pp. 14–35.

Sayer, Andrew (2014): *Why We Can't Afford the Rich*. London: Policy Press.

Scheff, Thomas (2000): Shame and the social bond: A sociological theory, *Sociological Theory*, 18(1): 84–99.

Snyder, Timothy (2018): *The Road to Unfreedom*. New York: Duggan Books.

Stavrakakis, Yannis and Katsambekis, Giorgos (2014): Left-wing populism in the European periphery: The case of SYRIZA, *Journal of Political Ideologies*, 19(2): 119–142.

Stiftung, Bertelsmann (2016): Fear not Values. Public opinion and populist vote in Europe. Eupinions 26/3 (edited by Catherine de E.Vries and Isabell Hoffmann).

Triandafyllidou, Anna, Wodak, Ruth and Krzyżanowski, Michał (Eds) (2009): *The European Public Sphere and the Media: Europe in Crisis*. Basingstoke: Palgrave Macmillan.

Wodak, Ruth (2013a): 'Anything Goes!' – The Haiderization of Europe. In: Ruth Wodak, Majid KhosraviNik and Brigitte Mral (Eds). *Right-Wing Populism in Europe. Politics and Discourse*. London: Bloomsbury, pp. 23–37.

Wodak, Ruth (2013b): The strategy of discursive provocation: A Discourse-Historical Analysis of the FPÖ's discriminatory rhetoric. In: Matthew Feldman and Paul Jackson (Eds). *Doublespeak. The Rhetoric of the Far Right Since 1945*. Stuttgart: ibidem Press, pp. 101–122.

Wodak, Ruth (2015a). *The Politics of Fear: What Right-wing Populist Discourses Mean*. London: Sage.

Wodak, Ruth (2015b): Normalisierung nach rechts: Politischer Diskurs im Spannungsfeld von Neoliberalismus, Populismus und kritischer Öffentlichkeit, *Linguistik Online*, 73(4): 27–44.

Wodak, Ruth (2017): The 'Establishment', the 'Élites', and the 'People'. Who's who?, *Journal of Language and Politics*, 16(4): 551–565.

Wodak, Ruth (2018a): Vom Rand in die Mitte–Schamlose Normalisierung, *Politische Vierteljahresschrift*, 59(2): 323–335.

Wodak, Ruth (2018b): The revival of numbers and lists in radical right politics. www.radicalrightanalysis.com/2018/06/30/the-revival-of-numbers-and-lists-in-radical-right-politics/ [6 August 2018].

Wodak, Ruth (2018c): Antisemitism and the radical right. In: Jens Rydgren (Ed). *Handbook of the Radical Right*. Oxford: Oxford University Press, 61–85.

Wodak, Ruth and Forchtner, Bernhard (2014): Embattled Vienna 1683/2010: Right-wing populism, collective memory and the fictionalisation of politics, *Visual Communication*, 13(2): 231–255.

Wodak, Ruth, KhosraviNik, Majid and Mral, Brigitte (Eds) (2013): *Right-Wing Populism in Europe. Politics and Discourse*. London: Bloomsbury.

Wodak, Ruth and Köhler, Katharina (2010): Wer oder was ist 'fremd'? Diskurshistorische Analyse fremdenfeindlicher Rhetorik in Österreich, *Sozialwissenschaftliche Studiengesellschaft*, 50(1): 33–55.

Wodak, Ruth and Krzyżanowski, Michał (Eds) (2017): Right-wing populism in Europe & USA: Contesting politics & discourse beyond 'Orbanism' and 'Trumpism'. Special Issue, *Journal of Language and Politics*, 16(4).

Wodak, Ruth, Pelikan, Johanna, Nowak, Peter, Gruber, Helmut, de Cillia, Rudolf and Mitten, Richard (1990): *"Wir sind alle unschuldige Täter!" Diskurshistorische Studien zum Nachkriegs-Antisemitismus*. Frankfurt/Main: Suhrkamp.

Wodak, Ruth and Pelinka, Anton (Eds) (2002): *The Haider Phenomenon in Austria*. New Brunswick: Transaction Press.

Zakaria, Fareed (1997): The rise of illiberal democracy, *Foreign Affairs*, 76(6): 22–43.

Zakaria, Fareed (2016): Populism on the March, *Foreign Affairs*, 85(11–12): 9–15.

3
ENVIRONMENTAL COMMUNICATION RESEARCH

Origins, development and new directions

Anders Hansen

Introduction

The 'environment' has become one of the key public and political concerns of our time. While concerns about nature, the natural environment and conservation go much further back in history, the public discourse on the environment – in its present ubiquitous, predominantly mediated, and familiar form – is of relatively recent origin dating back only to the 1960s.

In this chapter, I outline the historical development of research on media, communication and the environment, and its emergence into a broad body of research now generally referred to as environmental communication research. While very little – as noted by Forchtner (2019a) in his introduction to this volume – of the fast expanding body of research on environmental communication has touched directly or specifically on the far right and the environment, I indicate some of the linkages and research foci that may prove particularly productive in bringing the two bodies of research into dialogue with each other.

Starting with a brief reiteration of the importance of media and communication in the rise and development of environmental concern, I, first, delineate the disciplinary and theoretical context and some of the main emphases of environmental communication research. Second, I discuss key trends in research on news and other types of media representations of the environment, and, third, on the production and wider social and political implications of public communication about the environment. With reference to the implications of a rapidly changing media and communications landscape, I conclude with a summary of the major achievements of environmental communication research to date, and with suggestions for where future research emphases, including on far-right environmental communication, can usefully be focused.

The rise of environmental communication research

The study of mediated communication of the environment has come far since its emergence in the second half of the twentieth century. It has particularly developed and consolidated in exciting ways in the present century. This is evident not just in a marked increase in scholarly research on media, communication and the environment, but in the embedding of environmental communication research within university-level curricula and in sections and groups within national and international communication associations. Indeed, such is the growing interest in this field that new associations such as the International Environmental Communication Association (IECA, launched in 2011) have been formed. Sustaining the consolidation of 'environmental communication' as a recognised field of research is the growing body of book-length publications focused on environmental communication; for example, Pezzullo and Cox (2018), Hannigan (2014), Hansen (2010/2019), Lester (2010), Doyle (2011), Boykoff (2011), Anderson (2014), Tong (2015) and Priest (2016). In addition, there has been significant growth in journal articles across a range of science/environment/health and communications journals, including the establishment of academic journals specifically focused on environmental communication, notably the journal *Environmental Communication*. Perhaps the strongest indication of the increasing maturity of the field is the emergence of synoptic collections (Hansen 2014), handbooks (e.g. Hansen and Cox 2015; Sachsman and Valenti 2020), encyclopaedia (Nisbet 2018), and book series, such as the *Palgrave Studies in Media and Environmental Communication* launched in 2014.

As the public identification, construction and contestation of the environment as an 'issue' depends crucially on mediated public communication, it is no coincidence that the rise of the environment as a focus for social and political concern has gone hand-in-hand with, and been facilitated by, the growth in (mass) mediated forms of communication, including, perhaps particularly, the growth of visual media. Traditional broadcast and print media and newer forms of digital communication have been instrumental in defining 'the environment' as a concept and domain, and in bringing environmental issues and problems to public and political attention. The inherently long timescales and often low immediate visibility of many environmental changes mean that much of what publics know or recognise as 'the environment' or more particularly as 'environmental problems', is known or perceived through mainstream media and other mediated forms of communication. However, media and forms of communication are subject to multiple pressures, influences, and constraints – notably economic and technological, but also factors to do with the communicative resources and skills of those who contribute to, participate in, draw on and engage with communication about the environment.

Research on media, communication, public opinion and the environment draws from a broad range of both humanities and social sciences disciplines. Theoretically, dominant frameworks have included social constructionist theory, risk communication theory, Beck's work on the risk society, and Castells' work on communication, globalisation and the network society. Work in psychology has provided key models for understanding public perception of risk. Much work in recent decades on climate change communication and public perception similarly draws on psychology and social psychology for understanding how publics make sense of media and public communication about climate change and other environmental risks (Whitmarsh 2015).

The development of environmental communication research as a distinctive strand within media and communication studies generally is hardly surprising when considering the centrality of public media to drawing public and political attention to 'environmental problems'. Originally focused predominantly around the study of mainstream news journalism, media coverage of the environment and associated public opinion, environmental communication research has expanded to examine a much broader range of media, genres and forms of communication.

The social constructionist perspective in particular enabled communication research on environmental problems to focus on how the communicative practices of both journalists and their sources are circumscribed by and embedded within organisational, cultural and discursive structures (Hansen 2010; Anderson 2014). Within mainstream media and communication research, organisational and cultural perspectives on news production (Anderson 2015), agenda-setting research (Trumbo and Kim 2015), and framing-research (Nisbet and Newman 2015) have provided particularly productive frameworks for analysing environmental communication.

Given the centrality of mediated communication to the social, political and cultural construction of the environment, it is not surprising that the principal focus and point of departure for research has often been on *media representations* of the environment. It is also the case, however, that the analysis of media representation is almost invariably done with a view to understanding both the processes of its production and its wider social and political implications.

News media representation of the environment

The most frequent focus of research on mediated environmental communication has traditionally been that of news coverage (Hansen 2017), demonstrating what types of environmental issues receive media attention, why and with what implications for public and political action. Such studies have contributed significantly to mapping the processes involved in producing environmental communication, the nature and career of media representations of environmental issues, and the seemingly cyclical nature of environmental coverage and public environmental concern referred to as 'the issue-attention cycle' (Downs 1972).

As I have argued elsewhere (Hansen 2015a), it is particularly longitudinal studies of environmental news coverage that have contributed to our understanding of the complex factors influencing the extent and nature of news media attention to the environment. Mapping the significant ups and downs over time in media attention and drawing on the sociology of news, such research has begun to explain the complex roles of claims-making, news values, journalistic practices, issue-interaction, and other 'drivers' of news attention to the environment. Longitudinal studies have shown that, once introduced in the 1960s–70s, and despite the considerable fluctuations over time in media attention, the environment has remained firmly established on the media and public agenda (Djerf-Pierre 2013; Boykoff et al. 2015). In a comprehensive meta-analysis, Schäfer et al. (2014) show that key 'drivers' influencing the extent and framing of news coverage are political attention, the activities of campaigning groups and non-governmental organisations (NGOs), and major international summit events.

Environmental journalism and source influence

Research in the 1970s (Schoenfeld et al. 1979) showed that the extent and nature of news coverage of the then relatively novel notion of 'the environment' often depended closely on whether news organisations had a dedicated environment beat with specialist environmental journalists. It also showed the significant power of sources to influence the extent and nature of news definitions of environmental issues (Sachsman 1976).

The changing relationship between sources and journalists, and the role of sources in influencing the agenda and nature of news coverage of the environment, have continued to be central foci for research. Such research has also mapped how significant technological and economic changes have impacted media organisations and environmental journalism (Friedman 2015; Williams 2015).

Surveying the history of environmental journalism, Friedman (2015) thus notes how the rise of new digital media and forms of communication, increasing competition, and economic pressures have reduced or eliminated environmental beats within traditional news organisations. As a result, environmental news has often become dispersed across more mainstream news categories and covered by reporters, who would not necessarily have accumulated the breadth of insight or the range of reliable sources associated with specialist environmental or science reporters (Dunwoody 2015).

The contraction in environmental journalism has been matched by an expansion in sources' use of public relations strategies, resulting in an overall significant shift of power from journalists to sources in terms of ability to influence the agenda and nature of public debate about the environment (Williams 2015). Corporations/businesses, governments, research institutions, environmental pressure groups, NGOs, and other news sources have increasingly sought to influence and 'manage' public communication, including about the environment (Davis 2013; Williams 2015), by adopting promotional communication practices, now generally referred to as strategic communication (Frandsen and Johansen 2017). The ability of sources to influence public communication has also been significantly augmented by the exponential proliferation of communications modes, means, channels and formats characteristic of the new digital communications environment (Hansen 2018b).

Environmental pressure groups – along with many other types of source – have been quick to seize the greatly enhanced opportunities afforded by new digital and social media as a means to by-passing the control over information dissemination traditionally exercised by established media organisations. New digital media have been pressed into service – not least through collapsing of time and space barriers – to augment traditional campaigning and mobilisation strategies (Lester 2015).

Research charting the increasing professionalisation and strategic organisation of environmental communication by environmental pressure groups, government departments, NGOs and research institutions has been complemented with research on how corporations and big business deploy strategic environmental communication. Such studies have started to unveil some of the communicative strategies; for example 'casting doubt' and fuelling uncertainty about scientific evidence (Oreskes and Conway 2010), the use of 'front groups' (Beder 2002), think tanks and alliances of various sorts (Miller and Dinan 2015) skilfully deployed by big business and corporations to influence the public and political climate of opinion in ways advantageous to their vested interests.

There is a growing body of research showing how corporations, businesses and public institutions deploy environmental language, discourses and images in strategic communication aimed at enhancing their public environmental image and credentials. This includes communication strategies aimed at re-framing public discourse with regard to a particular industry's environmentally 'tarnished' image; for example, communication practices adopted by the coal industry to re-position coal energy as 'clean' (Schneider et al. 2016). And it includes research on the 'image repair' communication strategies pursued by companies and corporations in the wake of their involvement in environmentally damaging accidents and practices, for example, corporate communication following the Volkswagen emissions scandal of 2015 (Painter and Martins 2017) and the BP (British Petroleum) Deepwater Horizon oil spill of 2010 (Harlow et al. 2011; Schultz et al. 2012).

Studies of environmental journalism and source influence have confirmed that sources play an increasingly pro-active role in strategically influencing and shaping public communication about the environment. The key challenge for environmental communication research in this respect is to map and understand how the power to influence public communication about the environment is distributed, and to ask questions about why some constructions of environmental issues become much more successful than others in the public sphere, and to whose benefit.

The social and political implications of environmental news

Noting the tantalising parallels often observed between the ups and downs of media attention and public concern about the environment, a focus for environmental communication research has been on examining the relationship between media representations and public opinion, understanding and engagement with regard to the environment. Key frameworks for such research have included (Hansen 2019) agenda-setting, cultivation analysis, and quantity of coverage theory.

But one of the most productive developments in research on media representation of the environment has been the increasing use of the concept of framing. Research on framing draws attention to how the principles of 'selection' and 'salience' (Entman 1993; Nisbet and Newman 2015) in media content help structure audience responses by directing attention to: (1) what the issue/problem is, (2) who/what is responsible and (3) what the solution is (Ryan 1991).

Drawing on framing categories originally identified by Gamson and Modigliani (1989) in their analysis of nuclear issues and popular culture, studies of media and environmental issues have demonstrated how key interpretive packages ('progress', 'economic prospects', 'ethical/moral', 'runaway science', 'public accountability', etc.) are strategically deployed and manipulated by sources in public environmental debate. Deliberately and strategically chosen ways of framing environmental issues and controversies thus have been shown to influence both the way in which environmental issues are communicated in public media and, in turn, the mobilisation of public understanding, opinion and behaviour with regard to a broad range of issues, including nuclear energy, biotechnology, and climate change (Nisbet and Newman 2015).

The observed tendency of news reporting to balance – due largely to the journalistic values of objectivity and impartiality (Boykoff and Boykoff 2004; Dunwoody 2015; Hansen 2016) – opposing views in reporting of controversial environmental issues has the effect of influencing and perpetuating public perception of widespread scientific uncertainty long after scientific consensus has been reached. Studies of climate change news in particular have shown the influence of uncertainty-framing (Corbett and Durfee 2004; Philo and Happer 2013) on public understanding and perception. Uncertainty-framing in turn varies depending on whether environmental news is framed in political or in scientific terms. The framing of climate change in science news thus has been shown to reinforce public knowledge about climate change and beliefs that climate change is happening and needs addressing, while the framing of climate change in political news either contributes little to public understanding of climate change, or indeed reinforces doubt and scepticism about climate change (Hart et al. 2015; Nisbet et al. 2015).

Particularly promising indications of the significance of framing comes from work on how re-framing communication about climate change in terms of local – as opposed to global – concerns and in terms of a health-frame or a security-frame impacts on public understanding and, particularly, on public engagement. Thus, framing climate change in terms of its potential exacerbation of health issues that people are already familiar with enhances the effectiveness of communication. Likewise, 'replacing visuals of remote Arctic regions, animals, and peoples with more socially proximate neighbors and places across local communities and cities' (Nisbet and Newman 2015: 330) increases the likelihood of gaining local news media coverage, as well as impacting people's engagement with climate issues.

Key to understanding variations in how framing influences public engagement with mediated environmental communication is also the recognition that some frames resonate better than others with different publics or 'interpretative communities'. As Nisbet and Newman (2015: 337) explain, these are groups of 'individuals who share common risk perceptions about climate change, reflect shared schema, mental models, and frames of reference, and hold a common socio-demographic background'. While the notion of differentiated publics has been recognised in communication research since the mid-twentieth century, what is new and different in the present century is the increasingly fragmented and diverse digital media environment. That is, publics increasingly attend to and select the type of media and news content which confirm and reinforce, rather than change or broaden, existing views and beliefs characteristic of one's interpretative community. (see also Roser-Renouf et al. 2015).

Contrary to earlier public understanding of science perspectives, which tended to assume that the problem with public understanding of controversial scientific and environmental issues was insufficient or inaccurate communication of relevant evidence, framing theory stresses that public understanding (and associated perceptions and behaviour) relies on and is influenced by general and often ambiguous cues. Thus, on most issues we form our understanding and opinion on sometimes fleeting observations of key cues from our symbolic environment, notably as we become aware of them from the mediated content that we consume (Nisbet and Newman 2015).

Framing theory then adds to our understanding of public-mediated environmental communication and its influences by demonstrating how variations in prominence and absence (selection) and in context/perspective/framework/ discourse (e.g. science, health, security, politics, religion, etc.) influence how we take our cues from and make sense of public-mediated communication about the environment.

This body of literature in environmental communication research also arguably offers one of the most promising bridges to research on the far right, where studies have begun to uncover comparable framing strategies and discursive packages – and ultimately ideologies – underpinning and at work in far-right communication on the environment. Lockwood (2018) shows how right-wing populist parties' (RWPs) communication and ideology revolves around idealised/romanticised notions of national purity and sovereignty, anti-universalism and protecting the local/national environment against the forces of globalism. Referencing Forchtner and Kølvraa (2015) and Gemenis et al. (2012), Lockwood (2018: 724) summarises how such notions result in very different right-wing populist stances on climate change compared with stances on local and national environmental issues 'where there is often a strong element of conservation in RWP party manifestos and literature, framed by a romantic nationalism and hostility to immigration'. Further reference is made to the framing used to articulate discontent about the – broken – link between 'the people' and the political elite, and to how, for example, the RWP framing of energy independence and renewable energy is embedded in both material factors (whether countries have fossil fuel resources and a declining extraction industry) and in deep-seated notions about 'the people', 'the land' and national sovereignty.

Non-news media representation of the environment

While news media representations of the environment have historically been the most prominent focus for research, there is now also an expanding body of work on representations of nature and the environment in film (including animation and cartoons), advertising, and in documentary and entertainment television. Broadening the focus of research well beyond the news media is important in order to understand how public understanding and policy draw from and interact with the messages, images and ideologies about the environment that circulate in the wider cultural and symbolic environment. Studies of advertising, film and other popular culture representations have pointed to the prominent use of nature and the natural environment in ways that not only tap into deep-seated, often romanticised or nostalgic, views of nature, national identity and indeed other types of identity (race, class, gender, etc.), but also rework these as warrants (natural, authentic, genuine, ontological, god-given) for particular (ideological, political, commercial, etc.) purposes and views of what is right and wrong. Emerging work on the far right and the environment indicate similar trends (see work in this volume as well as, e.g. Richardson 2017: 163–173 and Forchtner 2019b).

Invoking nature/the natural is a key rhetorical component in which particular ideological views are constructed and communicated. A growing body of research on a range of genres, but perhaps most specifically advertising (Hansen 2010/2019), has demonstrated the prominent linking of a (romanticised) view of nature *with* a rural (idyllic) past *with* national identity as one of the most potent ideological uses in the modern age. This was used in the early parts of the twentieth century for naked political propaganda and mobilisation for war, and in the second half of the twentieth century for commercial purposes. Soper (1995: 194), in her discussion of the appropriation of nature and the 'rural imaginary' in commercial advertising, thus argues that in present times 'it is the marketing rather than the political propagandist potential of nature that is more exploited, and the clichés of nationalist rhetoric have become the eco-lect of the advertising copywriter'.

As I have demonstrated elsewhere (Hansen 2015b), the particular deployment and constructions of nature in advertising and other media have, broadly speaking, oscillated between, at the one extreme, a progress-package-driven view of nature as a resource to be dominated, exploited and consumed, and, at the other extreme, a romanticised – and often retrospective – view of nature as the (divine) source and embodiment of authenticity, sanity and goodness, to be revered and protected, 'not to be tampered with'. The latter construction of nature is also one which invokes a nostalgic view of an ideal and ordered past, with implications for the public construction of social class, gender, race and, not least, national identity.

Common to much of this work is the recognition that constructions of nature and the natural environment are central to the cultural, political and ideological power of communication about the environment. There are also exciting parallels – including some of the foundational literature drawn upon (e.g. Schama 1995) – between the wider research on images and uses of nature/the natural in advertising and other media genres and the appropriation of nature and the natural environment in far-right ideology. I would thus argue that this is one of the key areas where further cross-fertilisation between these bodies of literature holds the promise of deeper insights into the significant ideological power of constructions and appropriations of nature/the natural environment in environmental communication generally and in far-right communication more specifically.

Visual environmental communication research

While research on media representation of the environment has been predominantly focused on text, lexis and discourse in media coverage, a significant body of research is emerging – belatedly, it could justifiably be argued – into the mediated visual representation and construction of nature and the environment. And much of this, it would seem – with its attention to the visual linking of nature and the environment with national and other types of identity and with power – has direct applicability to research on visual representation in far-right communication.

Surveying the literature on visual environmental communication, I have noted elsewhere (Hansen 2018a) that such studies point to the visual representations of the environment and nature that tend to be decontextualised, aestheticised and symbolic in ways that enhance their flexible and versatile use across different genres of communication. Perhaps most significantly in terms of thinking about cross-fertilisation of research, including in relation to research on far-right visual communication, visual environmental communication research shows that images draw from historically and culturally resonant discourses of nature (including the tension between romantic and extractive/utilitarian views of nature) and re-work these to fit the needs of the communicators and the communicative context (e.g. advertising and the selling of products and ideas; or political communication and political persuasion). The decontextualised and generic nature of much visualisation of the environment point to the way that the 'meaning' of visuals is influenced or dominated by their textual anchoring: they mean what the text or voice-over *says* that they mean.

Studies that map historical changes in the visualisation of the environment and nature – for example in film (Mitman 1999), advertising (Kroma and Flora

2003; Howlett and Raglon 1992; Ahern et al. 2013) and news magazine covers (Meisner and Takahashi 2013) – are particularly useful for their demonstration of how changing visual constructions of nature and the environment both draw on and resonate with their wider political and cultural context. Research focused on mapping historical change in visual environmental communication helps demonstrate how types of imagery and visual representation are – like textual representations – carefully selected to play into wider 'stories of national identity, of culture, of gender, of race, and, most significantly, of power' (Peeples 2013: 205).

Other visual environmental communication research with potentially direct relevance to studying how the far-right visually constructs a right wing or populist representation of the environment comes from studies that focus on the use (or absence) of people in visual constructions of nature. Particularly instructive here are studies which have demonstrated the visual construction of 'sublime' nature and the concomitant symbolic annihilation of indigenous peoples (DeLuca and Demo 2000; Berger 2003).

Conclusion

Environmental communication research has come a long way in the past few decades. It has consolidated itself as a distinctive multidisciplinary field of inquiry drawing productively from both the humanities and social sciences. It has demonstrated the advantages of bringing diverse theoretical frameworks and analytical approaches to bear on the analysis not just of news media but a much wider range of media and communications forms and processes.

We thus now know a great deal about the publicity and campaigning practices of environmental claims-makers in the public sphere; about environmental journalism; about the organisational and economic pressures impinging on media organisations and their representation of the environment; about how nature and the environment are constructed – including visually – and appropriated for commercial and ideological purposes; and about the social, political and cultural implications of communication on the environment.

A growing body of research is beginning to show profound changes underway in how we engage with news and information about the environment, and, concomitantly, in how communication about the environment is manipulated for political and ideological purposes. The digital communications environment provides us with unprecedented and easy access to virtually any kind of information, 'evidence' and opinion, which is surely a positive thing. But at the same time, many of the traditionally trusted channels of communication (such as major news media organisations) have declined in importance as people's main sources of information. Similarly, many of the mechanisms of 'gate-keeping' and 'fact-checking' (including those traditionally performed by environmental journalists) and ways of establishing the credibility and validity of information, have been eroded significantly.

Lewandowsky and his colleagues (2017: 353), in a thoughtful examination of these trends and how to counter them, poignantly argue that what is now often referred to as the 'post-truth' world 'emerged as a result of societal mega-trends such as a decline in social capital, growing economic inequality, increased polarization, declining trust in science, and an increasingly fractionated media landscape'. A key characteristic of this new media and communications landscape, is the ability to choose (through custom-designed 'filter bubbles' and 'echo chambers') exposure to only, or mainly, information which conforms with and reinforces our pre-existing attitudes.

Another characteristic is the change in length, tone and rhetorical structure of newer digital communication forms, carrying with it the erosion or outright dismissal of evidence-based argumentation, and the concomitant rise of a simplistic public political discourse. This is a discourse where complex public issues, problems, policies and decisions are reduced to brief 'sound-bites' (rhetorically designed for maximum impact, using popular buzz-words, metaphors and slogans, rather than with a view to conveying information) and to what Lewandowsky et al. (2017: 359–360) characterise as a discourse of 'extreme incivility', fanaticism and outrage.

On the basis of the trends and changes in environmental communication and in environmental communication research surveyed and discussed in this chapter, I finish with the following suggestions for environmental communication research foci relevant to this book's focus on the far right, communication and the environment:

1 As the media and communications landscape changes (digital media, social media, user-generated content, citizen journalism, etc.), the manipulation of the mediated environment is becoming increasingly complex and diverse. Research needs to map and understand the skilful ways in which claims-makers of all types – including right-wing populist parties and the far right – are taking advantage of the affordances of the digital media landscape and influencing public communication about the environment. Some of the key challenges for environmental communication research concern the changing nature of how we interact with, assess and consume information. This involves moving from the focus on traditional media and genres, to examining how claims-makers are increasingly making use of a wide variety of digital/online media and media forms for promoting products, ideas, values and views regarding the environment and our relationship with it.
2 Environmental communication research focusing on the textual and visual construction and uses of nature/the natural in advertising, film and other media genres draws on much of the same literature as studies of the far right and its appropriation and framing notions of the people, the land, the nation, and indeed a romanticised and nostalgic view of the past. This is perhaps one of the most exciting and potentially fertile areas for further development and cross-fertilisation: exploring and comparing the frames and framing, discursive

cultural 'packages', culturally resonant themes, and visual representations deployed to promote particular world-views, values and political ideals.

3 As the media and communications environment changes and becomes increasingly diverse and differentiated, there is an urgent need for a much greater use of comparative research. In particular, environmental communication research – including research on right-wing populism and far-right communication on the environment – needs to examine how national environmental discourses are inflected by nationally specific material and ideological differences, and how such inflections draw from and re-work historically and culturally deep-seated views of nature and the environment.

4 Just as research is needed on the increasingly strategic use of environmental communication in the public sphere by multiple sources/communicators (government, NGOs, business, corporations, science institutions, political parties, environmental pressure groups, etc.), so too is research needed on how different publics interact with and act upon public information, debate and controversy regarding the environment. More research is needed on how diverse publics interpret and engage with environmental communication, how different publics (a) choose their information channels/media and sources, and (b) discern credibility and validity of information in an increasingly diverse media environment replete with 'echo-chambers', 'filter-bubbles', and a changed mode of – non–evidence-based – argumentation and rhetorical style.

If we acknowledge that studying and understanding mediated environmental communication in the public sphere is about understanding the processes by which politics, policy and cultural values concerning the environment are socially 'constructed', then the key task for environmental communication research should be to establish and understand how and why some definitions/views and 'solutions' regarding the environment become more prominent and successful than others in the public sphere, and – crucially – to determine whose interests are served (and whose are disadvantaged) by this.

References

Ahern, Lee, Bortree, Denise Sevick and Smith, Alexandra Nutter (2013): Key trends in environmental advertising across 30 years in *National Geographic* magazine, *Public Understanding of Science*, 22(4): 479–494.

Anderson, Alison (2014): *Media, the Environment and the Network Society*. Basingstoke: Palgrave Macmillan.

Anderson, Alison (2015): News organisation(s) and the production of environmental news. In Anders Hansen and Robert Cox (Eds). *The Routledge Handbook of Environment and Communication*. London and New York: Routledge, pp. 176–185.

Beder, Sharon (2002): *Global Spin: The Corporate Assault on Environmentalism*. Totnes: Green Books.

Berger, Martin A. (2003): Overexposed: Whiteness and the landscape photography of Carelton Watkins, *Oxford Art Journal*, 26(1): 1–23.

Boykoff, Maxwell (2011): *Who Speaks for the Climate? Making Sense of Media Reporting on Climate Change*. Cambridge: Cambridge University Press.

Boykoff, Maxwell T. and Boykoff, Jules M. (2004): Balance as bias: Global warming and the US prestige press, *Global Environmental Change-Human and Policy Dimensions*, 14(2): 125–136.

Boykoff, Maxwell T., McNatt, Marisa M. and Goodman, Michael K. (2015): Communicating in the anthropocene: The cultural politics of climate change news coverage around the world. In: Anders Hansen and Robert Cox (Eds). *The Routledge Handbook of Environment and Communication*. London and New York: Routledge, pp. 221–231.

Corbett, Julia B. and Durfee, Jessica L. (2004): Testing public (un)certainty of science: Media representations of global warming, *Science Communication*, 26(2): 129–151.

Davis, Aeron (2013): *Promotional Cultures: The Rise and Spread of Advertising, Public Relations, Marketing and Branding*. Cambridge: Polity.

DeLuca, Kevin Michael and Demo, Anne Teresa (2000): Imaging nature: Watkins, Yosemite, and the birth of environmentalism, *Critical Studies in Media Communication*, 17(3): 241–260.

Djerf-Pierre, Monika (2013): Green metacycles of attention: Reassessing the attention cycles of environmental news reporting 1961–2010, *Public Understanding of Science*, 22(4): 495–512.

Downs, Anthony (1972): Up and down with ecology: The issue attention cycle, *The Public Interest*, 28(3): 38–50.

Doyle, Julie (2011): *Mediating Climate Change*. London: Ashgate/Routledge.

Dunwoody, Sharon (2015): Environmental scientists and public communication. In: Anders Hansen and Robert Cox (Eds). *The Routledge Handbook of Environment and Communication*. London and New York: Routledge, pp. 63–72.

Entman, Robert M. (1993): Framing: Toward clarification of a fractured paradigm, *Journal of Communication*, 43(4): 51–58.

Forchtner, Bernhard (2019a): Far-right articulations of the natural environment: An introduction. In: Bernhard Forchtner (Ed). *The Far Right and the Environment: Politics, Discourse and Communication*. London: Routledge.

Forchtner, Bernhard (2019b): Nation, nature, purity: Extreme-right biodiversity in Germany. In: *Cultural Imaginaries of the Extreme Right. Special Issue of Patterns of Prejudice*, 53(2).

Forchtner, Bernhard and Kølvraa, Christoffer (2015): The nature of nationalism populist radical right parties on countryside and climate, *Nature and Culture*, 10(2): 199–224.

Frandsen, Finn and Johansen, Winni (2017): Strategic communication. In: Craig R. Scott and Laurie K. Lewis (Eds). *The International Encyclopaedia of Organizational Communication*. Oxford: Wiley-Blackwell, pp. 2250–2258.

Friedman, Sharon (2015): The changing face of environmental journalism in the United States. In: Anders Hansen and Robert Cox (Eds). *The Routledge Handbook of Environment and Communication*. London and New York: Routledge, pp. 144–157.

Gamson, William A. and Modigliani, Andre (1989): Media discourse and public opinion on nuclear power: A constructionist approach, *American Journal of Sociology*, 95(1): 1–37.

Gemenis, Kostas, Katsanidou, Alexia and Vasilopoulou, Sofia (2012): The politics of antienvironmentalism: Positional issue framing by the European radical right. Paper presented at the Paper prepared for the MPSA Annual Conference, Chicago, IL, 12–15 April 2012. https://ris.utwente.nl/ws/portalfiles/portal/6153509/MPSA. pdf [15 August 2018].

Hannigan, John A. (2014): *Environmental Sociology* (3rd ed.). London: Routledge.

Hansen, Anders (2010): *Environment, Media and Communication*. London: Routledge.

Hansen, Anders (Ed) (2014): *Media and the Environment* (Vol. 1–4). London: Routledge.

Hansen, Anders (2015a): News coverage of the environment: A longitudinal perspective. In: Anders Hansen and Robert Cox (Eds). *The Routledge Handbook of Environment and Communication*. London: Routledge, pp. 209–220.

Hansen, Anders (2015b): Nature, environment and commercial advertising. In: Anders Hansen and Robert Cox (Eds). *The Routledge Handbook of Environment and Communication*. London: Routledge, pp. 270–280.

Hansen, Anders (2016): The changing uses of accuracy in science communication, *Public Understanding of Science*, 25(7): 760–774.

Hansen, Anders (2017): Media representation: Environment. In: Patrick Rössler (Ed). *The International Encyclopedia of Media Effects*. Chichester: John Wiley & Sons.

Hansen, Anders (2018a): Using visual images for showing environmental problems. In: Alwin Fill and Hermine Penz (Eds). *The Routledge Handbook of Ecolinguistics*. London: Routledge, pp. 179–195.

Hansen, Anders (2018b): Environmental communication. In: Robert L Heath and Winni Johansen (Eds). *The International Encyclopedia of Strategic Communication*. London: John Wiley & Sons.

Hansen, Anders (2019): *Environment, Media and Communication* (2nd ed.). London: Routledge.

Hansen, Anders and Cox, Robert (Eds) (2015): *The Routledge Handbook of Environment and Communication*. London: Routledge.

Harlow, William Forrest, Brantley, Brian C. and Harlow, Rachel Martin (2011): BP initial image repair strategies after the Deepwater Horizon spill, *Public Relations Review*, 37(1): 80–83.

Hart, P. Sol, Nisbet, Erik C. and Myers, Teresa A. (2015): Public attention to science and political news and support for climate change mitigation, *Nature Climate Change*, 5(6): 541–545.

Howlett, Michael and Raglon, Rebecca (1992): Constructing the environmental spectacle: Green advertisements and the greening of the corporate image, *Environmental History Review*, 16(4): 53–68.

Kroma, Margaret M. and Flora, Cornelia Butler (2003): Greening pesticides: A historical analysis of the social construction of farm chemical advertisements, *Agriculture and Human Values*, 20(1): 21–35.

Lester, Libby (2010): *Media and Environment*. Cambridge: Polity.

Lester, Libby (2015): Containment and reach: The changing ecology of environmental communication. In: Anders Hansen and Robert Cox (Eds). *The Routledge Handbook of Environment and Communication*. London and New York: Routledge, pp. 232–241.

Lewandowsky, Stephan, Ecker, Ullrich K. H. and Cook, John (2017): Beyond misinformation: Understanding and coping with the "Post-Truth" era, *Journal of Applied Research in Memory and Cognition*, 6(4): 353–369.

Lockwood, Matthew (2018): Right-wing populism and the climate change agenda: Exploring the linkages, *Environmental Politics*, 27(4): 712–732.

Meisner, Mark S. and Takahashi, Bruno (2013): The nature of *Time*: How the covers of the world's most widely read weekly news magazine visualize environmental affairs, *Environmental Communication: A Journal of Nature and Culture*, 7(2): 255–276.

Miller, David and Dinan, William (2015): Resisting meaningful action on climate change: Think tanks, 'merchants of doubt' and the 'corporate capture' of sustainable development. In: Anders Hansen and Robert Cox (Eds). *The Routledge Handbook of Environment and Communication*. London and New York: Routledge, pp. 86–99.

Mitman, Gregg (1999): *Reel Nature: America's Romance with Wildlife on Film*. Cambridge: Harvard University Press.

Nisbet, Erik C., Cooper, Kathryn E. and Ellithorpe, Morgan (2015): Ignorance or bias? Evaluating the ideological and informational drivers of communication gaps about climate change, *Public Understanding of Science*, 24(3): 285–301.

Nisbet, Matthew (Ed) (2018): *The Oxford Encyclopedia of Climate Change Communication*. Oxford: Oxford University Press.

Nisbet, Matthew C. and Newman, Todd P. (2015): Framing, the media, and environmental communication. In: Anders Hansen and Robert Cox (Eds). *The Routledge Handbook of Environment and Communication*. London and New York: Routledge, pp. 325–338.

Oreskes, Naomi and Conway, Erik M. (2010): *Merchants of Doubt: How a Handful of Scientists Obscured the Truth on Issues from Tobacco Smoke to Global Warming*. London: Bloomsbury.

Painter, Christopher and Martins, Jorge Tiago (2017): Organisational communication management during the Volkswagen diesel emissions scandal: A hermeneutic study in attribution, crisis management and information orientation, *Knowledge and Process Management*, 24(3): 204–218.

Peeples, Jennifer (2013): Imaging toxins, *Environmental Communication: A Journal of Nature and Culture*, 7(2): 191–210.

Pezzullo, Phaedra C. and Cox, Robert (2018): *Environmental Communication and the Public Sphere* (5th ed.). London: Sage.

Philo, Greg and Happer, Catherine (2013): *Communicating Climate Change and Energy Security: New Methods in Understanding Audiences*. London: Routledge.

Priest, Susanna (2016): *Communicating Climate Change: The Path Forward*. Basingstoke and New York: Palgrave Macmillan.

Richardson, John (2017): *British Fascism. A Discourse-Historical Analysis*. Stuttgart: ibidem Press.

Roser-Renouf, Connie, Stenhouse, Neil, Rolfe-Redding, Justin, Maibach, Edward and Leiserowitz, Anthony (2015): Engaging diverse audiences with climate change: Message strategies for global warming's six Americas. In: Anders Hansen and Robert Cox (Eds). *The Routledge Handbook of Environment and Communication*. London and New York: Routledge, pp. 368–386.

Ryan, Charlotte (1991): *Prime Time Activism: Media Strategies for Grassroots Organizing*. Boston, MA: South End Press.

Sachsman, David B. (1976): Public relations influence on coverage of environment in San Francisco Area, *Journalism Quarterly*, 53(1): 54–60.

Sachsman, David B. and Valenti, JoAnn Myer (Eds) (2019): *The Routledge Handbook of Environmental Journalism*. London: Routledge.

Schäfer, Mike S., Ivanova, Ana and Schmidt, Andreas (2014): What drives media attention for climate change? Explaining issue attention in Australian, German and Indian print media from 1996 to 2010, *International Communication Gazette*, 76(2): 152–176.

Schama, Simon (1995): *Landscape and Memory*. London: HarperCollins.

Schneider, Jen, Schwarze, Steve, Bsumek, Peter K. and Peeples, Jennifer (2016): *Under Pressure: Coal Industry Rhetoric and Neoliberalism*. Basingstoke and New York: Palgrave Macmillan.

Schoenfeld, A. Clay, Meier, Robert F. and Griffin, Robert J. (1979): Constructing a social problem: The press and the environment, *Social Problems*, 27(1): 38–61.

Schultz, Friederike, Kleinnijenhuis, Jan, Oegema, Dirk, Utz, Sonja and van Atteveldt, Wouter (2012): Strategic framing in the BP crisis: A semantic network analysis of associative frames. *Public Relations Review*, 38(1): 97–107.

Soper, Kate (1995): *What Is Nature?* Oxford: Blackwell.

Tong, Jingrong (2015): *Investigative Journalism, Environmental Problems and Modernisation in China*. Basingstoke and New York: Palgrave Macmillan.

Trumbo, Craig and Kim, Se-Jin "Sage" (2015): Agenda-setting with environmental issues. In: Anders Hansen and Robert Cox (Eds). *The Routledge Handbook of Environment and Communication*. London and New York: Routledge, pp. 312–324.

Whitmarsh, Lorraine (2015): Analysing public perceptions, understanding and images of environmental change. In: Anders Hansen and Robert Cox (Eds). *The Routledge Handbook of Environment and Communication*. London and New York: Routledge pp. 339–353.

Williams, Andy (2015): Environmental news journalism, public relations and news sources. In: Anders Hansen and Robert Cox (Eds). *The Routledge Handbook of Environment and Communication*. London and New York: Routledge, pp. 197–205.

PART II
Western Europe

4

'PROTECTING OUR GREEN AND PLEASANT LAND'

UKIP, the BNP and a history of green ideology on Britain's far right

Emily Turner-Graham

Introduction

In a recent episode of the popular British detective series *Endeavour* (2018), a character closely modelled on the fascist Diana Mosley states that '[t]his house believes in an end to immigration and the repatriation of all settled immigrants to their ancestral lands' – 'This motion is not about the colour of one's skin, it is about national resources stretched to breaking point by an influx of immigrants to this green and pleasant land'. Likewise, in Max Schaefer (2011) novel *Children of the Sun*, which tells the story of a British skinhead in the 1970s and 1980s, a character modelled on former BNP leader Nick Griffin, showing prospective National Front members around his family farm, explaining that

> [y]ou're in Suffolk, which is part of East Anglia (…) named after the Angles, Anglo-Saxons who settled here from Germany around the fifth century (…). Down that hill is Ubbeston Wood, which is ancient forest, symbol of what the Front stands for. Other direction (…) is Sizewell nuclear power station. Symbol of what the Front is against (…) Breathe in. Go on. Look around you. This is what it's about. This is what we're fighting for (Schaefer 2011: 184–185).

Clearly the far right is regarded in the popular consciousness as identifying with the natural environment in defining its version of Britain. Yet, also in the popular consciousness, environmental awareness is often associated with left-wing politics. This chapter seeks to examine the far right's identification of Britain as a 'green and pleasant land' and how the populist (far) right has traditionally and currently interpreted that notion. First, two contemporary publications of the mainstream media which focus on Britain's natural environment and its connection to Britain's national identity, *Country Life* and *This England*, will be briefly examined in order to establish the essential tenets of how the countryside is now popularly understood. From there, I will consider how (neo-)fascist groups interpreted the countryside, followed by a look at the stance of two current groups on the populist (the United Kingdom Independence Party, hereafter UKIP) and extreme right (the British National Party, hereafter BNP). Although these two do not play a particularly significant role right now, they serve as exemplary actors who have voiced, in different ways, far-right views on the natural environment. As such, I attempt to make clear how the British natural environment is viewed in general and how the right has co-opted elements of those norms to increase their popularity and arrive at their current environmental positions.

Mainstream visions of a green and pleasant land

'And did those feet in ancient time', the evocative, well-known poem written in 1804 by the English Romantic poet William Blake, has long been a part of British culture, echoing from British churches, civic ceremonies and public schools, revering 'England's green and pleasant land'. Indeed, it has been referred to as Britain's alternative national anthem (Holmes 2013: 37). Its words link Britain with the divine, arguing that Britain's 'green and pleasant' land is such that it would make a suitable place for the very centre of the Christian faith (i.e. in Blake's time, the very core of humanity). In Blake's vision, Britain is a perfect rural idyll and thus the ideal realm of the divine. If nurtured as such, Britain has the potential to also be a springboard for a model people. The natural world and divinity thus connected, the 'dark Satanic mills' of modernity and urban life are wholly rejected. God is in the eternal natural environment, Satan in the new landscape of the modern world. But 'England's green and pleasant land' is not guaranteed, it must be fought for.

Today, while mainstream magazines like *Country Life* and *This England* are innocuous and in no way supporters of the far right, they nonetheless express long-held desires of middle England – 'little England' – to leave behind the apparent negative effects of urban life to claim their slice of the 'real' Britain. This is reminiscent of the historical 'intelligentsia and the middle classes' yearning for '[t]he countryside and the "natural" life, as an antidote to the materialism and competitive individualism of city existence' (Forchtner and Kølvraa 2015: 205). In a manner, they also articulate the notional wellspring from which parties like UKIP and the BNP have been able to pitch their environmental policies. In 1939, *Country Life* set down its mission:

> [i]t is our purpose to continue to produce a paper that shall reflect the substance of England that exists in moor and downland, in mansion and hamlet, in the ploughed field and the swift-running beck (…). It will continue to illustrate and describe the homes, and the sports and pastimes of country-loving folk, the way of the bird in the air, the life of the field and forest.
>
> *(quoted in Strong 1999: 126)*

It is important to note that even within the largely mild and picturesque world of *Country Life*'s countryside, a strong connection was made between an ideal landscape producing an ideal Briton. Even in 2017, writer and *Country Life* editor Mark Hedges (2017: 40) wrote in the magazine that 'our countryside (...) [is] in the British DNA' for it has long 'created feelings of patriotism and nostalgia and pulled at the heartstrings'. Indeed '[o]ur countryside is a new religion' and how the countryside was and is perceived and developed 'will define what it (...) mean[s] to call yourself a Briton' (ibid.: 41). And in the 2016 documentary, 'Land of hope and glory – British country life' (BBC 2016), it is observed in its introduction, the following narrative:

> *Narrator*: British are known for their enduring love affair with our landscape. Some would say that our obsession with the rural dream comes with a question. If we could, would we choose to live in it? (...)
>
> *Interviewee*: You smell them cutting the grass, the farmers working, you're just in a part of merry old England and it's lovely. For almost one hundred and twenty years, *Country Life* magazine has been aspiring to capture the elusive soul of the British countryside.
>
> *Interviewee*: This is the place of dreams. This is the place where people just sit back with their cup of tea and dream that they could live in a place like this.

In *Country Life*, Britain's natural world is depicted as one intrinsically intertwined with British identity. In an article by the renowned nature writer John Lewis-Stempel on walking in an ancient wood, for example, it is pronounced that '[g]athering kindling (...) aligns you with all your ancestors, ever' (Lewis-Stempel 2018). But as well as lauding Britain's environment, *Country Life* also now focusses on the realities (and difficulties) of farming as well as the impact of climate change and poor governmental environmental planning. For example, in a soon-to-be-post-Brexit Britain, there is

> [a]lready, a landscape scarred with disused petrol stations and emptying hypermarkets is being defaced with wind turbines, distribution warehouses and once-banned advertising hoardings. The Briton's 'right to roam' has been transferred to the bulldozer.
>
> (Jenkins 2018)

While we shall see similar discourse arise from UKIP and the BNP, another publication, *This England*, similarly focuses not only on Britain's landscape as a key characteristic of British national identity, but also links it to traditional values and customs of British people: Christian, conservative, pro-British and anti-European Union (EU). *This England*, though, is infinitely more jingoistic than *Country Life*. It looks back nostalgically at selected aspects of Britain's past through trenchantly rose-coloured glasses. Amidst inoffensive celebrations of Gilbert and Sullivan and Paddington Bear, poetry ties Britain's (pictured) landscape firmly with its glorious (human) history: 'Of those who still love England deep and true/and keep alive traditions that made us great/ (…) Be not afraid to capture moments filled with beauty/Our first Elizabeth and that golden age/ And Nelson's victory when men performed their duty'. The tone then turns clearly political with phrases like 'Don't let Europe rule Britannia' oft repeated (*This England* 2008).

In the 1990s, *This England* even supported the far-right New Britain party. The title, *This England*, does reference John of Gaunt Shakespeare's King Richard II: 'This royal throne of kings, this sceptred isle (...) This blessed plot, this earth, this realm, this England'. Its masthead reads 'For all those who love this green and pleasant land'. Amongst its mostly mild articles focussed on Britain's landscape and past popular culture, Union Jack and St. George flags, for example, are liberally displayed and sold in various permutations in the *This England* shop. While these flags are now mostly benign in and of themselves, within context it must also be noted that the red and white St. George flag in particular has been profusely employed as territorial markers by the English Defence League and much is made of the 'red, white and blue' of the Union Jack by the BNP.

Historical context

With a contemporary, mainstream notion of representations of the British countryside now established, our examination of far-right approaches to the environment must begin with a brief consideration of rightist environmental thought within the twentieth century. While those on the British far right during the interwar period cannot be simply dismissed as pale imitators of their European confrères, their focus did at least partially reflect the *völkisch* concerns of the German National Socialists and their ideological ancestors (for more on the German context, see Chapter 13 in this volume by Forchtner and Özvatan 2019). Heinrich Himmler (quoted in Turner-Graham 2011: 71) encapsulated best the notion of *völkisch* thought as it was interpreted by the National Socialists when he said '[c]owards are born in towns; heroes in the country', but it is important to emphasise that for both Germans and the British, *völkisch* thought covered a considerable spectrum of ideas. At one end in nineteenth and twentieth century Germany was the comparatively innocuous *Lebensreform* (Life Reform) Movement and proto-Green movements like the *Heimatschutz* (Homeland Protection); the former focussed on the importance to the *Volk* of a healthy diet and exercise in the outdoors, the latter espousing similar ideas, along with the need to preserve (and return to) vernacular art, literature and architecture. Fierce anti-Semites like Georg von Schönerer and Theodor Fritsch were at the extreme end of *völkisch* thought (Turner-Graham 2011: 73).

Taking their cue from the German Romantic writers of the early nineteenth century, the *völkisch* movement at large understood a strong link between the environment in which people lived and the resulting character of those people. Nature, therefore, was no mere backdrop to human activity. It shaped history by leaving an indelible mark on the population that lived within it. As such, the people of the countryside, peasants as they were then regarded, were seen as key ingredients in a healthy nation. Similar concerns marked the arguments of the British Romantics and the myriad of the 'neo-Romantic, anti-industrialist' thinkers who can be seen at least in some respects as their ideological successors (Dietz 2008: 809).

As such, modernity and its milieu, the city, represented the very antithesis of a healthy living space. Rampant industrialisation, individual dislocation (rootlessness) and a subsequent lack of national unity had resulted in a national disassociation from the all-important natural environment and its people. The essential nobility of the peasantry, those products of the idyllic countryside for which the *völkisch* ideologue and Third Reich Minister for Food and Agriculture Richard Walter Darré so strongly advocated, had been compromised. As with all polarised views, there was (and is) a shadow self acting as the inversion of the vital peasant. The Jew, taken up as such by not only the Germans but also branches of British fascism and neo-fascism, was regarded as the epitome of the rootless, unhealthy city dweller. Literally and symbolically, the Jews, as a 'desert people', sought out the environment and soul of the dark European forests.

With the artificial, unnatural social constructions of cities and thus civilisation, *völkisch* thinkers argued for 'Culture, Not Civilisation' (Turner-Graham 2011: 78). Oswald Spengler, the noted philosopher in whom many on the British far right were to find inspiration, argued that 'A culture has a soul (...) [whereas] civilisation is the most external and artificial state of which humanity is capable (ibid.). The virtuous peasant was honest, with an uncomplicated integrity and a love of family. Yet it was acknowledged that the measures needed to maintain the *Volk*, to support *völkisch* culture, may be cruel and brutal. That is, all means of defending one's *Volk* were acceptable; and that the natural world, the ideal world, could be a pitiless one.

Similarly, in Britain, there was a wide breadth of belief amongst the interwar 'back-to-the-landers', from 'rural revivalists such as Harold John Massingham and Rolf Gardiner to fully-fledged fascists such as Viscount Lymington and Robert Saunders' (Stone 2004: 183). Seemingly benign agricultural matters, such as 'mechanization, the industrialization of agriculture and the increasing use of chemicals', carried threats of 'racial degeneration, rather than (...) purely health and nutrition issues' (Stone 2004: 183). Under the broad umbrella of a 'search for 'Englishness' and 'English heritage', England's 'rural revival' took on both comparatively mild but also radical complexions (Dietz 2008: 802) – and it is this aspect that is still present on the right. For example, Massingham (1938: 8), the renowned writer on ruralism, asked in 1938, 'how can this tatterdemalion England patch-up the windowed raggedness of her past so that she may still wear something of that rural dress which the smart new fashions set by progress are fast tearing off her back?' Massingham (ibid.: 29) then lauded the connection between (country) man and his (natural) environment.

> Ritual, founded by the ancestral form of the village community and swayed by the periodic rhythms of nature, had entered so deeply into the subconscious mind of the peasant that it integrated his whole life. It gave dignity to his labours, depth and joy to his celebrations of them and a sense of something universal in his partnership both with nature and his fellows. How appalling to him must have been the loss by which he became a lonely and landless wage-earner.

As much as the focus of these figures was the actual landscape and the manner with which it was dealt (as continues to be the case today), there was a symbolic landscape in focus too. In the interwar period, and, as we shall see in the late twentieth and early twenty-first centuries, there was 'a movement that derived its political strategy and symbolic action from (putatively) real landscapes', '[A]n organo-fascist vision of a culturally homogeneous nation or race, dependent on the soil and deriving identity from it' (Stone 2004: 183). As in Germany and many other parts of Europe, the turn 'back to the land' in Britain was a response to the cultural despair that flooded the region in the first half of the twentieth century. From the First World War, the Bolshevik Revolution and the widespread crisis of democracy, the appeal of 'defending tradition' and returning to the apparent 'natural order of things' as symbolised through 'blood, soil, and (family) tree' (Stone 2004: 184) is clear.

But it must be emphasised once more that Britain's back-to-the-landers were no mere imitators of the *völkisch* movement, particularly as it was perceived and depicted by the Nazis. In England, the 'reactionary-modernist synthesis' was never achieved (Stone 2004: 185). Still, environmental scientist Reginald George Stapledon argued that 'unless rural England is provided with the amenities and facilities necessary rural England and rural psychology are doomed – and then the driving force behind the English character would be lost' (Stone 2004: 186). 'Degeneration' was the watchword of Britain's rural revivalists. Degeneration in the countryside meant 'the deterioration of the English race and its national values, resulting in the decline of the British Empire' (Dietz 2008: 810).

Stapledon was an associate of the Kinship in Husbandry group, founded by the likes of Gardiner, Massingham and Gerard Wallop (Viscount Lymington) and focussed on organic farming. Amongst a number of others, Jorian Jenks also became a member of the group (for an extensive biography, see Coupland 2017). Jenks was firmly connected with the British Union of Fascists (hereafter BUF) and advised them on agricultural matters. Along with Gardiner, Massingham and Lymington, Jenks went on to become a part of the Soil Association, an organisation which argued that 'out of the soil are we fashioned, and by the products of the soil is our earthly existence maintained' (Balfour quoted in Stone 2004: 187). According to Moore-Colyer (2004: 354), Jenks' 'anti-modern, neo-romantic and quasi-sacramental approach to rekindling the mythical rural roots' informed his advice to the BUF. The BUF agricultural policy focussed on ridding the countryside of pests and disease, educating the public about agriculture, preventing the interference of field sports with farming and rural recreational activities and of the general public with rural communities so that 'the natural vigour of productivity' might be maintained (ibid.: 360).

While it cannot be reasonably argued that mere association with these groups equalled fascist belief, in context, the 'quasi-mystical reverence for the soil' brought to mind blood-and-soil arguments, especially when espoused by men such as Jenks and Lymington. To quote Lymington:

> [i]t is blood and soil which rule at last; but if they fail only anarchy and slavery succeed (...) if we serve our soil we can bring back the fertility of the strong breeds that will people the Empire with desired men and women who could hold it against the tides of yellow man and brown.
>
> *(Dietz 2008: 807)*

Beyond Jenks and well beyond Massingham, Viscount Lymington took these arguments one step further. Beyond simple organic farming, he put forward that the land is in fact 'not only a reservoir for health but for leadership (...) [since it was] still the nucleus of a true aristocracy' (Stone 2004: 189).

Positive landscapes could of course be found in the health-giving countryside, peopled by the very best of Britain, firmly rooted, salt-of-the-earth countryfolk. Cities, by contrast, were dirty, health-sapping spaces, populated by the worst elements of British society – foreigners, and here too particularly Jews, who took the form of both 'rootless refugees and internationalist financier-cosmopolitans' (Stone 2004: 182). Gardiner (quoted in Dietz 2004: 806), the rural revivalist and organic farmer, said '[o]ur cities, un-English, centres of sheer barbarism, harbour our population; the countryside is empty, although any culture worth the name flourishes in our villages rather than in our towns'. Within this wide breadth of belief lay a core understanding that the British countryside and its inhabitants represented the 'true' Britain; and it is this core understanding which can still be seen in rightist political belief and even in popular culture today.

The environmental position of the contemporary far right

In 2009, the then-leader of the BNP, Nick Griffin, claimed that 'global warming is essentially a hoax (...) being exploited by the liberal elite as a means of taxing and controlling us' (Hickman 2009). In 2015, the then-leader of UKIP, Nigel Farage, posited 'If [*the Greens*] won the election, we'd all be living in caves (...) the whole thing is based on a fallacy: that our fossil fuels are going to run out and therefore we have to adapt the way we live' (King 2015). These two examples of right wing positioning on environmentalism fit into long-heard arguments from the right avowing climate change scepticism and outright denial. Yet, as has been already suggested, both parties place great emphasis upon the natural environment in their all-important definition of Britain and British identity. How does this seeming dichotomy work? To consider the question, some background on each party followed by an examination of their environmental stance is required.

The UKIP can be identified as a right-wing populist political party. At the time of writing, it has three representatives in the House of Lords and seventeen members of the European Parliament. It has five assembly members in the National Assembly for Wales and two members in the London Assembly. It began as the Anti-Federalist League, a single-issue Eurosceptic party, in 1991. In 2006, Farage officially became leader, and under his direction, UKIP widened its platform to also encompass rising immigration. This focus sought to particularly tap into the concerns of the white British working class, and it resulted in significant electoral successes from 2013 to 2016. Following a 'successful' Brexit campaign, Farage stepped down and UKIP's fortunes have subsequently dwindled. Still, its communication can still illuminate our understanding of what 'the environment' signifies for a particular ideological segment.

The UKIP promotes a unitary British identity, opposing Welsh and Scottish nationalisms, largely equating Britishness with Englishness. It has called for lower immigration, a rejected multiculturalism, and opposed what it calls the Islamification of Britain. Drawing its lineage from the right wing of the Conservative party, it is economically libertarian and socially conservative, but it draws heavily on populist discourse, referring to its supporters as the 'People's Army'. In this way, UKIP places itself within the Europe-wide milieu of populist right-wing parties.

In 2017 the UKIP spokesperson for the environment, Julia Reid, updated their 2015 policy statement to 'protect this green and pleasant land'. It clearly politicises the environmental debate, putting the relationship between the environment, the EU and Brexit at the top of its concerns. 'The idea that our membership of the EU has been only good for our environment is quite simply false. In some ways we have benefited, but in others our natural environment has suffered as a consequence of EU policy' (Reid 2017). Further, Reid (ibid.) stated that '[t]he EU's Common Agricultural Policy has damaged our countryside. The Common Fisheries Policy has devastated fish stocks around our coastline'. Reid's statement nurtures the populist right-wing image of the EU as an omnivorous tyrant, preying on individuals in lone nations. It then takes a swipe at climate change science, stating, 'UKIP will promote *evidence-based* environmental schemes [*my italics*]' (Reid 2017). The UKIP clearly continues to see England as a green and pleasant land, calling for scope for 'the human need to breathe in open, green spaces'. Their notion of (an ideal) England is a rural one, calling for a 'safeguard protection for Britain's wildlife, nature reserves, areas of outstanding natural beauty, countryside, and coastlines in a new Environmental Protection Act, prioritising policies to protect our precious countryside for future generations' (ibid.). Equally, UKIP pledges to 'protect (…) our ancient woodlands' (Reid 2017), prioritising the countryside over urban space. For example, brownfield (instead of greenfield or agricultural land) is supposed to be used for new housing and grant schemes should help to preserve natural habitats.

Looking at a slightly different source, James Delingpole wrote *The Little Green Book of Eco-fascism: The Plan to Frighten Your Kids, Drive Up Energy Costs and Hike Your Taxes!* in 2013. It was published by Biteback Publishers, a right-wing publisher whose authors include Nigel Farage. In it, Delingpole finds in climate science and arguments for climate change a plethora of social and ultimately national ills. It encapsulates well the politicised environmental stance of many on Britain's far right. Delingpole is a writer who has contributed to various mainstream British newspapers and is executive editor for the London branch of the far-right *Breitbart News*. He describes himself as a 'libertarian conservative' (Delingpole 2012) and has been described as a 'prominent voice of the right' (Gander 2015). With a phrase typical of the far right, he has described himself 'as a member of probably the most discriminated against subsection in the whole of British society – the white, middle-aged, public-school-and-Oxbridge educated middle-class male' (Harris 2007). In addition to that, and of particular relevance here, Delingpole is a climate sceptic who has regularly published articles expressing his disbelief in the existence of significant anthropogenic global warming and his opposition to wind power. And yet, he has no expertise in the area; as Painter and Gavin (2016: 440) have pointed out, climate sceptic articles have been especially marked by writers of no 'relevant expertise', often in the form of 'uncontested' opinion pieces. British public opinion on climate change has been especially exposed to this form of reporting, along with America and Australia. Delingpole (2013: 44) also refers to Rachel Carson as a 'mass murderer' while Adolf Hitler was a '[v]egetarian; anti-smoker; environmentalist; keenly aware of overpopulation issues, land use and scarce resources (aka *Lebensraum*)' (ibid.: 131). Although not espousing allegiance to Hitler, Delingpole is in fact employing the well-known stereotype of the over-zealous green 'Nazi'.

Significantly, of UKIP, Delingpole (ibid.: 292) states, that 'how come of all the vaguely mainstream parties in Britain they're the only one with an energy and climate policy which actually accords with the science, the economy and the views of the majority of the electorate?'.

Nigel Farage, the long-time leader of UKIP, has often described himself as an environmentalist (Stringer 2013), but he too firmly places that connection with the environment within a rural context. Depicting himself as a countryman, a keen fisherman and 'fanatical lepidopterist' with a wide knowledge of flora and fauna, Farage too politicises Britain's connection with the environment and the attitude towards environmental change, when he, like Reid, suggests that the true Britain is in the country, while the majority of current environmentalists have their focus on the city (Stringer 2013). Added to this, Farage has been clear about who should populate England's 'true' environment. In the manner of numerous figures on the further reaches of the right, he has repeatedly referred to England's apparent over-population and the need to dramatically slow immigration. This reached its zenith during the 2016 Brexit campaign when Farage controversially appeared with a billboard showing a long line of Syrian refugees entering Slovenia and captioned with the words 'Breaking point – the EU has failed us all'. This blatant call to a territorial ethno-nationalism can be foreseen; this position will only become more entrenched under UKIP's current leader Gerard Batten, who has pledged to take the party further to the right. Indeed, this is certain if a social media account entitled 'Make UKIP great again' (2019) (referencing Donald Trump's successful presidential campaign), which lauds the leadership of Batten, is anything to go by. In posting a clip entitled 'I am an Englishman', the view within this political milieu of 'true' English people and their rightful environment is made clear. Using old British Pathé News footage, spliced with shots of the controversial conservative post-war politician Enoch Powell and scenes of the British landscape, the text in part runs:

> Mr. Powell once spoke of the destruction of ancient Athens and the miraculous survival in the blackened ruins of that city of the sacred olive tree; the symbol of Greece, their country. And he also spoke of us, the English, at the heart of a vanished empire, seeming to find within ourselves that one of our own oak trees, the sap rising from our ancient roots, and he said perhaps, after all, we who have inhabited this island fortress for an unbroken thousand years, brought up, as he said, within the sound of English bird song under the English oak, in the English meadow, beneath the red cross of St. George, it is us who know most of England.
>
> *(Make UKIP great again 2017)*

Similarly, a photograph of a hand with a small green branch lying over the top of it is captioned 'Nationalism and environmentalism go hand in hand: it is pride in your people, pride in your nation and pride in the very soil of the land'. A photograph of green rolling hills is superimposed with the message 'Every year 15,000 acres of this' followed by a cramped housing estate and the words 'becomes this due to mass immigration' (ibid.).

As Bernhard Forchtner and Christoffer Kølvraa argue (2015: 200), British nationalist groups in the twenty-first century approach the issue of environmentalism by highlighting the local and the national in the face of the transnational issue of environmental crisis. This is something that is certainly seen in the firmly British focus of both UKIP and, to a great extent, the BNP – nationally focussed to the point of hostility towards the world at large.

Further, Forchtner and Kølvraa (2015: 201) break the right's perception of nature into three main groups – the aesthetic, the symbolic and the material. The aesthetic, how the right interprets the natural environment itself; the symbolic, how the right connects a nation's inhabitants to that environment; and the material, how that territory is best managed. As they (ibid.: 209) state, nature for Denmark's People's Party is much as we will see it for UKIP and the BNP:

'Not just ecologically or geographically Danish, but also culturally and historically, because the landscape is a canvas on which the history and traditions of the nation can be performed, traced, and revisited'.

Turning now to the BNP, is an extreme right party, neo-fascist, it has no elected representatives at any level of UK government. It was founded in 1982 by the well-known extreme right figure John Tyndall and other members of the neo-fascist National Front. Up until the 1990s, it focussed on street politics, but it was the ascension to the leadership of another long-term traveller on the extreme right, Nick Griffin in 1999, which saw the BNP's electoral base studiously broadened. In the 2000s, the party reached its peak when it had over fifty seats in local government, one seat on the London Assembly and two members of the European Parliament (Forchtner 2015).

The BNP broadened its electoral appeal by targeting feelings of disenfranchisement within the white working class. The party zeroed in on concerns about rising immigration rates and emphasising localised community campaigns, those foreign to Britain's environment were enemies and those within it required protection. As such, it became the most electorally successful extreme-right party in British history. However, infights (ultimately resulting in the ousting of Griffin) and the rise of the BNP splinter group, Britain First, and of the English Defence League saw the BNP falling back into political irrelevance.

Under Tyndall's leadership, it was more specifically regarded as neo-Nazi. The party is ethnonationalist and has called for an end to non-white migration into the United Kingdom and for non-white Britons to be stripped of citizenship and removed from the country. From its initial position of compulsory expulsion for non-whites, since 1999 it has advocated voluntary departures, persuaded by financial incentives. The BNP advocates a biological understanding of race as well as a belief in the white genocide conspiracy theory. Tyndall promoted an anti-Semitic stance which has turned to Islamophobia under Griffin, adopting the extreme right's current enemy number one. The BNP also stands for economic protectionism, Euroscepticism, and a strident social conservatism, opposing feminism and greater societal inclusion of Lesbian, Gay, Bisexual, Transgender, Intersex and Queer (LGBTIQ) community, the latter of which is a firm leitmotif of the contemporary far right as a whole.

As for the environment, 'our green and pleasant land' is represented as the English shire: green, orderly fields separated by hedgerows and including a few trees' (Forchtner and Kølvraa 2015: 208). Yet efforts to protect these environments are regarded with suspicion by these further reaches of the right. The BNP declares that [i]nvestigation of the topic of global warming appears to be less about climate than about developing international political power structures aimed at eroding the sovereignty of the nation state (Forchtner and Kølvraa 2015: 214). There is concern about possible 'dependency' on foreign resources with 'renewables hailed not for their environmental benefits but for being homegrown' (ibid.: 216) and a call for 'state sustainability' (ibid.: 218).

The BNP led their 2018 environmental policy with a connection between Britain's environment and those who inhabit it:

> The British National Party is this nation's only true Green party which has policies that will actually save the environment. Unlike the fake 'Greens' who are merely a front for the far left of the Labour regime, the BNP is the ONLY party to recognise that overpopulation – whose primary driver is immigration, as revealed by the government's own figures – is the cause of the destruction of our environment. Britain is one of the most densely populated countries in the world and our population is increasing – due entirely to immigration – which necessitates the building of ever more homes, which in turn places a strain on our infrastructure such as transport and water supplies. Independent environmental organisations believe that Britain's population needs to be significantly reduced. Our immigration policies will achieve this. [*capital letters in original text*]
>
> (BNP 2018)

Other BNP (2018) environmental policies can be seen to echo those put forward by UKIP, including a call 'to stop building on green land', but to build 'on derelict brown land' and to push 'supermarkets to supply more local and seasonal produce'. The party even argues in favour of a 'rapid switchover to organic and low fossil fuel farming' (ibid) – though yet, once again, these policies take on additional complexions within the context of the far right. Rural land ('green land') is prized above all and supermarkets are encouraged to keep their stock British.

While a particular kind of care for the environment is thus clearly present and longstanding, the BNP has seemingly moved away from a strongly climate sceptic one (as reported in Forchtner and Kølvraa 2015).

> Finally, the BNP accepts that climate change, of whatever origin, is a threat to Britain. Current evidence suggests that some of it may be man-made; even if this is not the case, then the principle of 'better safe than sorry' applies and we should try to minimise the emission of greenhouse gases and other pollutants.
>
> (BNP 2018)

So, while it is by no means a full acceptance of anthropogenic global warming, it *does* accept that measures need to be put in place to protect the green and pleasant land. But that land is viewed through a very specific lens.

In the BNP's ideology-focussed publications *Identity* and *Voice of Freedom*, a stronger, more recognisably *völkisch* message comes through. An entirely unique and white Britain is put forward as the prize which must be conserved, thus bringing the landscape and its population firmly together and ensuring that the environmental debate is seen as not only a battle for the land but for the white race. It is thus no surprise that one of the books advertised by the BNP's

Excalibur Press is entitled *The Killing of the Countryside* (VoF 2008: 11). A number of articles, for example, focus on the connection of the invasion of the natural environment and that of the 'indigenous' population.

> Hot on the heels of British jobs being taken by Polish immigrants, we now have British garden plants under threat from East European species. Our traditional Clematis plants are now being rapidly displaced by Polish varieties such as the Lech Walesa, General Sikorski and the Jan Pawell II. In addition, most trees sold as native species now actually originate from East Europe. Included in these examples are so-called English Oaks, which have mostly been grown from acorns gathered in Poland and other East European countries.
>
> *(VoF 2008: 6)*

Along these lines, migrant workers in Britain are portrayed as not worthy of the environment they find themselves in: rather than admiring London's native swans, they allegedly eat them (VoF 2008: 11) and yet, not all seems to be lost. Concerning the rural life and landscape, the BNP stated in 2002:

> Throughout the British Isles there is a small but growing band of people who are determined to defend our British way of life from those who seek to destroy it (...). We support all peaceful protests and non-violent direct action aimed at curbing the tyranny of the out-of-touch, Politically Correct, urbanite Government, and at preserving the traditional identity, freedoms and independence of rural communities.
>
> *(quoted in Richardson 2017: 168)*

However, when they attempted to attach their flag to the mast of bona fide, mainstream countryside protection groups like the Countryside Alliance and broaden their support base they had little success. The BNP encouraged their members to spread its 'patriotic, pro-countryside message to the huge contingent of radicalised Middle Britain who will flood central London' when the Countryside Alliance held their Liberty and Livelihood rally in 2002. The Countryside Alliance responded by saying that '[e]verything we stand for is the opposite of what they believe in. We repudiate any links between themselves and ourselves (...) [but] [i]t's a public demonstration and there nothing we can physically do to stop people coming along' (Parrone 2002).

Conclusion

Thus, it can be seen that, there is a long history of a very specific type of green consciousness within the spectrum of Britain's far right. Although now facing the political wilderness themselves, UKIP and the BNP provide strong contemporary examples of the manner in which this ideological path continues.

A traditional vision of what Britain's landscape must be and how that vision connects with Britain's national identity continues to be maintained by both parties. Given the more mainstream character of one (UKIP 2018) and the more extreme nature of the other (BNP), that vision is depicted with varying emphases while consistently recalling this core notion.

Both parties have downplayed the extent of the issue of climate change and have failed to move far away from the traditional far-right position on the environment. Just as there is and always has been a wide range of thought under the *völkisch* and right-wing environmental umbrella, so too can we see graded degrees of thought and interpretation in the populist radical right UKIP and the extreme right BNP. The rise of Islamophobia and the upheaval of recent international events has provided new grist to their traditional mill, bolstered already by an appeal to long held, conventional views of the British natural environment.

References

BNP (2018): Environment. https://bnp.org.uk/policies/environment/ [19 August 2018].
VoF (2008): Rural Britain faces cuts to fund African villages, *Voice of Freedom*, 94: 11.
Coupland, Philip M. (2017): *Farming, Fascism and Ecology: A Life of Jorian Jenks*. Oxon: Routledge.
Delingpole, James (2012): About James Delingpole. https://web.archive.org/web/20120529123609/http://jamesdelingpole.com/wordpress/about/ [19 August 2018].
Delingpole, James (2013): *The Little Green Book of Eco-fascism: The Plan to Frighten Your Kids, Drive Up Energy Costs and Hike Your Taxes!* London: Regnery Publishing.
Dietz, Bernhard (2008): Countryside-versus-city in European thought: German and British anti-urbanism between the wars, *The European Legacy*, 13(7): 801–814.
Forchtner, Bernhard (2015): Longing for communal purity: Countryside, (far-right) nationalism and the (im)possibility of progressive politics of nostalgia. In: Christian Karner and Bernhard Weicht (Eds). *The Commonalities of Global Crises*. Basingstoke: Palgrave, pp. 271–294.
Forchtner, Bernhard and Kølvraa, Christoffer (2015): The nature of nationalism: Populist radical right parties on countryside and climate, *Nature and Culture*, 10(2): 199–224.
Forchtner, Bernhard and Özvatan, Özgür (2019): Beyond the 'German forest': Environmental communication by the far right in Germany. In Bernhard Forchtner (Ed). *The Far Right and the Environment: Politics, Discourse and Communication*. London: Routledge.
Gander, Kashmira (2015): Lord Ashcroft's Cameron biography: Source James Delingpole defends alleged cannabis revelations, *The Independent*, 27 September. www.independent.co.uk/news/uk/politics/lord-ashcrofts-cameron-biography-source-james-delingpole-defends-alleged-cannabis-revelations-a6669566.html [19 August 2018].
Harris, John (2017): The Cameron club, *The Guardian*, 17 February. www.theguardian.com/commentisfree/2007/feb/16/takingthebullingdonbytheh [19 August 2018].
Hedges, Mark (2017): It's in our DNA, *Country Life*, 4 October.
Hickman, Leo (2009): 'Global warming is hoax': The world according to Nick Griffin, *The Guardian*, 9 June. www.theguardian.com/environment/blog/2009/jun/09/climate-change-oil [19 August 2018].
Holmes, Richard (1994): 'Lord of Unreason' a review of E.P. Thompson, Witness Against the Beast: William Blake and the Moral Law, *The New York Review of Books*, 12 May. www.nybooks.com/articles/1994/05/12/lord-of-unreason/.

Jenkins, Simon (2018): The new British countryside is being not planned, but plonked, *Country Life*, 20 June. www.countrylife.co.uk/nature/simon-jenkins-new-british-countryside-not-planned-plonked-179662 [19 August 2018].

King, Ed (2015): Nigel Farage on climate change: In his own words, *Climate Change News*, 11 March. www.climatechangenews.com/2015/03/11/nigel-farage-on-climate-change-in-his-own-words/ [19 August 2018].

Lewis-Stempel, John (2018): A walk in the woods: Tranquillity, beauty, and a 500,000-year-old connection to our ancestors, *Country Life*, 18 March. www.countrylife.co.uk/nature/walk-woods-tranquility-beauty-500000-year-old-connection-ancestors-175058 [19 August 2018].

Make UKIP Great Again (2017): I'm an Englishman. Post on 26 July 2017. www.facebook.com/MakeUKIPGreatAgain/?ref=page_internal [19 August 2018].

Make UKIP Great Again (2018): Make UKIP Great Again. www.facebook.com/MakeUKIPGreatAgain/?ref=page_internal [19 August 2018].

Massingham, Harold John (1938): Our inheritance from the past. In: Clough Williams-Ellis (Ed). *Britain and the Beast*. London: J. M. Dent & Sons, pp. 8–31.

Moore-Colyer, Richard (2004): Towards 'Mother Earth': Jorian Jenks, organicism, the right and the British Union of Fascists, *Journal of Contemporary History*, 39(3): 353–371.

Painter, James and Gavin, Neil T. (2016): Climate skepticism in British newspapers, 2007–2011, *Environmental Communication*, 10(4): 432–452.

Parrone, Jane (2002): BNP urges members to attend countryside march, *The Guardian*, 20 September. www.theguardian.com/uk/2002/sep/20/ruralaffairs.thefarright [19 August 2018].

Reid, Julia, http://juliareid.co.uk/ukip-manifesto-2017-protecting-our-environment/.

Richardson, John E. (2017): *British Fascism: A Discourse-Historical Analysis*. Stuttgart: ibidem Press.

Schaefer, Max (2011): *Children of the Sun*. London: Granta Books.

Stone, Dan (2004): The far right and the Back-to-the-Land movement. In: Julie V. Gottlieb and Thomas P. Linehan (Eds). *The Culture of Fascism: Visions of the Far Right in Britain*. London: I.B. Tauris, pp. 182–198.

Stringer, Leigh (2013): UKIP's Nigel Farage on wind farms, global warming and Charles Darwin, 4 June. www.edie.net/library/UKIPs-Nigel-Farage-on-wind-farms-global-warming-and-Charles-Darwin/6342 [19 August 2018].

Strong, Roy (1999): *Country Life 1897–1997: The English Arcadia*. London: Boxtree.

This England (2018): www.thisengland.co.uk/ [19 August 2018].

Turner-Graham, Emily (2011): '*Never forget that you are a German*': Die Brücke, 'Deutschtum' and National Socialism in interwar Australia. Frankfurt/Main: Peter Lang.

UKIP (2018): UKIP's manifesto. www.worthingukip.org/ukips-manifesto [19 August 2018].

5

FROM BLACK TO GREEN

Analysing *Le Front National*'s 'patriotic ecology'

Salomi Boukala and Eirini Tountasaki

Introduction

On 26 January 2017, Marine Le Pen, president of *Le Front National* (FN, *National Front*)[1] announced the party's environmental agenda during the first of a series of so-called 'presidential conferences' held in parallel with the FN's campaign for the presidential race. As she mentioned: 'I love the products of the French forests and seas as I love the monuments and those works that forge our national memory'.[2]

The above extract from her speech constitutes her vision for a national ecology that synthesises French natural sources and the national identity. By emphasising in her speech environmental issues, such as climate change, pollution, energy policies, resources depletion and genetically modified food, Marine Le Pen attempts to present an environmental agenda that aims not only for the protection of public health and security, but also the French nation, territory and national identity.

According to Peter Davies (1999), the FN's emphasis on the nation as a key ideological principle dates from its formation in 1972 and has been enriched by references by the party's leadership to national identity, heritage and ecology. Moreover, the FN's conception of nationhood is linked to a synthesis of blood, culture and territory built upon the idyllic icon of a rural France and a dichotomy between the FN's imagery of a pure France and cosmopolitanism, as mainly represented by French socialists (1999: 68–69). Hence, Marine Le Pen's turn to a national ecology, the French nation and its identity and heritage, is not an innovative communication strategy. Quite the contrary, it can be seen as a hybrid campaigning plan drawing on the party's ideological tenets, a green politics tradition embracing issues of European integration, environmental sustainability, global security and grassroots democracy (Bomberg 1998) and European far right parties' recent environmental communication (Douglas 2002; Lockwood 2018; see also the Introduction of this volume by Forchtner 2019a). Hence, Le Pen's presidential campaign and the party's emphasis on green politics transcend the technique of scapegoating immigrants and Islam that has been used by right-wing populist parties looking through the prism of a politics of fear (Wodak 2015) and illustrate that green imagery is not limited to green parties and movements. Following the example of European right-wing populist and extreme right parties that emphasise climate change and turn to environmental communication (Gemenis et al. 2012; Forchtner et al. 2018; Lockwood 2018), Marine Le Pen utilises her party's agenda for the environment, industry and agriculture to explicate how '*patriotic ecology*' (*écologie patriote*) could contribute to the protection not only of the environment, but also the French people, their culture, heritage and identity.

In the 2017 French presidential election, Marine Le Pen won almost 22% of the vote, a historical highpoint, and then participated in the final round where she was beaten by the current President of France Emanuel Macron. Le Pen's electoral success can be seen as a consequence of the FN's attempt to broaden its appeal within mainstream politics (Shields 2011, 2013; Almeida 2013; Stockemer and Barisione 2017). Moreover, the results of the presidential election manifest the FN's growth on the French political scene and its transition to right-wing populism. We claim that the party's environmental agenda provides evidence of its transformation, while an in-depth study of its environmental discourses unmasks its ultranationalist tradition and its links to the French far right.

Hence, in this chapter, we aim to identify and examine how the FN's '*new ecology movement*' has adopted green politics, while at the same time being linked to the party's populist and nationalist values. In the following, we first briefly review the FN's representation on the French far-right spectrum, from the Vichy regime and the *Nouvelle Droite* to Marine Le Pen's leadership. By using the slogan '*in the name of the people*' [*au nom du peuple*] and the underlying patriotic character of her party, we argue that Marine Le Pen has sought a 'de-demonisation' of the FN to win the popular vote. In addition, we claim that nativism and Euroscepticism morphed into populism and patriotism and have been linked to everyday life concerns through the FN's environmental policies.

In a second step, after summarising the Discourse Historical Approach (DHA) to Critical Discourse Studies and drawing on argumentation strategies and the Aristotelian concepts of topoi and fallacies, we illustrate the discursive resurgence of the French nation and nation-state through an analysis of the anti-Europeanisation and populist anti-elitism that dominate the French presidential candidate's rhetoric on 'patriotic ecology'. More specifically, we focus on the FN's '*new ecology movement*' website[3] and analyse Marine Le Pen's speech on '*France durable*' that was given in Paris on 26 January 2017.[4]

Thus, this chapter is an attempt to trace the discursive revival of the French nation-state as an imagined community with its national identity, and nationalism masquerading as something of interest to common people through a systematic analysis of the FN's environmental vision. We believe that the usage of the

DHA on the basis of theoretical approaches to national identity, patriotism and nationalism can reveal silent strategies that challenge the multicultural character of French society and pursue nationalist and populist discourses in times of crisis and with Euroscepticism on the rise across the whole of Europe.

Le Front National and the French far right: In defence of the nation and its natural heritage

As Goodliffe notes:

> At root, in both its historic and contemporary guise, the French radical Right must be considered the product of the central conflict underlying post-Revolutionary French political history: namely, the conflict opposing the republican defenders of the philosophical and political legacy of 1789, the champions of political and economic liberalism, to the opponents of that legacy, the proponents of antiliberalism and antirepublicanism.
>
> *(Goodliffe 2012: 31)*

In 1940, those proponents of antiliberalism and antirepublicanism triumphed over the French Resistance and the National Assembly and created the Vichy regime, led by the First World War hero Phillippe Pétain. During the next 4 years (1940–1944), Pétain administrated a satellite regime of Nazi Germany that was based on anticommunism, anti-Semitism and antimodernism, and which emphasised nationhood, 'pure Frenchness' and French heritage and territory (Jackson 2000; Shields 2007). Although the Vichy regime is not characterised as fascist (Jackson 2000; Shields 2007), it fed those 'collaborationists' who wanted to fight against communism, democracy and civil rights and the return to the bases of '*Liberté, Egalite, Fraternité*' of the Fourth Republic (Shields 2007).

After World War Two, the movement that raised the hopes of Vichy nostalgia was Poujadisme. The movement was launched in 1953 as a localised protest by small shopkeepers against the tax system. Pierre Poujade, who was a stationer and municipal councillor from the small town of Saint-Céré, created the Defence Union of Shopkeepers and Craftsmen (UDCA) to organise protests. In a climate of political instability and division and as a political movement with 400,000 members, the UDCA participated in the 1956 national election; its campaign was built on the idea of defence of the common man against a corrupt democracy and elites. Poujadism was also opposed to industrialisation, urbanisation and modernisation – which were all perceived as a threat to French identity and rural France – and introduced the idea of a national ecology that emphasised the protection of local products and natural sources as part of the national heritage (Milza 2002; Shields 2007). On 2 January 1956, the newly constituted Poujadist party (UFF) elected 52 members of Parliament. The youngest MP was the then leader of the UDCA youth branch, Jean-Marie Le Pen. Although the Poujadisme movement was founded on the basis of a micro-economy and nationalism, Le Pen's entry to the party and Parliament was accompanied by the introduction of anticommunism and xenophobia, which characterised far-right power (Shields, 2007). Poujadisme proved that the French extreme right was not capable of becoming a mass movement and it faded from view in 1958; however, we claim that it left a political stamp of populism, national ecology and ultranationalism on the French political scene, as Shields (2007: 89) also maintains:

> In their failure to fill the doctrinal void of Poujadism and convert its revolt into revolution, elements within the French extreme right discovered not a new horizon of opportunity but the confirmation of their continued marginality a decade after Vichy.

In opposition to Poujadisme, which did not succeed to transform into an international movement and was limited to the French far right spectrum, the movement that has concerned or even divided scholars of the extreme right (Griffin 2000), influenced various European far-right parties and shaped a transnational impact in the European far right (Bar-On 2011), is that of the New Right (*la Nouvelle Droite*). It is not possible to refer to the complicated ideological considerations of *la Nouvelle Droite* within the scope of this chapter. However, we consider that a brief presentation of the intellectual movement that shaped and continues to frame the ideological basis of the extreme right, and was created in opposition to and thereby incorporated the political heritage of the left, should not be absent from the following analysis of Marine Le Pen's discourse. In 1968, Alain de Benoist created the Research and Study Group for European Civilisation (GRECE). Benoist had a long experience of right-wing groups and was influenced by the Vichy regime, while at the same time adopted the tactics and ideology of Marxism, and especially Antonio Gramsci, in an attempt to weaken the leftist ideological foundations of French society. The GRECE and the *Nouvelle Droite* in general emphasised the importance of European culture, the French nation and its heritage, and opposed multiculturalism and egalitarianism (Milza 2002; Shields 2007; Bar-On 2011). Benoist asserted race diversity and hierarchy in individual and social groups. Moreover, he and his supporters reacted to the criticism of being racist by presenting immigrants' integration into French society as a synonym of ethnocultural genocide (ibid.). Under the mask of intellectualism, Benoist's thinking was disseminated to a wider audience in the 1970s, but started to wane in the 1980s, especially after the Socialists' victory in the presidential election of 1981 (Griffin 2000; Shields 2007).

Hence, *Nouvelle Droite* revised the political agenda of the extreme right in key areas by opposing a multicultural society through the introduction of homogenous ethnic-cultural communities within the framework of a federalist European 'empire'. The Western form of democracy had to be replaced by the democracy of an 'organic community'. Cosmopolitanism had be to overcome on the basis of the defence of authentic cultures and cultural difference. Finally, the *Nouvelle Droite* focused on Judeo-Christianity and in the end on left-right division (Griffin 2000). Moreover, as Bar-On (2001) also notes, the *Nouvelle Droite* was challenged the Western industrial civilisation, liberal democracy and global capitalism along with the Greens. However, *Nouvelle Droite's* ecological agenda was shaped on the basis of ethnic purity and anti-modernity. The return to the nature was important for the protection of the French psyche and cultural heritage against globalisation and multiculturalism (Milza 2002). The FN's electoral success in 1984 might have been the end of the public attention paid to the *Nouvelle Droite*, but it also illustrates the transition of an intellectual movement into a political party and the *Nouvelle Droite's* political impact.

The FN was founded in 1972 to unite the diverse far right and was located on the margins of extreme-right ideological radicalism (Shields 2007, 2011; Reynie 2016). In 1984, in the European Parliament election, the FN won 11% of the vote. The FN's electoral success marked the resonance of a radical, xenophobic and anticommunist political discourse with French voters, and continued through Jean-Marie Le Pen's successes in the 1988 and 1995 presidential elections. Through the prism of the *Nouvelle Droite's*, the FN's, political tradition, focused on French heritage and racial difference, anti-immigration and anti-globalisation, and together with Le Pen's discourse on the superiority of the French nation the French nation and its sovereignty, led in the 2002 election to Le Pen having a run-off against Jacques Chirac. Le Pen also integrated Pujadisme's populism into his discourse by putting the 'people' first in his political campaigns; however, as Reynie (2016: 51) explicates, 'under Jean-Marie Le Pen the FN was a protest party. It could make noise, but it had no real plan for gaining power'.

The FN has experienced a political revival under Marine Le Pen's leadership (2011) by following the path of heritage populism (Reynie 2016). Marine Le Pen attempted to lead the party from far-right to mainstream politics (Balent 2012) by adopting a populist discourse that obscures the radical-right component (Stockemer and Barisione 2017). She has focused on the people and their needs, French nationality, heritage and identity, while Euroscepticism has replaced her father's antiglobalisation (Shields 2011, 2013; Almeida 2013; Stockemer and Barisione 2017). Following the FN's green tradition that calls for a return to the local and opposes industrialisation, globalisation and modernity and drawing upon the *Nouvelle Droite's* anti-modernity and its emphasis to the protection of rural France and the French heritage (Milza 2002), Marine Le Pen has attempted to broaden her appeal, and remove the 'curse' of racism, by proceeding to a *dédiabolisation* (Almeida 2013) of the FN via a patriotic agenda that underlines issues of ecology, national identity, territory and heritage. She opposes the EU and multiculturalism and asks for a return to a 'Europe of nations'. Ecology plays an important role in her discourse and political agenda, insofar as she develops her Euroscepticism and nationalism through the FN's environmental values that highlights the natural environment, the purity of the French landscape and the uniqueness of the French culture and identity. She suggests economic and environmental protectionism in the name of the French people and via a new ecology movement, a faction inside the party that deals with nature and society, and illustrates FN's interest to environmental communication. Hence, via the triangulation of nation-heritage-ecology, Le Pen shapes a holistic green profile that is relevant to the FN's ideological tenets, which we attempt to analyse by focusing on discourse and via the usage of the DHA.

Methodological framework

The DHA links discursive practices, social variables, institutional frames, and sociopolitical and historical contexts and aims to explore how discourses, genres and texts change in relation to sociopolitical change. As Wodak further explains:

> The DHA attempts to integrate a large quantity of available knowledge about the historical sources and the background of the social and political fields in which discursive 'events' are embedded. Further, it analyses the historical dimension of discursive actions by exploring the ways in which particular genres of discourse are subject to diachronic change. Lastly, and most importantly, this is not only viewed as information. At this point we integrate social theories to be able to explain the so-called context.
> (Wodak 2001: 65)

We assume the DHA as/being an adequate theory and methodology to analyse, understand and explain the intricate complexities of national identity and the FN's green agenda. Our aim is to illustrate the tradition of 'eco-naturalism' (Olsen 1999) that signifies an identity between nature and society that presents national culture as an ecosystem.

In our analysis, we draw upon its discursive strategies (Reisigl and Wodak 2001) and especially argumentation strategies and the concepts of topoi and fallacies. The DHA's argumentation schemata, illustrate silent, prejudiced and racist discourses and contribute to the in-depth analysis of the fallacious arguments usually employed by politicians in positive self and negative 'other' presentations (Reisigl and Wodak 2001, 2009). In the analysis below we introduce the usage of Aristotelian topoi and false reasoning and intend to elucidate the links between the DHA and the Aristotelian tradition in the establishment of a discursive opposition between 'us' and 'them' by examining topoi and fallacies in Le Pen's rhetoric, as we believe these concepts are salient for the comprehension of prejudiced and racist discourses.

According to Aristotle, topoi are 'search formulas' that examine *endoxon*, or common knowledge, and comprise fallacious reasoning (topoi of fallacious enthymemes) (1992, 100a: 25–27). Hence, Aristotelian dialectic topoi examine *endoxa* and select accepted opinions that can develop dialectic arguments and lead to the solution of dialectical problems and the pursuit of a 'truth' that can always be challenged, while topoi in Aristotle's *Rhetoric* (2004) are means of persuasion. In *Rhetoric* B23, Aristotle categorises the topoi that apply to all subjects in common. These topoi are, as they are all devices for arriving at a certain conclusion about a case. While they do not all have universal applicability, they can be applied to every rhetorical case (Rubinelli 2009: 84). Aristotle (2004) provides a holistic classification of *topoi* that can be used by interlocutors to persuade an audience, though they might be named differently, in relation to their arguments. For this reason, *topoi* can be useful in a systematic analysis of various discourses. Aristotle also distinguishes between 'topoi of probative/ real enthymemes' and 'topoi of fallacious enthymemes'; and as he explains, via a number of examples, topoi are usually expressed by the proposition: 'if one … then the other' (Rubinelli 2009). Thus, *topoi* are central to the analysis of seemingly convincing fallacious arguments, which are widely adopted in prejudiced and discriminatory discourses (Reisigl and Wodak 2009; Wodak 2015; Boukala 2016), and can be useful in the systematic analysis of right-wing populist discourses, such as Le Pen's environmental agenda.

In the first lines of *Sophistical Refutations* (1994), Aristotle provides an implicit definition of a fallacy by explaining that, in his work, he intends to discuss 'arguments that appear to be logical refutations, but in fact they are not; they are fallacies' (1994, 164a: 19–21). Various scholars and disciplines, such as Pragma-dialectics, Argumentation and the DHA, have focused on fallacies as means that serve the justification of discrimination and illustrate the diachronic character of Aristotle's thought that could contribute to the understanding of contemporary political discourse. Van Eemeren and Grootendorst (1987) developed a set of norms for the identification of fallacies and the development of critical discussion. In this way, an argumentative tactic that violates any of the rules of defence of a standpoint[5] can be considered a fallacy. Moreover, the discourse-historical approach shares with Pragma-dialectics an interest in analysing critical discourse and describing strategies of argumentation. As Reisigl and Wodak (2009) claim, the line between reasonable or fallacious argumentation cannot be drawn clearly in any case. Following the DHA's and Pragma-dialectics' emphasis on false reasoning, we argue that false reasoning cannot be ignored in our analysis. Furthermore, we assume that a systematic analysis of fallacious arguments on discrimination that dominate right-wing populist and far-right discourses requires an in-depth study of Aristotelian fallacies and topoi, which we intend to utilise in the next section in order to analyse Marine Le Pen's 'patriotic ecology'.

Towards a durable France – Marine Le Pen's green agenda

The aims of the FN's '*nouvelle écologie*' become visible on a first reading of the movement's website https://rassemblementnational.fr/communiques/communique-de-presse-du-collectif-nouvelle-ecologie/. The majority of '*collectif pour une écologie patriote-nouvelle écologie*' comments refer to issues such as nuclear power, pollution, nutrition, health and European policies, and all of them are developed on the basis of a criticism of the party's political opponents and the constitution of an environmental view that accuses the Anthropocene, the French state and the EU of impairing the quality of life of the French people. However, scientific opinions are absent from the FN's rhetoric on environmental issues and most of the comments are built on contradictions. The most characteristic examples of the FN's contradictory arguments are those comments in which the FN criticises the Minister for Ecological and Inclusive Transition, and environmental activist, Nicolas Hulot, for his decision to reduce the use of oil and gas in France (6 September 2017) and the then President Nikola Sarkozy for the privatisation of nuclear power stations (30 April 2016), respectively https://rassemblementnational.fr/author/collectif-nouvelle-ecologie/. The FN explicates that although it agrees that the use of fossil fuels can be noxious and nuclear power stations are not safe, the party's leadership comprehends that both of them are important for France's autarky and independence in energy. Hence, sovereignty and autonomy are central in the Front National's environmental discourse and linked to the nature-nationalism tradition of the far right (Olsen 1999: 78).

According to Mudde (2007, 2017) the populist radical right shares an ideology that combines three features: nativism, authoritarianism and populism. We claim that the FN affirms Mudde's view through the party's introduction of the '*nouvelle écologie*' and that the above three features can be illustrated via the fallacy of *argumentum ad ignorantiam* (Reisigl and Wodak 2001: 72), an appeal to ignorance, which means that from a failed defence that does not combine scientific elements, but is justified through the party's interest in the French people, the FN's spin doctors attempt to prove their 'knowledge' on the issue and to challenge the party's political opponents' decisions. Nativism, authoritarianism and populism also dominate Marine Le Pen's presidential campaign and her environmental agenda. Following Eatwell's (2017) approach of Jean-Marie Le Pen's 'charisma', we argue that the communication, rhetoric skills and ability to be presented as an effective leader have been demised from father to daughter and contributed to the constitution of the above three features that characterise the FN's ideological basis.

In the first of the 'presidential conference' speeches, Marine Le Pen described in detail her vision for a green France and a patriotic ecology. As she maintained at the beginning of her speech:

> I love France ... France is not an idea. In my view, France is a living reality of men and women, lands and seas, trees and birds, rivers and forests, flavours and words. I love everything about France ... because France is one of the lands (territories) on the planet most favourable, benevolent and friendly to human life.
>
> *[J'aime la France ... La France n'est pas une idée. Pour moi la France est une réalité vivante d'hommes et de femmes, de terres et de mers, d'arbres et d'oiseaux de fleuves et de forêts, de saveurs et des mots. J'aime tout de la France ... parce que la France est un des territoires sur le planète le plus favorable, le plus bienveillant, le plus amicaux à la vie humaine].*[6]

Through an appeal to emotions, Marine Le Pen explicates that France is not a vague idea; it has a territorial substance that include humans, flora and fauna. Moreover, the FN's leader notes the country's natural superiority by ascertaining that 'France is one of the lands on the planet most favourable, benevolent and friendly to human life', and presupposes a distinction between a superior France and the 'Others'. In this way, she reintroduces the synthesis between nature and nationalism that has appeared in the far right green agenda (Olsen 1999; Forchtner et al. 2018; Forchtner 2019b). Thereafter, Marine Le Pen refers to those factors that pollute France and threaten the French people's quality of life. The following extract summarises Le Pen's concerns vis-à-vis the climate and the French nation.

> The threats to our health, our life expectancy, our heritage, due to climate change, air, water and soil pollution and the great extinction of species, have to be faced ... all policies regarding independence and national security will be policies that favour a durable France ... the President has to lead this policy. I will lead it.
>
> [Face aux menaces sur notre santé, sur notre espérance de vie, sur notre patrimoine qui font poser le dérèglement du climat, la pollution de l'air, de l'eau et de sol et la grande extinction des espèces ... toute politique d' indépendance et de sécurité nationale sera une politique en faveur d' une France durable ... c'est au Président qu'incombe de conduite cette politique. Je la conduirai].

Here, Marine Le Pen does not refer to threats specifically, but rather to vague threats that stem from the consequences of pollution and the overexploitation of natural resources, described in her speech, and which are implicitly linked to international enterprises, globalisation and European integration. Moreover, she does not omit referring to the climate change that is a central issue of the far-right parties' environmental agenda (Forchtner et al. 2018; Lockwood 2018). She does not explicate the concept of the climate change further, although she underlines it as a threat to the French nation and nature. Hence, she builds an invisible and vague enemy via the Aristotelian *topos of the consequential* or the DHA *topos of threat* (Boukala 2016), which here relies on the conditional 'if there are specific threats to our heritage, health and environment, then the French authorities must do something about them', threats that have to be faced by a capable leader. She also presents herself as a future president through an appeal to authority linked to the *fallacy of authority* or *argumentum ad verecundiam* (Reisigl and Wodak 2001: 72), which explicates the false presentation of oneself as authority. Furthermore, she emphasises her leadership profile by employing the Aristotelian *topos of analogue consequence* (B23, 1399b) or the *topos of responsibility* (Reisigl and Wodak 2001: 78), which here is developed through the conditional 'if Marine Le Pen as FN's leader and potential president of France is responsible for the French then she has to find solutions to French environmental problems'.

Another important part of the above extract is Le Pen's usage of the first-person pronoun '*Je*' (I), in contrast to utilisation of the plural form '*Nous*' (We) that is mainly used in populist rhetoric (Wodak 2015). We assume that this differentiation is linked to FN far-right roots that impose the leader's authority (Shields 2007), Le Pen's 'charisma', and her attempt to present herself as a capable leader during the presidential campaign. Moreover, Marine Le Pen synthesises the environment and French heritage and underlines the necessity of national security and independence. In this vein, she illustrates the nationalist character of FN, insofar as her strategy is not innovative, quite the contrary, it is also followed by many European far-right parties that have developed their environmental agenda based on the combination of the nation and its natural environment (Olsen 1999; Douglas 2002; Lockwood 2018; Forchtner 2019b).

Thereafter, the FN leader depicts her vision of a durable France and the French economy in a new age by referring to how the economy should not be.

> This new age redefines the mission of the government and the state ... an economy that focuses only on numbers ... a growth which measures ignore the destruction of resources and overvalue, that serves the capital cannot be considered a political action that can guarantee to the French the essentials, health and security.
>
> *[Ce nouvel âgé redéfinie la mission du gouvernement et de l'Etat ... une économie qui ne connait que les chiffres ... une croissance dans la mesure qu'ignore la destruction des ressources et survalorise, qui services d'argent ne peut être juge d'une action politique qui doit d'abord garantir aux Français l'essentiel, la santé et la securité].*

In the above extract Marine Le Pen implicitly criticises French governments and the EU that built upon an economy that focuses on numbers and ignores the French people, its health and security via the Aristotelian *topos of induction*, which here can be labelled the *topos of the FN's vision of France* and paraphrased through the conditional: 'if some policies do not emphasise the French people, then they cannot ensure the French people's health and security'. The French people and the French nation dominate Le Pen's rhetoric and illustrate the populist and nationalist features of the party's character that question France's Europeanisation through a silent critique of environmental and financial policies that have been applied in France. Hence, following the European far-right green tradition (Olsen 1999; Douglas 2002; Forchtner 2019b) and by implicitly criticising the EU, European policies and globalisation, Marine Le Pen's environmental rhetoric leads to a discursive reconstruction of the French nation-state.

Cornelius Castoriadis (1981) claims that the imaginary institution of society is an invented representation of reality which is cultivated to assuage social consciences and becomes acceptable via symbols, which are considered the only truth. Consequently, the nation has an imagined character because it is based on symbols, such as a national flag, anthem and emblem, which its members accept as common symbols, and an invented common history, which members of the nation accept as the absolute truth (1981: 240–244). The imaginary institution of the nation was also studied by Benedict Anderson (1983 [2006]), who considered the nation to be 'an imagined political community; and imagined as both inherently limited and sovereign' (2006: 6). Thus, the nation unifies its members and therefore excludes the 'Others', i.e. non-members. It creates a sense of belonging and solidarity for its members and for this reason can be considered an 'imagined community'. Marine Le Pen stresses a discursive dichotomy between 'Us', the FN and defenders of the French nation, and 'Them', the politicians who do not care about France via an environmental agenda, which highlights the threats to the French quality of life, its territory and people, and also draws upon national symbols and French natural resources. In particular, she repeats her plans for 'blue development' and highlights the importance of France's 'marine heritage'. The emphasis on the sea and the colour blue is implicitly linked to the blue of the national flag that represents the notion of liberty. Hence, Marine Le Pen reveals the party's Euroscepticism and provides the idea of a powerful and free French nation-state that unifies those who are integrated into the imagined community of the French nation based on its natural resources and heritage. Two other concepts that dominate Le Pen's speech and *nouvelle écologie* rhetoric in general, and are relevant to the nation-state's foundations, are those of national identity and sovereignty. These two concepts dominate the European far-right environmental agenda and are connected to nationalism and the importance of the natural environment in the creation of national imaginaries (Olsen 1999; Douglas 2002).

National identity, like the nation and nation-state, is a complex construction that is based on a number of political, economic and territorial components. There are many different approaches to national identity, and many different definitions of it. However, the majority of scholars agree that national identity is composed of a number of features (Gellner 1983). As Anthony Smith (1991) claims, the fundamental features of national identity are:

1 An historic territory, or homeland
2 Common myths and historical memories
3 A common, mass public culture
4 Common legal rights and duties for all members
5 A common economy features which are referred to Marine Le Pen's environmental discourse (1991: 14)

As Marine Le Pen maintains:

> The fight for national sovereignty is a fight for the diversity of people, agriculture and civilisation. The fight for French identity is a fight to keep our gardens, mountains, companions, flowers, birds, butterflies … it is a fight for a fair and sustainable economy. The fight for the borders is a fight that respects each nation and does not reduce men, companies and territory to

the level of market prices. It is the fight against a European administration that has never intended to protect European populations.

[Le combat pour la souveraineté nationale, c'est le combat pour la diversité des peuples, des agricultures et des civilisations. Le combat pour l'identité de la France c'est le combat pour demeurent dans nos jardins, les montagnes, les compagnes, ces fleurs, ces oiseaux, ces papillons ... c'est le combat pour une économie équitable et durable. Le combat pour les frontières, c'est le combat qui respecte chaque nation et ne reduit pas les hommes, les enterprises et les territoires a des prix du marché. C'est le combat contre une administration européenne qui n'a jamais voulu protéger les populations européennes].

Here, the FN's leader explicates her vision for ecology by utilising the conceptual metaphor of a fight. According to Le Pen, France has to fight for its national sovereignty, identity, flora and fauna. France has to fight against an enemy that is openly referred here through the concept of 'European administration'. The EU is also represented as being indifferent to European populations and based on an economy that intensifies the markets against people's interests. This argument is further supported by the Aristotelian *topos of induction* or the *topos of the FN's vision of France* (see above), which here extends beyond the limits of the economy and includes national sovereignty, territory, culture and identity. In addition, the French people are represented as being deceived by the EU, an appeal to compassion that is connected to the *argumentum ad misericordiam* fallacy (Reisigl and Wodak 2001: 72), which replaces relevant arguments by emphasising empathy and compassion and leads to the victimisation of the French people in a vein of populism and nativism.

FN's *'nouvelle écologie'*, also refers to the duty of the French people and its leaders to protect French territory, heritage, culture and local products. In her speech, Marine Le Pen explicates that:

> To defend the national identity of France means to affirm our preference for a culture rich in history and national legends and for the symbols which make the community of French people ... it is to defend the biological quality of our products and it is to proceed towards this new mode of production and distribution of energy and human and rural mobility which will give a new meaning to local life and responsibility to territorial actors.

> *[Défendre l'identité nationale de la France, c'est affirmer notre préférence pour une culture riche d'histoire de légendes nationales et de ses symboles qui font la communauté des Français ... c'est défendre la qualité biologique de nos produits et c'est avancer versce nouveau mode de production et de distribution d'énergie de mobilité humaine et rurale qui donneront un nouveau sens à la vie locale et à la responsabilité des acteurs territoriaux].*

The FN's ecology, thus, synthesises national identity and protection of the French environment and production, and highlights localisation and the French nation in contrast to the EU and international markets. The fight against globalisation and a return to the nation, as well as localisation against modernity, are traditional ideological axioms of the FN (Davies 1999; Olsen 1999; Shields 2007), which this time build upon the party's new environmental agenda and signify its rhetorical transmission to less radical, although nationalist, tenets (Almeida 2013; Mayer 2013; Shields 2013).

The national identity that dominates FN's rhetoric and shapes Marine Le Pen's ecological vision also divides people. It is based on strategies of inclusion and exclusion and separates the members of a nation from 'Others' (Wodak et al. 2009), especially supranational organisations such as the EU and the French authorities that imposed European policies. In other words, via its 'patriotic ecology' linked to national identity and protection of the French nation, the FN seeks a discursive resurgence of the nation-state against the EU, which is presented together with French governments as 'Others' that are excluded from the imagined community of the culturally and naturally superior French nation and territory. Thus, the FN's antiglobalisation and populism (Bar-On 2001) has been recently transformed into an anti-Europeanism that, together with the party's emphasis on an environmental communication that formed through the prism of nature-nationalism tradition (Olsen 1999; Shields 2007), led to the propagation of ultranationalism in the name of a 'patriotic ecology'.

Conclusion

On the basis of the far-right environmental communication (Olsen 1999; Forchtner 2019b) natural environment, national identity and heritage were all employed in Marine Le Pen's 2017 official campaign; images of French seashores and Le Pen sailing provided a sense of liberty and continuation that are linked to her agenda for a secure and prosperous France. The *'nouvelle écologie'* movement of the FN constitutes a de-demonisation of the FN, insofar as the party's rhetoric turns to environmental issues and notions of national heritage and national identity. A turn that become a cliche for FN when Marine Le Pen presented her vision of a 'Europe of nations and people' and the 'world's first ecological civilisation' based on localism during the party's 2019 European Parliament election campaign'.[7] However, the discursive (re)construction of national identity, especially in an age of multiple European crises, cannot be politically neutral. Wodak et al. (2009: 33–42) distinguish the different strategies that may be employed in the discursive formation of national identity. The form that can be discerned in the above examples of Le Pen's environmental agenda is one of *constructive strategies* that focus on unification, identification and solidarity, as well as differentiation, which is used by the FN leader when she refers to the EU.

The discourse-historical approach and the argumentation schemata of fallacies and topoi that we utilised in our analysis reveal that Mudde's (2007) three features of populist radical-right parties – nativism, populism and authoritarianism – coexist with Euroscepticism and discrimination between French patriots and 'Others' in the FN's '*nouvelle écologie*'. It draws upon a green agenda that emphasises not only pollution and health issues, but also climate change and the concepts of national sovereignty, identity and heritage. Hence, Marine Le Pen's and the FN's '*nouvelle écologie*' movement's references to the superiority of French territory and heritage, and the protection of local products and French culture, heritage and identity illustrate that the FN's environmental agenda is based on the party's ideological patterns that oppose globalisation, the nature-nationalism synthesis (Olsen 1999) and the far right concerns over the natural environment, the purity of the nation and the national heritage (Forchtner 2019b).

Notes

1. On 1 June 2018, Marine Le Pen announced that the FN changed its name to National Rally (*Rassemblement National*). As we investigate a period in which the party was still called FN, we keep using its old name.
2. www.youtube.com/watch?v=cRe74tIsviQ&t=2s.
3. https://rassemblementnational.fr/communiques/communique-de-presse-du-collectif-nouvelle-ecologie/
4. www.youtube.com/watch?v=cRe74tIsviQ [14 August 2018].
5. See Van Eemeren and Grootendorst (1987), Van Eemeren et al. (2009) for a detailed presentation of the ten rules.
6. All the extracts that we analyse here are included in Marine Le Pen's speech on the party's environmental agenda that took place in Paris, 26 January 2017 and opened her presidential campaign. See: www.youtube.com/watch?v=cRe74tIsviQ&t=2s.
7. See https://rassemblementnational.fr/videos/pour-une-europe-des-nations-et-des-peuples-allocution-de-marine-le-pen/.

References

Almeida, Dimitri (2013): Towards a post-radical Front National? Patterns of ideological change and dediabolisation on the French radical right, *Nottingham French Studies*, 52(2): 167–176.
Anderson, Benedict (2006 [1983]): *Imagined Communities*. London: Verso.
Aristotle (1992): *Topics*. Athens: Kaktos.
Aristotle (1994): *Sophistical Refutations*. Athens: Kaktos.
Aristotle (2004): *Rhetoric*. Thessaloniki: Zitros.
Balent, Magali (2012): *Le Monde Selon Marine: La Politique International du Front National*. Paris: Armand Colin.
Bar-On, Tamir (2001): The ambiguities of the Nouvelle Droite 1968–1999, *European Legacy*, 6(3): 333–351.
Bar-On, Tamir (2011): Transnationalism and the French Nouvelle Droite, *Patterns of Prejudice*, 45(3): 199–223.
Bomberg, Elizabeth (1998): *Green Parties and Politics in the European Union*. London: Routledge.
Boukala, Salomi (2016): Rethinking *topos* in the discourse historical approach: Endoxon seeking and argumentation in Greek media discourses on 'Islamist terrorism', *Discourse Studies*, 18(3): 249–268.

Castoriadis, Cornelius (1981): *The Imaginary Institution of Society*. London: Polity Press.
Davies, Peter (1999): *The National Front in France: Ideology, Discourse and Power*. London: Routledge.
Douglas, Mary (2002): *Purity and Danger*. London: Routledge.
Eatwell, Roger (2017): The rebirth of right-wing charisma? The cases of Jean Marie Le Pen and Vladimir Zhirinovsky. In: Cas Mudde (Ed). *The Populist Radical Right: A Reader*. London: Routledge, pp. 223–237.
Forchtner, Bernhard (2019a): Far-right articulations of the natural environment: An introduction. In: Bernhard Forchtner (Ed). *The Far Right and the Environment: Politics, Discourse and Communication*. London: Routledge.
Forchtner, Bernhard (2019b): Nation, nature, purity: Extreme-right biodiversity in Germany, *Patterns of Prejudice*, 53:3, 285–301.
Forchtner, Bernhard, Kroneder, Andreas and Wetzel, David (2018): Being skeptical? Exploring far-right climate change communication in Germany, *Environmental Communication*, 12(5): 589–604.
Gellner, Ernest (1983): *Nations and Nationalism*. London: Blackwell.
Gemenis, Kostas, Katsanidou, Alexia, and Vasilopoulou, Sofia (2012, April 12–15): The politics of anti-environmentalism: Positional issue framing by the European radical right. *Paper prepared for the MPSA Annual Conference*, Chicago.
Goodliffe, Gabriel (2012): *The Resurgence of the Radical Right in France: From Boulangisme to the Front National*. Cambridge: CUP.
Griffin, Roger (2000): Plus ça change! The fascist pedigree of the Nouvelle Droite. In: Edward J. Arnold (Ed). *The Developments of the Radical Right in France*. London: Macmillan, pp. 217–252.
Jackson, Julian (2000): Vichy and fascism. In: Edward J. Arnold (Ed). *The Developments of the Radical Right in France*. London: Macmillan, pp. 153–171.
Lockwood, Matthew (2018): Right-wing populism and the climate change agenda: Exploring the linkages, *Environmental Politics*, 27(4): 712–732.
Mayer, Nonna (2013): From Jean-Marie to Marine Le Pen: Electoral change on the far right, *Parliamentary Affairs*, 66(1): 160–178.
Milza, Pierre (2002): *L'Europe en chemise noire: Les extremes droites européennes de 1945 à aujourd'hui*. Paris: Anthème Fayard.
Mudde, Cas (2007): *Populist Radical Right Parties in Europe*. Cambridge: CUP.
Mudde, Cas (2017): Introduction to the populist radical right. In: Cas Mudde (Ed). *The Populist Radical Right: A Reader*. London: Routledge, pp. 1–10.
Olsen, Jonathan (1999): *Nature and Nationalism: Right-Wing Ecology and the Politics of Identity in Contemporary Germany*. London: Palgrave.
Reisigl, Martin and Wodak, Ruth (2001): *Discourse and Discrimination: Rhetorics of Racism and Antisemitism*. London: Routledge.
Reisigl, Martin and Wodak, Ruth (2009): The Discourse Historical Approach. In: Ruth Wodak and Michael Meyer (Eds). *Methods of Critical Discourse Analysis*. London: Sage, pp. 87–121.
Reynie, Dominique (2016): 'Heritage Populism' and France's National Front, *Journal of Democracy*, 27(4): 47–57.
Rubinelli, Sara (2009): *Ars Topica: The Classical Technique of Constructing Arguments from Aristotle to Cicero*. Berlin: Springer.
Shields, James (2007): *The Extreme Right in France: From Petain to Le Pen*. London: Routledge.
Shields, James (2011): Radical or not so radical? Tactical variation in core policy formation by the Front National, *French Politics, Culture & Society*, 29(3): 78–100.

Shields, James (2013): Marine Le Pen and the 'new' FN: A change of style or substance? *Parliamentary Affairs*, 66(1): 179–196.

Smith, Anthony (1991): *National Identity*. Nevada: NUP.

Stockemer, Daniel and Barisione, Mauro (2017): The 'new' discourse of the Front National under Marine Le Pen: A slight change with a big impact, *European Journal of Communications*, 23(2): 100–115.

Van Eemeren, Frans and Grootendorst, Rob (1987): Fallacies in Pragma-dialectical Perspective, *Argumentation*, 1: 283–301.

Van Eemeren, Frans, Gerssen, Bart and Meuffels, Bert (2009): *Fallacies and Judgements of Reasonableness: Empirical Research Concerning the Pragma-Dialectical Discussion Rules*. Berlin: Springer.

Wodak, Ruth (2001): The discourse historical approach. In: Ruth Wodak and Michael Meyer (Eds). *Methods of Critical Discourse Analysis*. London: Sage, pp. 63–94.

Wodak, Ruth (2015): *The Politics of Fear: What Right-Wing Populist Discourses Mean*. London: Sage.

Wodak, Ruth, de Cillia, Rudolf, Reisigl, Martin *and Liebhart, Karin (2009): The Discursive Construction of National Identity*. Edinburgh: EUP.

6
ENVIRONMENTAL POLITICS ON THE ITALIAN FAR RIGHT
Not a party issue?

Giorgia Bulli

Introduction

Environmental issues were never a major concern for the most important Italian parties during the so-called First Republic era (from the end of World War Two to the late 1980s). The growth of an environmental awareness took place later in Italy than in the United States and other European countries (Menichini 1983). However, such awareness was never able to give rise to a successful and long-lasting green party. The failure of the institutionalization of environmentalist parties in Italy has been explained as a consequence of the particular characteristics of the Italian party system after the end of World War Two and the impossibility to make ecological issues a new cleavage (Rokkan 1970; Diani 1988: 200–201). In contrast to, for example, Germany (Müller-Rommel 1985), where Green parties entered the Bundestag in the early 1980s, environmental awareness expressed itself in Italy mainly through activism by environmental associations established in the 1960s and 1970s, and in their lobbying from the 1980s onwards (Lodi 1988).

Empirical research (Biorcio and Lodi 1988) conducted among environmental political activists evidences that their involvement started in the mid-1970s following the oil crisis of the early 1970s (Lodi 1988: 18). The peak of environmentalist mobilization was reached in the movement against the use of nuclear energy (Diani 1988: 81). In 1987, one year after the Chernobyl nuclear disaster, three issues regarding the production of nuclear power were submitted to referendum. Nearly 80% of the Italian citizens who participated (the turnout was around 65%) voted against the use of nuclear energy, which was eventually abandoned.

Despite numerous attempts to convey this success into the establishment of a unitary environmentalist party, the main attitude of environmentalist associations in Italy was clearly oriented towards cooperation with institutions, not with parties (Diani 1988: 167), an aspect mirrored by the resistance of the mainstream parties to opening up an internal political space for the representation of environmental issues, which were considered not sufficiently profitable in electoral terms.

The underrepresentation of environmental issues in mainstream political terms persisted even after the collapse of the so-called First Republic. Starting from the 1990s, only one party contained explicit reference to environmentalist issues in its name and symbol. This was *Sinistra Ecologia Libertà* (Left, Ecology Freedom), a left-wing party founded in 2009 as a merger among various small left-wing parties.

Indeed, environmentalism is not considered a profitable electoral issue by most of the Italian political parties. However, this does not seem to be true for some of the principal political movements of the far right. My aim in this chapter is to describe the positions of important far-right political actors on environmentalist issues and, more specifically, to evidence the role played by ideological and organizational factors in the coverage by far-right movements and parties of environmental issues. In so doing, I will distinguish between those organizations that are still attached to a classic extreme-right ideology and those parties and/or movements (Caiani et al. 2012) that have embraced a post-industrial (Ignazi 2000) interpretation of extreme-right ideology. I will also make reference to the right-wing populist positions of the *Lega Nord* (LN – Northern League),[1] showing that the extreme right interpretation of the environment reflects a different political culture if compared to the right-wing populist agenda of environment protection.

Following this introduction, I will first focus on the evolution of the concept of 'environment' and on the role of the Italian representatives of the *Nouvelle Droite* in this field. These thinkers were particularly active within the youth section of the post-fascist party *Movimento Sociale Italiano* (MSI), promoting the renovation of the ideological repertoire in post-fascist Italy.[2] Second, the article will describe the position of one of the most successful extreme-right movements in Italy, CasaPound Italia (CPI) (Di Nunzio and Toscano 2011; Albanese et al. 2014; Rosati, 2018), in relation to environmental issues.

The environment and post-fascism

Italian Fascism cannot be compared to the National Socialist ideology in terms of the latter's features of a 'religion of nature' consisting of 'primeval Teutonic nature mysticism, pseudo-scientific ecology, irrationalist anti-humanism and a mythology of racial salvation through a return to the land' (Staundenmaier 2011: 26). As von Hardenberg (2014: 275) states, nature conservation was 'not a central issue in the activities of Mussolini's regime'. The relationship between the Italian fascist regime and nature must more correctly be ascribed to the notion of 'regeneration' of the lost unity between people and nature (Armiero 2014: 241).

For the Fascist regime, regeneration implied an active relationship between humankind and nature. The reclamation policy of the Fascist regime is cited by Ben-Ghiat (2011, quoted in Armiero 2014: 242) as an example of a broader cultural reclamation typical of totalitarian efforts to create the '*uomo nuovo*' (Bernhard and Klinkhammer 2017). Despite these interesting considerations, one must acknowledge that academic reflection on the relationship between the Fascist regime and nature has been limited (Armiero 2014: 244). This is matched by a general lack of attention to environmental history in Italy and the abovementioned underrepresentation of environmental issues by the Italian parties, the far right included.

However, the attempt by the youth sector of the far right to initiate an ideological renewal during the late 1970s is an important exception to this general trend. The history of Italian post-fascism was marked since the beginning of the Republic by the MSI militants feeling of ghettoization (Germinario 1999: 71). Despite its immediate political and parliamentary representation, the MSI was always excluded not only from participation in governing coalitions but also from playing any significant role in Italian institutions. Being part of this '*Polo Escluso*'[3] was differently perceived by older and younger generations active in the party.

The frustration of younger people who had joined the MSI without a living memory of the Fascist regime started to become visible in the 1970s due to, for example, experiences of polarization provoked by the cultural movement of 1968 and impatience with the rigid organization of political activity within the party. The tensions between the MSI and its youth wing, the *Fronte della Gioventù* (FdG), were visible and growing in that period. The FdG was much more involved in cultural matters than the central organization. The 'quasi monopoly' expressed by the younger generation over the intellectual production of the far-right party (Ignazi 1989: 118) was frowned upon by the strict and authoritarian organization of the MSI.

The cultural debate was particularly active among younger people inspired by the political figure of Pino Rauti, one of the founders of the MSI. Rauti's influence was due to his position on the social wing of the Fascist regime (Rao 2006: 244–246) and to his closeness to the Evolian doctrine (Ignazi 1989: 119) and its mysticism.[4] It is not by chance that the first explicit reference to a new model of ecology emerged in this subcultural milieu. However, this concept of 'ecology' did not refer to a generic protection of the environment but included a more complex understanding of political ecology (Giovannini 1987: 10). In the late 1970s the biologist Alessandro Di Pietro, a leading figure in the youth organization of the MSI, founded the GRE – *Gruppi di ricerca ecologica* (ecological research groups) – an organization devoted to ecological issues (Baldoni 2009: 238).[5] The GRE's exploration of the relationship between human beings and the environment went beyond the protection of nature. With its focus on anticonsumerism, it also went beyond the traditional concept of anti-communism and the old repertoires of Italian post-fascism action. In that period, traditional forms of activism in the far right subculture coexisted with new forms of political activism. 'Alternative' issues were approached, including environmentalism.

One of the main cultural innovations during those years was the magazine *La voce della fogna* (*The voice of the sewer*), whose title ironically referred to a famous anti-fascist slogan '*Fascisti, carogne, tornate nelle fogne*' (Fascist rats, go back to the sewers). The subtitle of the magazine was *Giornale differente* (*Different magazine*). It pointed not only to a new form of aesthetics centred on the use of comic strips but also to innovative political contents, where environmental issues soon were represented. The magazine was published for the first time in 1974 and appeared regularly until 1983. These years coincided with the penetration of the ideas of the French *Nouvelle Droite* into the Italian far right. Overall, publications like *La voce della fogna*, small and independent bookshops, and the 'creation of adversarial ecological associations' expressed an underground culture that challenged the official party line.

The development of these forms of political activism culminated in the organization of the three Hobbit Camps in 1977, 1978 and 1980. These camps were cultural festivals, responding to the needs for new forms of activism by those embracing the metapolitics of the *Nouvelle Droite*. The festivals were centred around concerts by far-right bands, theatre performances and public debates covering 'heretical issues' such as the condition of women in contemporary society, the contradictions of capitalist development and the protection of nature. For example, the leaflet advertising the third Hobbit Camp stated '*Campo Hobbit 3. Musica cultura ecologia, grafica poesia cinema teatro cabaret*' (Hobbit Camp 3. Music, culture, ecology, graphics, cinema, theatre, cabaret). The name Hobbit is a clear reference to *The Lord of the Rings* saga by J.R.R. Tolkien. The novel depicts the evocative scenario of a new form of community and the symbolism of a nature consisting of myths very distant from the traditional fascist imaginary which was perceived as oppressive and frustrating by the younger generation.

At these camps, environmental issues were present not only in terms of specific debates but also, implicitly, in the choice of places where the festivals were held and in their symbolic evocation. The first camp was organized in the small village of Montesarchio in the rural south, far from the cities and political tensions of the north. In the northern regions, the organization of a similar event would have provoked an immediate counter-reaction by extreme left militants. The third camp took place in Castel Camponeschi, an abandoned village in the southern region of Abruzzo, whose topography represented the ideal stage for the festival's cultural performances.[6] The symbolic repertoire of the Hobbit Camps proposed a mode of 'living together' in between individualism and collectivism, which animated the cultural activities as well as the public debates. The idea that 'the private is political' (Tarchi 2010) ceased to be a catchphrase for only the extreme left and started to gain currency on the far right. In environmental terms, this resulted in an individual commitment to the environmental cause that implied an explicit critique of the capitalist model of production. The Hobbit Camps represented an important but isolated phase in the long history of Italian post-fascism, which was not even significant for all young members of the MSI. The following years were marked by the passage to the terrorist underground on the one hand (Telese 2006) and by realignment with the party discipline on the other. In the meantime, however, important features of the above-described ideological innovation continued to play a role in the evolution of post-fascist political culture.

The Hobbit Camps posed a challenge to the extreme left with regard to the latter's claimed unique representation of the revolutionary ambitions of youth in cultural terms. Allegations of cultural plagiarism were made by political opponents as well as by the mainstream media, which tried to discredit the novel contribution represented by the Hobbit Camps. This is an important element if one considers that the same accusations would be directed at the most recent forms of far-right activism in the second decade of the following century.

The commitment to environmental issues by extreme-right youth in this period should therefore not be evaluated with regard to their genuine environmentalism. The scant importance that the Fascist regime gave to the environment – both in policy making and in the construction of a symbolic understanding of nature – obstructed the structuring of a far-right ecology. Protection of the environment was considered to be a consequence of human primacy over technological development and not an aim *per se*; the original criticism concerning what would later become an open attack on globalization pointed to the role of the human being in the chain of wealth production and redistribution; and, finally, the unlimited exploitation of natural resources conducted by the international powers was rejected. This accusation was addressed to both sides of the then bipolar world (Tarchi 2010).

These elements were not completely abandoned in the following years. However, the combined effect of the final phase of the terrorist period, the marginalization of the youth faction drawing on the ideological innovation of the *Nouvelle Droite* within the party, and the rigid leadership of Giorgio Almirante contributed to interrupting this cultural innovation. Although this also affected reflections on environmental issues, it did not prevent the birth of an ecological association, *Fare Verde* (Go Green), which was founded in 1987 as a reaction to Chernobyl. *Fare Verde* was the first organized grouping directly linked to the MSI after the aforementioned *Gruppi di Azione Ecologica* was created in 1978. Similarly, to the latter, *Fare Verde* interpreted its ecological commitment in terms of communitarian belonging (Loreti 1999) and in the context of criticism towards consumerism and capitalism (Amorese 2014: 188–190). Crucially, however, the association subsequently lost its far-right ideological dimension, and the organisation's president, Paolo Colli, together with one of the leading figures of the environmental movement, Alex Lange, contributed to a substantive debate on the status of environmental issues since the early 1980s (De Meo and Giovannini 1985) beyond the left and the right (De Benetti 1995: 72). This specific element would play a crucial role in the subsequent development of far-right coverage of environmental issues in the following decades.

Beyond parties and movements: CasaPound Italia and environmentalism

After the successful referendum campaign against nuclear power, environmentalist associations continued their lobbying, but their impact on the ecological debate was limited. This also applied to organizations on the far right within the context of the end of the First Republic. The restructuring of the Italian party system due to the corruption scandals and the end of the Cold War significantly influenced the MSI as well, which went through a process of ideological transformation that resulted in a change of name and party logo (Tarchi 1997).

The last party congress of the MSI, held in Fiuggi, led to the birth of *Alleanza Nazionale* (AN) and a split in the party (Tarchi 1997: 121–124). Pino Rauti was among those who refused to agree with the new course. He founded the *MSI Fiamma Tricolore* (Tricolour Flame) party. In contrast, the new leader of AN, Gianfranco Fini, directed the party into the government (headed by Silvio Berlusconi). His later definition of fascism as the 'absolute evil' (Scaliati 2010) severed all remaining ties with the post-fascist political culture of the MSI. During those years of dramatic confrontation within the far right, no political space was available for the thematization of environmental issues. In the newly structured party coalitions that emerged in the 1990's, environmental issues were represented by the Federazione dei Verdi (Federation of the Greens). The FdV was a constitutive member of the center-left coalition headed by Romano Prodi called L'Ulivo (the Olive tree). During the years 1996-2008, the Ministry of the Environment was held four times by FdV politicians (Edoardo Ronchi 1996-1998; 1998-1999; 1999-2000. Alfonso Pecoraro Scanio 2006-2008) within Centre-left government coalitions.

Also within the split parties of *MSI Fiamma Tricolore* and the later-founded *Forza Nuova*, the ideological tensions were mostly concentrated on the internal fight for hegemony within what was left of the post-fascist political culture (and post-fascist electorate). The youth organization of the MSI was transformed into *Azione Giovani* in 1996 (Antonucci 2011: 77) and followed the transformation of AN. It replicated the model of 'allegiance' that had marked the relationship between MSI and *Fronte della Gioventù* without evident traces of an ideological debate concerning the relationship between politics and the environment which had characterized the above-described phase (Tarchi 1997: 339–340). Dissatisfaction among the most active members of AN was evident (Ignazi and Bardi 2006: 52–54), and in smaller far-right parties, too, a sense of stagnation prevailed among younger activists (Antonucci 2011).

In reaction to this wider development, CasaPound Italia was officially founded as a cultural association in 2008 (Albanese et al. 2014: 21). Abandoning *MSI Fiamma Tricolore* precisely because of the party's ideological and symbolic immobility, a group of young militants formed around the leader of the popular 'identitarian rock' (Di Giorgi 2008) band Zeta Zero Alfa, Gianluca Iannone. The splinter group had already started a few years earlier, making use of an action-repertoire which made it well-known in the political subculture of Rome's far right. Along with the opening of pubs and other places devoted to cultural and leisure, from bookstores to gyms where one could engage in martial arts or boxing (Antolini 2010: 50). CPI's activities included the squatting of buildings.[7]

This was the beginning of a long and composite set of practices that has made CPI a point of reference for other national and international far-right organizations (Koch 2013). CPI's combination of practices of action typical of a social movement and activities distinctive of a political party enables the group to gain attention in different environments, ranging from the young cohorts of high school students to disenchanted political activists of the far right and young militants with no previous political experience. The feature that makes CPI attractive is its invocation of the primacy of action(ism). Environmentalism finds a place in this celebration of political action.

The organization of CPI is built around its leader group based in Rome and structured into a large number of collateral organizations that promote the ideals and the programme of CPI in specific subcultural milieus. One of these milieus focuses on environmentalism and protection of nature (Castelli et al. 2013: 242). In the following section, I will describe the positioning of CPI with regard to the environment contained in the last electoral programme and environmental campaigns developed by related cultural associations.

CPI declared its intention to participate in the administrative elections in Rome during its official 2012 summer festival. Subsequently, it also took part in national elections in 2013 and 2018 – although with very limited electoral success.[8] The programme for the 2018 general elections was issued under the title *Una Nazione* (One Nation) and began by making references to an organic and inclusive state whose aim should consist in the reaffirmation and reconquest of national sovereignty. The discourse on nationalism and sovereignty is an important dimension of the extreme right parties 'common ideological core' in Europe (Mudde 2000: 176–178). An organicist understanding of the State instead tends to coexist with a 'holistic perspective according to which organisms should be understood in terms of their embeddedness in an interdependent system' (Forchtner 2019).

The organization's programme continues with a description of what CPI considers as 'enemies'. The list includes all elements that provoke the disfigurement of peoples, of persons and of cultures, and everything which is '*enemy of the form*'. The list goes on with the attribution of 'friend' to all those who 'operate in the interest of the Italian people, and care about its destiny, *beauty* and social justice'

(CPI electoral programme 2018: 2. Author's italics). Along with the classic opposition between friend and foe typical of an authoritarian understanding of politics, the reference to the concept of beauty associated with the nationalist and ethnic defence of the Italian population is in line with an extreme-right concept of aesthetics widely analysed in recent literature (Forchtner and Kølvraa 2017). This ideal of aesthetics is also connected with CPI's romanticized narrative of the Fascist regime (Castelli Gattinara and Froio 2014: 158). The centrality of 'form' is of key importance in CPI's understanding of the relationship between humankind and the environment as continuous pursuit of an organicist ideal (Scianca 2011: 73). Within this framework, the words of Gabriele Adinolfi – considered a leading figure in the ideological and cultural domain by CPI's activists (Caldiron 2013: 131) – are indicative of this relationship.

> Environmental protection as a segregated ideology is a nonsense. To recover the linkage that ties us to the ancestral is a completely different thing, and so is re-establishment of the correct relationship between nature and culture.
>
> *Adinolfi (2008: 99; Author's translation)*

Chapter 8 ('For an energy sovereignty') and Chapter 11 ('For a non-compliant ecology') of the CPI 2018 electoral programme directly address environmental issues. The proposal to reintroduce nuclear energy seems to contradict the call for public financing of renewable energy. However, the main ideological focus of CPI is the sovereignty of the nation. Consequently, the desire for energy autarchy – a feature shared by other far right movements in Europe (Forchtner et al. 2018: 4) – ultimately justifies the proposal of abolishing one of the greatest successes of the Italian ecological movement.

CPI ecology differs in many respects from the contemporary ideal of environmental protection. Indeed, it is closely linked to the idea of the sacred union between 'idea and action' that is one of the main keys to the interpretation of CPI's success (Albanese et al. 2014). The following passage is evocative of this relationship.

> The environment is us and vice versa. The real problem is not how 'not to pollute' but how to breathe together with the cosmos. We propose an enlightened ecology that does not criminalize humans but wishes to exploit all their abilities to build and give an order to the world. We oppose this vision against a gloomy green fundamentalism and its often anti-ecological taboos. For us, there is more ecology in the redeemed soil of a city under construction than in a 'extremely natural' stagnant swamp.
>
> *(CPI 2018: 13)*

In CPI's electoral programme, there is no trace of a nostalgic feeling for a collective purity (Forchtner 2016: 291) realized in the natural setting of the countryside, as in the case of other extreme right actors in Europe. The references to the *terra redenta* (redeemed soil), as well as the action of man who imposes his will on nature are indicative of an extreme right understanding of the relationship between humankind and the environment typical of the Italian tradition. Here, there is no trace of the holistic understanding of interdependence among humans in their environment cited above (Forchtner 2019). On the contrary, an anthropocentric vision prevails, and the exaltation of *azionismo* in futuristic terms is coherent with the Italian fascist glorification of action, virility and struggle (Traverso 2008: 304). This attitude combines historical Fascism's approach that conceals its autarchy aims behind a rhetoric of nature protection with a vague reference to the mystique of the German ecologism of the 1930s. Interestingly, no reference is made to the concept of 'beauty' if not in the sense of the protection in nationalist terms of the Italian people. The 'respect for harmony and beauty of nature' (Champetier and de Benoist 2000: 10) addressed by the New Right is not a point of reference for the Italian extreme right movement. This is in line with the interpretation of CPI's political action as distant from the intellectual understanding of the *Nouvelle Droite* which animated the cultural phase of the 1980s described above.

Furthermore, the proposal to 'carry out an analysis and recovery of *Italian biodiversity*' (CPI 2018: 13. Author's italics) does not simply identify the desire to protect natural species, but indirectly hints at the ideals of anti-egalitarianism expressed by the *Nouvelle Droite* and cultural differentialism (Taguieff 1994: 112–116). The reference to biodiversity can be read in terms of the first article of the chapter devoted to immigration ('Stop immigration, no to the ius soli, repatriation') which declares that mass migration is 'one of the main causes of deracination and social, cultural and existential deprivation to the detriment of all involved populations, hosts and guests' (CPI 2018 4). From this point of view, CPI seems to offer a notion of biodiversity that diverges from the reference to the differentiation of the natural species and their protection in the ecological system (Forchtner 2019) and indirectly points to the protection of *Italian biodiversity* in nationalistic terms.

The references to the only visible policies of the Fascist regime directly aimed at preservation of the environment – that is, the institution of national parks (von Hardenberg 2014) – features also in CPI's political programme with the proposal to enhance the foundation of parks and nature reserves, and to realize a culture of 'participation in parks'. This last part of the proposal is consistent with CPI's emphasis on individual health and well-being that passes through the ideal of aesthetics of the (predominantly male) body. Here, CPI's suggestions include the proposal to increase the number of hours devoted to physical education in the school curriculum by 150%[9] (CPI 2018, 13) and the introduction in public schools of excursions to mountain and sea camps. The recommendation to financially support the 'alpinist, maritime, parachuting associations and all other activities directly linked to nature' is another reference to how the organization CasaPound Italia is structured on the co-existence of diverse grassroots organizations. Indeed, environmentalist associations play a major role in CPI's self-representation as a lively social movement. There are three main associations that promote environmental issues: *la Foresta che Avanza*, *la Salamandra* and *la Muvra*.

La Foresta che Avanza (The Advancing Forest) describes itself as the 'environmentalist movement of CasaPound Italia'.[10] The official programme of the organization lists the proposals of the group, ranging from opposition to the 'industry of the meat' to resistance against the practices of vivisection and the use of animals in circus performances. The programme ends with the movement's slogan '*Per un REGIME della natura*' ('For a REGIME of nature'; capital letters in the original) which is a play on words with the historical concept of 'regime'. On exploring the website, one gains the impression that *La Foresta che Avanza* can count on many local branches. However, in most cases, the content of the respective webpage is the same for all Italian regions.[11] It represents two major campaigns of the organization centred on opposition against the Harlan Laboratories and on the replanting of trees on Monte Giano.

Let me close this section by describing these two campaigns in some detail. The first concerned opposition to the Harlan Laboratories which delivered animals for research purposes.[12] The movement against the Harlan Laboratories developed in 2013 and 2014, mobilizing thousands of participants belonging to diverse pro-animal associations. Members of the movement distanced themselves from the participation of the far right and developed counter-information to reveal the presence of extreme right associations in environmentalist and animalist actions, something which happened in other European countries as well. Similar reactions surfaced because of the involvement of CPI in solidarity actions put in place by the organization *La Salamandra* – a civil protection association whose members carry out food collection, food distribution to poor families, and organize blood donations.[13]

The second campaign, the Monte Giano campaign, was organized by CPI in coincidence with the 'Day of the Tree' (21 November). Its purpose was to raise funds to replant trees in the forest of Monte Giano that spelled out the word D U X in huge letters on the side of the mountain – an evident tribute to Benito Mussolini that could be seen from the city of Rome. The website of the environmentalist organization of CPI dedicated to the campaign[14] explains the origins of the 'Day of the Tree' (established in 1923) and calls for the raising of funds to replant the trees. Their slogan, '*La storia non si cancella*' ('One cannot erase history') reaffirms the relationship between CPI and Italian fascism and ultimately indicates the use of a symbolism more interested in evocation of the past than in commitment to nature.

If we briefly compare CPI's interpretation of ecology and environment to the positions expressed on the same issues by the Lega, we find surprising differences. In the document that served as an internal guide for elaboration of the programme on the occasion of the 2013 local elections at the municipal level, environment protection was understood as 'the means by which it is possible to develop sustainable development integrated with economic improvement' (Lega Nord 2013: 19). This quotation confirms the 'tradition' of limited attention paid by the Italian parties to environmental issues and confirms their limited efforts to define ecology outside the wide concept of sustainability.

The attention paid to environmental politics grew in the Lega 2018 party manifesto issued for the general elections. Here, environmental proposals are contained in the chapter 'Environment, green economy and quality of life' (Lega 2018: 35–42). The chapter ranges from generic support of the green economy to the development of a new hydro-geological plan for the country and it opens with the party's commitment to support the 'green economy, through the funding of innovation, education on environmental actions, and industrial competitiveness in our industrial system' (Lega 2018: 35). The main feature of the pages devoted to the environment in the party programme (8 out of 72) evidences the Lega's understanding of environmentalism as linked to the promotion of economic development.

The headline of the chapter 'Humankind and environment are two faces of the same coin. Those who do not respect the environment do not respect themselves' (Lega 2018: 35) is a generic declaration of intents that is not followed by a more specific description of the environmentalist conception of the party. In its proposals in the protection of the environment, the party mainly refers to public policies – garbage management, water management plans, support to the investments in the field of electric cars – to be implemented both at the national and local level. In the description of these policies, the main focus lies in the interaction of the institutions involved (the state, the regions, and the public administration) in the decision-making process and in the consequent benefits of the citizens from an economic point of view. The party does not deny the existence of a climate change, and proposes the implementation of sustainable models of economy as well as the support of technical and political plans to promote the production of renewable forms of energy. The description of the environment proposals of the Lega reflect, therefore, the preoccupations of a party used to act at the level of the policy making. No symbolic reference to the environment is present in the party's description, as in the case of the above described positions of CPI.

Conclusion

One would be tempted to say that environmentalism is not a party issue when looking at the failed institutionalization of environmentalist parties in Italy. The results of the 2019 European elections – where the party Europa Verde (Green Europe) obtained 2,3% and no seats at the European Parliament - testifies of the limited investment in environmentalism by the Italian party system. This is due not only to the difficulties of the Italian Green parties described in the first section of this chapter. The hopes for an environment commitment of the Movimento Cinque Stelle - Five Star Movement – the party created in 2009 by the comedian Beppe Grillo with a clear emphasis on ecological issues (Corbetta and Gualmini 2013, 38) – declined with the difficulties encountered by the party in the government coalition signed in 2018 with the Lega. However, this statement appears too absolute when analysing the discourse on environmental issues by the Italian far right. Without taking account of the rich debate on the consequences of the 'silent revolution' (Inglehart 1977) and its influence on the interests of left-wing parties and movements in ecological issues (Diani 1988), the relationship between far-right organizations and the environment is more complex than one might assume.

This chapter has focused on two major distinctions: the organizational and the ideological dimensions. This distinction has been developed with a particular focus on the early 1980s and early 2010s. Following the organizational dimension, the chapter has evidenced that on the far right, no defining line can be drawn between successful movements and unsuccessful parties. The political phase marked by youth activism around the thinkers of the *Nouvelle Droite* coincided with a lively debate on the essence of ecologism. This debate originated within the main party of Italian post-fascism, even if it was limited to its youth organization. The experience of the *Gruppi di azione ideologica* and of *Fare Verde* can only be understood by considering it from within the political milieu of MSI. Their contribution to environmental issues, connected to the consequences of capitalism, consumerism, cultural and economic globalization, as well as on the impact of colonialism, went beyond the rigid distinction between far right and far left.

After a long break imposed by the restructuring of the Italian political system at the beginning of the 1990s, environmental issues were mainly reactivated on the far right by CasaPound Italia. This particular type of organization consisting of a soft structure, a limited number of members and a strict hierarchy facilitated spectacular campaigns to capture the attention of the media (Castelli and Froio 2017). At the same time, CPI's participation in the elections forced it to elaborate on its core issues – environmentalism included. The proposals contained in the electoral programme reflect, however, the desire of the organization to capture the addressees' imagination much more than containing profound reflection on the concept of ecology in the current national and international context. Most references to ecology are made by authors considered to be CPI ideologues and whose political activities go back to the 1970s.[15] CPI's connections to many organizations (environmental groups included) is instrumental to its territorial ramification and its recruitment. In terms of impact and a new environmentalist conscience and behaviour, the campaigns organized by associations like *La Foresta che avanza* or *La Salamandra* respond much more to the principle of '*azionismo*' than to an intellectual and ideological set of proposals. This is confirmed by CPI's propensity to invest in issues like immigration and law and order much more than in environmentalist themes (Castelli Gattinara and Froio 2017: 67–68).

In ideological terms, the coexistence of a rhetoric that denies the validity of the categories of right and left (Adinolfi 2008: 82; Iannone, quoted in Antolini 2010: 65) and a symbolic repertoire made of a continuous recall of the Fascist regime (Albanese et al. 2014: 49–51) in CPI produces an ambiguity in the organization's proposals on environmental issues. The exaltation of sovereignty generates contradictions, e.g. when both call for the reintroduction of nuclear power and demand higher investments in renewable forms of energy, with no clear indication on how to reconcile these sources of energy in the movement's ecological framework. No clear position is taken on the tension between industrialization and respect for the environment. The criticism of the globalization focuses much more on its economic and cultural consequences in terms of financial crisis and migration than on the environment. The impact of CPI's environmentalist campaigns are therefore very limited, not only on public opinion at large, but also within the political milieu of the far right. This is due, as I have shown, to a mixture of organizational and ideological factors. These suggest that the popularity of CPI depends much more on the repertoire of actions adopted than on the organization's disposition to strengthen its links within the far-right milieu through ideological incentives, which go beyond the organization's focus on hierarchy, sovereignty and actionism. The environment is not included.

Notes

1 The Northen League abandoned the reference to the 'North' since the 2018 general elections, where the party was present in the official lists under the name 'Lega'.
2 The MSI was created immediately after the end of World War Two with a clear reference to the *Fascismo Movimento* (Ignazi 1994: 11) and obtained an uninterrupted parliamentary representation during the First Republic.
3 *Polo Escluso* (The excluded Pole), is the title of one Piero Ignazi's (1989) books devoted to the organization and political culture of MSI. The expression refers to the *cordon sanitaire* adopted by 'constitutional parties' towards the post-fascist MSI.
4 On the complex relationship between the Italian post-fascist right and Evolian reception, see Germinario (2005: 47–63).
5 In 1988, 10 years after the founding of the organization, the *Gruppi di ricerca ecologica* counted more than 15.000 members (Diani 1988: 61).
6 The second camp is considered the less successful due to the attempt by the *Fronte della Gioventù*, loyal to the MSI secretary, to control over organization and political content. It was also held in the southern region of Abruzzo.
7 The first squatting in 2002 gave birth to the experience of 'Casa Montag'. An insider description is available in Di Tullio (2006).
8 0.14% in 2013 and 0.9% in 2018.
9 The importance devoted by the Fascist regime to the spiritual, sportive and pre-military education of the youth is underlined by, e.g. Gentile (1995: 192).
10 www.facebook.com/pg/laforestacheavanza/about/?ref=page_internal [20 March 2018].
11 www.laforestacheavanza.org/search/label/Puglia [13 March 2018].
12 www.casapounditalia.org/2013/03/vivisezione-harlan-deve-chiudere-500.html [16 April 2018].
13 The association was particularly active following the earthquakes that happened in 2016 in central Italy.
14 www.laforestacheavanza.org/2017/11/il-monte-giano-riavra-la-sua-scritta.html [13 March 2018].
15 Gabriele Adinolfi for instance has been the funder of the organization *Terza Posizione*, an extra-parliamentarian, extreme-right organization created in 1978.

References

Adinolfi, Gabriele (2008): *Tortuga, l'isola che (non c'è). Pensieri non conformi di lotta e di vittoria.* Milano: Società Editrice Barbarossa.
Albanese, Matteo, Bulli, Giorgia, Castelli, Pietro and Froio, Caterina (2014): *Fascisti di un altro millennio? Crisi e partecipazione in CasaPound Italia.* Catania: Bonanno.
Amorese, Alessandro (2014): *Fronte della Gioventù. La destra che sognava la rivoluzione. La storia mai raccontata.* Massa Carrara: Eclettica.
Antolini, Nicola (2010): *Fuori dal cerchio. Viaggio nella destra radicale italiana.* Roma: Elliot.
Antonucci, Maria Cristina (2011): *La cultura politica dei movimenti giovanili di destra nell'era della globalizzazione.* Milano: Franco Angeli.
Armiero, Marco (2014): Introduction: Fascism and nature, *Modern Italy*, 19(3): 241–245.
Baldoni, Adalberto (2009): *Storia della destra. Dal postfascismo al Popolo della libertà.* Firenze: Vallecchi.
Ben-Ghiat, Ruth (2011): *Fascist Modernities: Italy, 1922–1945.* Berkeley: University of California Press.
Bernhard, Patrick and Klinkhammer, Lutz (2017): *L'uomo nuovo del fascismo, La costruzione di un progetto totalitario.* Roma: Viella.
Biorcio, Roberto and Lodi, Giovanni (Eds) (1988): *La sfida verde. Il movimento ecologista in Italia.* Padova: Liviana.
Caiani, Manuela, della Porta, Donatella, and Wagemann Claudius (2012): *Mobilizing on the Extreme Right Germany, Italy, and the United States.* Oxford: Oxford University Press.
Caldiron, Guido (2013): *Estrema Destra.* Roma: Newton Compton.
CasaPound Italia (CPI) (2018): Electoral Program. *Una Nazione.* www.docdroid.net/Bg8qGdw/programma-casapound-2018.pdf [24 September 2018].
Castelli Gattinara, Pietro and Froio, Caterina (2014): Discourse and practice of violence in the Italian extreme right: Frames, symbols, and identity-building in CasaPound Italia, *International Journal of Conflict and Violence*, 8: 154–170.
Castelli Gattinara, Pietro and Froio, Caterina (2017): Comunicazione del terzo millennio? La politica mediatizzata di CasaPound Italia, *Comunicazione Politica*, 1: 55–76.
Castelli Gattinara, Pietro, Froio, Caterina and Albanese, Matteo (2013): The appeal of neo-fascism in times of crisis. The experience of CasaPound Italia, *Fascism*, 2: 234–258.
Champetier, Charles and de Benoist, Alain (1999): Manifeste: La Nouvelle Droite de l'an 2000, *Elements*, 94: 11–23.
Corbetta, Piergiorgio and Gualmini, Elisabetta (2013): *Il partito di Grillo.* Bologna: il Mulino.
De Benetti, Lino (1995): *Verde scuro, verde chiaro. L'ecologia politica: etica, sviluppo sostenibile e democrazia.* Genova: Le Mani.
De Meo, Massimo and Giovannini, Fabio (1985): *L'onda verde. I verdi in Italia: la storia, il dibattito, gli indirizzi, i risultati elettorali.* Roma: Alfamedia.
Di Giorgi, Cristina (2008): *Note alternative. La musica emergente dei giovani di destra.* Roma: Edizioni Trecento.
Di Nunzio, Daniele and Toscano, Emanuele (2011): *Dentro e fuori Casapound. Capire il fascismo del Terzo Millennio.* Roma: Armando Editore.
Di Tullio, Domenico (2006): *Centri sociali di destra. Occupazioni e culture non conformi.* Roma: Castelvecchi.
Diani, Mario (1988): *Isole nell'arcipelago. Il Movimento ecologista in Italia.* Bologna: Il Mulino.
Forchtner, Bernhard (2016): Longing for communal purity: Countryside, (far-right) nationalism and the (im)possibility of progressive politics of nostalgia. In Christian Karner and Bernhard Weicht (Eds). *The Commonalities of Global Crises: Markets, Communities and Nostalgia.* Basingstoke: Palgrave, pp. 271–294.

Forchtner, Bernhard (2019): Nation, nature, purity: Extreme-right biodiversity in Germany, *Patterns of Prejudice*, 53(2): 285–301.
Forchtner, Bernhard and Kølvraa, Christoffer (2017): Extreme right images of radical authenticity: Multimodal aesthetics of history, nature and gender roles in social media, *European Journal of Cultural and Political Sociology*, 4(3): 252–281.
Forchtner, Bernhard, Kroneder, Andreas and Wetzel, David (2018): Being skeptical? Exploring far-right climate change communication in Germany. *Environmental Communication*, 12(5): 589–604.
Gentile, Emilio (1995): *La via italiana al totalitarismo. Il partito e lo stato nel regime fascista*. Roma: Carocci (quoted in the text: 2nd edition 2001).
Germinario, Francesco (1999): *L'altra memoria*. Torino: Bollati Boringhieri.
Germinario, Francesco (2005): *Da Salò al governo. Immaginario e cultura politica della destra italiana*. Torino: Bollati Boringhieri.
Giovannini, Fabio (Ed) (1987): *Le culture dei verdi. Un'analisi critica del pensiero ecologista* Edizioni Dedalo: Bari.
Ignazi, Piero (1989): *Il polo escluso*. Bologna: Il Mulino.
Ignazi, Piero (1994): *Post-fascisti? Dal Movimento Sociale Italiano ad Alleanza Nazionale*. Bologna: Il Mulino.
Ignazi, Piero (2000): *L'estrema destra in Europa*. Bologna: Il Mulino.
Ignazi, Piero and Bardi, Luciano (2006): Gli iscritti ad Alleanza nazionale: attivi ma frustrati. *Polis*, 17: 31–58.
Inglehart, Ronald (1977): *The Silent Revolution*. Princeton: Princeton University Press.
Koch, Heiko (2013): *CasaPound Italia. Mussolinis Erben*. Münster: Unrast Verlag.
Lega (2018): *Elezioni 2018. Programma di governo*. www.leganord.org/component/tags/tag/programma-elettorale [20 February 2018].
Lega Nord (2013): *Electoral program municipal elections*. www.leganord.org/index.php/component/phocadownload/category/7-comuni-al-voto?download=789:programma-elezioni-amministrative-2013 [20 February 2018].
Lodi, Giovanni (1988): L'azione ecologista in Italia: dal protezionismo storico alle Liste Verdi. In Roberto Biorcio and Giovanni Lodi (Eds). *La sfida verde. Il movimento ecologista in Italia*. Padova: Liviana, pp. 17–47.
Loreti Andrea (1999): *Gli amici di Gaia. Storia, atteggiamenti e comportamenti degli ambientalisti italiani*. Roma: La Sapienza.
Menichini, Stefano (Ed) (1983): *I Verdi. Chi sono, cosa vogliono*. Roma: Savelli.
Mudde, Cas (2000): *The Ideology of the Extreme Right*. Manchester: Manchester University Press.
Müller Rommel, Ferdinand (1985): The greens in Western Europe. Similar but different, *International Political Science Review*, 6: 483–499.
Rao, Nicola (2006): *La Fiamma e la Celtica*. Milano: Sperling and Kupfer.
Rokkan, Stein (1970): *Citizens, Elections, Parties: Approaches to the Comparative Study of the Processes of Development*. Oslo: Universitetsforlaget.
Rosati, Elia (2018): *CasaPound Italia. Fascisti del Terzo Millennio*. Milano-Udine: Mimesis.
Scaliati, Giuseppe (2010): *Il Male assoluto. Da Fiuggi al PDL*. Catania: Bonanno.
Scianca, Adriano (2011): *Riprendersi tutto. Le parole di CasaPound: 40 concetti per una rivoluzione in atto*. Cusano Milanino: Società Editrice Barbarossa.
Staudenmaier, Peter (2011): Fascist ecology: The 'Green Wing' of the Nazi Party and its historical antecedents. In: Janet Biehl and Peter Staudenmaier (Eds). *Ecofascism Revisited: Lessons from the German Experience*. Porsgrunn: New Compass Press, pp. 13–42.
Taguieff, Pierre-André (1994): *Sur la nouvelle Droite. Jalons d'une analyse critique*. Paris: Descartes. Italian translation, *Sulla nuova destra. Itinerario di un intellettuale atipico*. Firenze: Vallecchi.

Tarchi, Marco (1997): *Dal Msi ad An*. Bologna: Il Mulino.
Tarchi, Marco (2010): *La rivoluzione impossibile. Dai Campi Hobbit alla Nuova Destra*. Firenze: Vallecchi.
Telese, Luca (2006): *Cuori neri*. Milano: Sperling and Kupfer.
Traverso, Enzo (2008): Interpreting fascism: Mosse, sternhell and gentile in comparative perspective, *Constellations*, 15(3): 303–319.
Von Hardenberg, Wilko (2014): A nation's parks: Failure and success in Fascist nature conservation, *Modern Italy*, 19(3): 275–285.

PART III
Nordic countries

7

WOLVES IN SHEEP'S CLOTHING? THE DANISH FAR RIGHT AND 'WILD NATURE'

Christoffer Kølvraa

Introduction

The far-right Danish People's Party (DPP) have often been accused of being 'wolves in sheep's clothing' by left-wing opponents, insinuating that their immigration-hostile populism carries fascist elements at heart. However, when a heated political debate about immigration of real wolves into Denmark erupted in 2018, the DPP immediately sided with the sheep and their farmers, and in doing so demonstrated more clearly than usual their core idea of nature and its relationship to human societies. The fact is while the environment is not a core political concern of the DPP (Hervik 2011; Forchtner and Kølvraa 2015; DPP 2018a), it does not mean that the party is devoid of what might be called an 'environmental imaginary'. As argued by Forchtner and Kølvraa, an understanding of far-right actors need to go beyond a concern with their explicit political priorities, and also take an interest in how their wider (cultural) imaginaries more implicitly – through iconographic imagery, popular culture preferences and consumption patterns – construct and perpetuate a number of ideals pertaining to both collective and personal life (Forchtner and Kølvraa 2017). As argued in the introduction to this volume (Forchtner 2019a), and as demonstrated by Forchtner and Kølvraa elsewhere (Forchtner and Kølvraa 2015), nature is a core element in various far-right imaginaries; something partly due to the central place that both the aestetics, symbolism and materiality of the national countryside (Palmer 1998; Cosgrove 2004), as well as a range of biological/racial metaphors (Olsen 1999; Olwig 2003), have traditionally held the ideology of nationalism (Forchtner 2019b).

In the following, I continue this line of thought by further exploring the DPP's 'environmental imaginary', by which I mean a wider, ideologically tinted but not necessarily explicitly politically formulated, idea of nature and a set of ideals or assumptions as to its proper relation to human societies. I argue that while actors at the extreme end of the far-right spectrum, often have a more *völkish-organicist* 'ecological imaginary' (Forchtner 2019), the DPP's environmental imaginary and thus their view on nature and wildlife – as illustrated in the 'wolf-debate' – remains within a populist-anthropocentric frame.

I will first sketch the general features of the DPP's environment imaginary by examining the party's statements and positions on climate change. While these statements allow one to outline the overarching dynamics of the DPP's stance on 'nature', the 'wolf-debate' in the spring of 2018 – to which I next turn – affords a narrower discursive context in which to reveal some of the deeper ideological tenets of the DPP's environmental imaginary.

Roland Barthes, *Mythologies* and 'vertical analysis'

In his early work *Mythologies*, Roland Barthes demonstrated how seemingly marginal or limited cultural or political phenomena – from the new Citroën to Poujadism – could be analysed to shed light on wider ideological structures (Barthes 2013). Even 'innocent objects' – such as an image of an African boy saluting – could be shown to entail a much wider ideological connotation, and thus became a 'myth'. Barthes insisted that myths had to be analysed by unfolding both the wider semiotics of the object, and the ideological use to which it was being put. In myth, language could be both political and innocent because its function was to naturalize certain meanings – to hide the fact that certain meanings carried a certain ideological perspective (Barthes 2013: 215–230).

But what is also first established in *Mythologies* – and which Barthes later develops in his work with movie stills and photographs (Barthes 1977; Barthes 1981) – is the analytical strategy of engaging with a limited 'fragment', rather than with more extensive corpora of (ideological) discourse. As Barthes argued such 'vertical' analysis, entailing a highly focused and deep exploration of, for example, a single character in a novel or a single movie still, serves to bring to light new aspects, implications or meanings, that would not necessarily emerge if a more 'horizontal' analysis – relating to the entirety of the novel, or the movie plot – was undertaken (Barthes 1977: 64; White 2012: 38–53).

I approach the 'wolf-debate' as a fragment in relation to DPP discourse about nature. Thus, while the DPP's general political discourse about the environment (emerging for example around the issue of climate change) is not only sparse, but often pragmatically bent to accommodate political alliances or other priorities. And for this reason not necessarily consistent over time, their discourses and performativity in the limited context of the wolf-debate constitute an ideological 'prism' through which a wider environmental imaginary might be seen. I employ a corpus of media and social media texts, political manifestos, the party magazine, official statements, party website content, and TV and radio shows containing statements and performances by the DPP and its members. The corpus stretches from 2009, when the DPP first came out forcefully as climate change sceptics (Forchtner and Kølvraa 2015) and up to the 'wolf-debate' in the spring of 2018.[1]

The nature of the Danish Peoples Party

The DPP started as a breakaway group from the Progress Party, which first emerged in the early 1970s as a populist party with a strong anti-tax agenda. But while DPP quickly shifted towards a neo-nationalist and far-right brand of populism, it is important to note that the party does not have its roots in, or are associated with, neo-fascist or overtly racist groupings or milieus. However, the DPP's shift from the Progress Party's classical poujadist low-tax populism to a more overtly nationalist and at times certainly culturally xenophobic discourse, has certainly proven a successful formula with the party securing 21.1% of the vote in the 2015 elections (Juul Christiansen 2016).

The academic literature on the DPP is still somewhat sparse and is almost exclusively focused on the conventional far-right tenets of their ideology (Hervik 2011; Siim and Meret 2016). Work on the DPP's environmental stance or ideas about nature seem much more rare allthough the party is often mentioned in comparitive discussions about attitudes to climate change on the far-right (Lockwood 2018). More extensively Forchtner and Kølvraa's exploration of neo-nationalist ideas of 'countryside' and 'climate change' compares the DPP and the British National Party (Forchtner and Kølvraa 2015). They argue that climate change is a phenomenon at odds with the aesthetic, symbolic, and material frames through which nature is understood in nationalism. Thus, climate change scepticism is in fact ideologically meaningful, (Forchtner et al. 2018; Lockwood 2018). Thus, while the DPP consistently uses the aesthetic depiction of the Danish countryside in its public communication, and while its political programme does contain a short ambition about caring for nature for future generations (DPP 2018b), thereby symbolically evoking the temporal eternity of the imagined national community (Anderson 1983), it nonetheless came out as strongly sceptical towards the anthropogenic nature of climate change in the context of the 2009 COP15 United Nations Climate Change Conference in Copenhagen. Its spokesman on energy policy, Morten Messerschmidt, developed a 'scientific' line of argument, seeking – as is typically the case on the far right – to discredit mainstream climate change scientific consensus, by resorting to and elevating as the 'people's scientists', those relatively few voices in the scientific community who doubt the anthropogenic nature of climate change (McKewon 2012; Forchtner et al. 2018). The DPP even arranged an alternative climate conference alongside COP15 (Forchtner and Kølvraa 2015). Both Messerschmidt and the party leader Pia Kjærsgaard would simultaneously employ a wider mode of political discourse informed less by 'alternative scientific facts', and more by a general suspicion towards domestic elites and foreign actors and institutions. Here the national and international 'climate change lobby' were depicted either as a transnational conspiracy designed to rob nation-states of their sovereignty, or as a dogmatic 'thought police' geared to squash any ideas but their own. The future party leader Kristian Thulesen Dahl, for example, compared himself to the little boy in the Hans Christian Andersen fairy-tale 'The Emperors New Clothes' (Ritzau 2009) and the party leader Pia Kjærsgaard continously referred to the scientific consensus as 'hysterical', 'a religion' or even 'totalitarian' (Kjærsgaard 2009a, 2009b),[2] Such modes of denial is in fact conventional on the far right when it comes to climate change denial (Forchtner et al. 2018).

However, besides these elements of what might be identified as 'attribution' and 'process' scepticism, the DPP has also voiced a 'response skepticism' (Rahmstorf 2004, Van Rensburg 2015). This 'response skepticism' ironically ends up implicitly admitting the link between climate change and human activity because it it moves from what Kemp and Nielsen has called the 'fatalism' of rejecting any possible action at all, to the 'insignificance complex' of arguing that given Denmark's small size it does not matter what the Danes do (Kemp and Nielsen 2009). For example, energy spokesman Jørn Dohrmann argued in 2011 that since countries such as China, India and Brazil would most likely increase their CO_2 emissions in coming years, it made no sense for Denmark to spend money lowering its emission levels (Wolfhagen 2011). Beyond these occasional statements in the context of climate policy, and despite the fact that formally – in the long periods where the DPP has supported liberal governments – it has partaken in environmental legislation initiatives, the DPP's public communication very rarely focused directly on environmental issues. In fact, Pia Kjærsgaard caused a minor scandal in 2015 when she bluntly admitted in a radio-interview that nature was simply not a priority for the DPP (DR-Radio 2015). Nonetheless, this first exploration allows for some overarching points. Firstly, it is clear that the rejection of either the truth of anthropogenic climate change or Denmark's responsibility in relation to it is premised on a strictly national understanding of sovereignty, agency and normative subjectivity. One almost gets the impression that anthropogenic climate change is rejected *because* its solution would imply transnational forms of political agency, and the resultant 'sacrifice' of national sovereignty. Indeed, what quickly comes to dominate the DPP's stand is the ability to shift the discussion from one about the 'reality' of climate change, to one about the democratic (if not heroic) legitimacy of doubting and questioning all accepted or authoritative knowledge – including the reality and anthropocentric nature of climate change. Thereby, the DPP could move away from the increasingly weaker 'scientific' argument against anthropogenic climate change and towards a more classic populist self-conception as persecuted and ridiculed by powerful (international) elites brutally enforcing their ideological dogma.

While this 'horizontal' analysis, sweeping the party's environmental communication across the past decade, has helped us sketch the overarching principles, it is through the examination of the DPP's performance in the wolf-debate – to which I now turn – that one is able to flesh out the DPP's environmental imaginary beyond this initial ideological outline.

Wolves at the door: The origins of the Danish wolf-debate

Recents debates about wild wolves are certainly not an exclusively Danish phenomonen. Indeed in European countries where wolf populations are much bigger (Sweeden, Norway, Germany, Spain and Austria) similar debates are ongoing. In the Danish case, however, a number of factors has served to put far-right voices – i.e. the DPP – at the centre of the debate.

In fact, it was the DPP in the form of their well-known foreign policy spokesman Søren Espersen who unwittingly initiated the debate. On the day before Christmas 2017, he tweeted a link to an otherwise little noticed news article, which informed that 2017 had the highest number of wolf attacks on sheep (24) since the reimmigration of wolves into Denmark had been confirmed in 2012 (Miles 2017). Espersen's tweet commented: 'Wasn't it about time that we got around to shooting these ravenous beasts?' (Miles 2017). Almost immediately something of a Twitter and social media storm ensued and soon the debate expanded into the mainstream media, seemingly dividing the country between left wingers and right wingers, town and country, the capital of Copenhagen and the countryside in Jutland, the ordinary people and the experts.

What is interesting about the wolf-debate is not only how it positioned different political actors in relation to different ideas of (wild) nature – how it illustrated different environmental imaginaries – but also that it became so heated at all. After all, at the time of the debate, experts estimated a population between 8 and 11 wolves, and although 24 sheep attacks in 2017 was a rise, the problem remained limited, both economically and geographically. More than half of these attacks occurred at the same sheep farm, and the compensation – which the state pays to farmers having lost sheep to wolves – amounted to only about 100.000 dkr (ca. 15.000 USD). It is also worth noting that in 2017 95 attacks on sheep were reported, while wolves could only be made responsible for 24 of these attacks – many of the remaining being attributed to domestic dogs (Ritzau 2017).

Not only did the actual problem of wolves in Denmark thus seem strangely inadequate in relation to understanding the intensity of the debate, but also the actual 'policy suggestions' were somewhat vacuous. While relatively few political actors outright supported Espersen's recommendation of eradication (by shooting), many (including the DPP's animal welfare spokesperson Karina Due [see Kynde and Hartung 2018b]) favoured control through sterilization but soon moved on to advocating either incarceration (see Stie 2018) (catching and fencing in the wolves) or expulsion (sending them to other countries with more extensive national park systems, such as France) (see Byrne 2018). The irony was that none of these solutions were practically possible or, in the case of shooting, legal. While one could make the easy point that this spectrum of measures is somewhat recognizable from discussions about human immigration undertaken by the far right, what actually sustained the debate, and gave it its intensity, seem to have been neither the extent of the actual problem nor the quality of the suggested solutions, but instead the very fact that the wolf – perhaps to an extent, unique among European wildlife – was the subject of a wide cultural semiotics, a multifaceted mythology ready for ideological use.

'Ravenous beast': The cultural semiotics of the wolf

In surveying what Tønnesen calls the *zoo-semiotics* of the wolf, it seems almost as though the wolf was always a cultural stereotype first and a zoological species second (Tønnessen 2016). There is an abundance of wolves, not just in fairy-tales and folklore ('Little Red Riding Hood', the 'Three Little Pigs', the 'Boy Who Cried Wolf' etc.), but also in mythology and religion (Romulus and Remus fed by a she-wolf, the Fenris wolf in Norse Mythology and the Bible imagery of the wolf and the lamb resting together as a sign of the blissful harmony in God's kingdom) and in philosophy and psychology – here often utilizing the powerful imagery of Lycanthropy, where the wolf comes to signify the untameable, wild, aggressive and asocial side of man (Arnds 2015). Hobbes mobilized this meaning when drawing on the proverb '*Homo Homini Lupus*' to describe a pre-societal state of nature, Freud borrowed it from Hobbes to describe the unstable stand-off between the drives and civilization (Freud 1962), and Bettelheim has proposed in the same vein that 'Little Red Riding Hood' should be psychoanalytically read as a cautionary tale about aggressive male sexuality (Bettelheim 1976).

The classical 'myth' of the wolf is quite simply to reduce it to a ravenous and terrible appetite, to make it a sign of that wild nature – 'red in tooth and claw' – which supposedly stand in direct opposition to human societal order and security. This dichotomy is what is metaphorically signified by the wolf often appearing in contradistinction to animals of human husbandry (most often sheep), as is evident, for example, in the imagery of Christ as a shepherd tending his flock (Tønnessen 2016). It is important to note however that the 'wildness' of the wolf can be both an object of fearful condemnation and of fascinated appropriation. Indeed, it is worth pointing out – especially as we are dealing with a political party often accused of harbouring elements of fascism – that among the modern ideologies, national-socialism had perhaps the most extensive use of wolf-semiotics (Arnds 2015: 122–125), which entailed fascination and an ideal of emulation rather than rejection of the wolf. In Nazi ideology the 'wildness' and the merciless struggle for survival identified with 'untamed' nature became a racial and eugenic social ideal. While the Nazis actually pioneered a number of groundbreaking conservationist laws, 'wild nature' was also brutally incorporated into the wider national socialist ideology (Uekötter 2006; Chapoutot 2012).

The 'ecological imaginary' of Nazism is certainly complex and cannot be fully unfolded here. It contained a tension between technological modernism and ideological fascination with 'nature' and 'natural struggle' (Fritzsche 1996; Griffin 2007), as well as between the idealization of untamed 'wildness' and the *Blut und Boden* celebration of agriculture (Chapoutot 2012). But the wolf nonetheless remained a favourite metaphor for the 'life of struggle' which would forge the national-socialist 'New Men' in the image of Nietzsche's 'blond beast' (Arluke and Sax 1992). The wolf in Nazism came to represent animal instinct as a 'rebellion against culture and intellectualism' (Arluke and Sax 1992: 10), the antidote to the softness and the effeminate degeneration of modern urban life. Thus, not only has Hitler's personal fascination with wolves been noted by several biographers (he choose 'Wolf' as his party code name, and even supposedly asked his sister to change her name to 'Frau Wolf' [Langer 1972: 83, 217–218; Roberts 2009: 136]), but he also disseminated this fascination into the official ideological and military mythologies of the Third Reich. For example, he referred to the *Schutzstaffel* (SS) and the Hitler Youth respectively as his 'pack of wolves' and 'wolf cubs' (Victor 1998: 89), and named his various headquarters *Wolfschlucht* and *Wolfschanche* (Felton 2014: 123). Indeed, the planned Nazi resistance after the war was named 'Werewolf' in order to play 'on the primal anxieties of the savage and relentless ferocity of the wolf' (Fritz 2004: 195). Even today, depictions of wolves remain among the most popular animal symbols for neo-fascist and extreme right groups (Caiani and Parenti 2013: 88; Forchtner and Kølvraa 2017).

The classical mythology of the wolf is certainly echoing in many of the DPP's claims in the wolf-debate – from Espersen's designation of them as 'ravenous beasts' to his subsequent claim that 'wolves target children and the elderly' (Thiis 2017). Indeed, the authority of these statements seems linked to the plots of narratives like 'Little Red Riding Hood', not to zoological fact. It is equally clear, however, that in spite of the accusation of being 'wolves in sheep's clothing', there is no trace of the 'fascist' fascination with the wolf in the DDP's performance in the wolf-debate, or in their discourses on the environment more generally. In fact, their 'myth' of the wolf is to an astonishing degree one which sides with the sheep, and which ultimately unfolds an environmental imaginary informed by an anthropocentric hegemony so extensive that it ultimately questions even the wolf's cultural privilege as the feared representative of a wild untameable nature.

The politics of lupophobia

From the very onset of the debate, the DPP's rejection of the wolf never entailed painting it as an 'invasive species'; that is, as a foreign element diluting, destroying or degenerating the unique particularity of Danish nature. The wolf was always granted at least the fact that it was 'returning'. Its rejection rested not on a claim that it was a 'cultural stranger', but on the argument that it constituted a risk and a danger to the rural populous. The fundamental argument was thus one of 'space'. The DPP claimed above all that there was simply 'no room' for a large predator in a country as relatively densely populated and heavily farmed as Denmark (see Thiis 2017). Wolves in Denmark was as such a dangerous proposition simply because they were bound to encounter people, unlike – it was pointed out with an eye to the 'expulsion' solution – in countries such as Canada, Norway or even France.

However, the factual basis of this argument could be relatively easily challenged. It was not only that the 'problem' in fact amounted to a maximum of 11 animals, which zoologists assured the public were inherently timid and shy towards humans, or that it had (at least in comparison with other environmental or societal challenges [see Esbjerg and Schleicher 2018]) a very limited geographical and economic extent, but also that the fundamental idea of a lack of room could be questioned, both as it related to human society and to the ecosystem. Regarding the former, Northern Spain with an unproblematic population of about 600–800 wolves, has a comparable human population density; regarding the latter, Denmark actually struggles with an overproduction of deer exactly because there are no large predators to keep this stock in check (DR-TV 2018).

However, just as the DPP in the debates on climate change developed two distinct lines of argument, and could as such shift away from the increasingly untenable position of 'scientific' climate change denial towards one of ideological suspicion of the 'climate change lobby', so the party in the wolf-debate proved equally successful in pivoting away from a factual space where they were confronted by voices of scientific knowledge (biologists, zoologists and other 'wolf'experts) into a more hospitable ideological home ground. It is not that the cultural semiotics of 'ravenous beasts' defeated the discourses of scientific knowledge, but rather that the former enabled a shift into the subjunctive mode. This left behind the dreary 'expert knowledge' of 'what has happened' (almost nothing) and 'what is statistically likely to happen' (actually nothing), for the much more flexible space of imagining 'what might happen', which, given the infinity of the imagination and the rich material of the cultural semiotics of the wolf, was quite a lot. The DPP's Pia Adelsteen already had imagined a scenario of 20 kindergarten children on a picnic accidentally separating a she-wolf from her cubs (Adelsteen 2017). The horrifying results needed not even be articulated, but justified her determination not to 'wait for a child to get bitten before doing something' (Pedersen 2017) and her later conviction that it was 'only a matter of time before something violent happens' (Andersen 2018). Similarly, Søren Espersen offered the prospect of babies being attacked in their trolleys or old people tripping over branches in the forest only to be immediately fallen upon by wolves they did not even know had been stalking them. He even argued that we might unwittingly be preparing our own doom, since, if we successfully barred the wolves from feeding on our sheep, this would only facilitate them turning to human prey instead (Jyske Bank TV 2018).

The effect of such subjunctive discourse was to shift the object of the political debate from the likelihood of being attacked by wolves on the basis of what is known, to the legitimacy of harbouring a fear of wolves on the basis of what might be imagined. While the former could be challenged, the latter seems unassailable. The problem which now demanded a political solution was no longer the actual wolf but rather the 'popular lupophobia' of which the wolf was simultaneously statistically innocent and culturally guilty. In other words, it was no longer crucial that wolf attacks have been all but unknown in Europe in the past half decade, because the mere horrific fantasies so readily (and culturally) available when it came to the wolf meant that people were nonetheless adversely affected by its presence. While nobody was actually being stalked by wolves, the fear of wolves could be portrayed as in itself something that stalked and hampered the lives of those affected by it. As Karina Due reported: 'we are already now hearing that there are many who no longer dare to let their children go out alone' (Kynde and Hartung 2018a). And once the problem was 'psychologized', it immediately became a much more expansive category. The psychological burden suffered, hinged not just on the fear of attacks on humans, but equally regarded the trauma of finding one's sheep attacked, or ultimately even imagining such a calamity befalling one's husbandry. Karina Due personally enacted this trauma as she announced on TV – filmed in her empty sheepfold – that she had sold her five sheep, not because of an attack, but because the fear itself had been too much for her. She added as a general dictum that 'people are worried about the wolf, and we have to accommodate that worry' (TVMidtvest 2018). 'Popular lupophobia' could thus – quite divorced from actual wolves – be assigned its own very real and socially detrimental effects. Effects that in and of themselves demanded a political response. What the DPP primarily articulated across the debate was in fact not claims to the actual dangerousness of wolves – indeed Karina Due admitted that there was no real risk to people (DR-TV 2018) – but rather the demand that 'the politicians', 'the experts' or 'those in favour of the wolves', should 'take seriously', 'recognize' or 'accommodate' the worry and anxiety which supposedly plagued the population in the wolf-inhabited part of the country (see Bergmann 2018; Dohrmann 2018; Ritzau 2018). Crucially however, the fact that the problem was now psychological rather than zoological did not prevent the DPP from advocating a solution which targeted the latter rather than the former. Indeed, unlike other ideologies fond of 'the people', populism alone harbours no ambitions to shape, better, educate or realize the populous into its true and worthy form (the responsible citizen, the fascist new man, the self-aware working class) (Mudde 2004). It is certainly significant that there is no idea that lupophobia – as a factually mistaken belief –might be eliminated in the realm of 'belief', i.e. by changing the signification of the wolf through information or education. Indeed, such a privileging of 'expert knowledge' over and against the people's own ideas would be fundamentally at odds with core ideological tenets in populism. The solution to lupophobia is thus not to attack its mistaken signification, but to rid oneself of the referent which gives occasion for its articulation – however innocent this creature might be in the semiotics afforded it by human societies. The wolf ultimately becomes an unfortunate collateral damage accepted as a necessary prize to pay in the fight against popular lupophobia.

But one might even question if this so noisily declared fear is really what animates the DPP's discourse and explains its affective intensity. As Barthes argued in relation to his idea of the 'third meaning', something done to excess – overdone if you will – reveals its own artifice and should be interrogated as a 'disguise' which as such inadvertently points to that which it seeks to cover

over (Barthes 1977; Owens 1980; Oxman 2010). Indeed, Barthes distinguishes between a loudly proclaimed 'sticky' emotionality and the kind of deeper emotional dispositions implicitly revealed in and behind the artifice of the third meaning; 'Caught up in the disguise, such emotion is never sticky, it is an emotion which simply designates what one loves, what one wants to defend' (Barthes 1977: 59). This might be taken to mean exactly that to explore the actual affective dynamics behind a certain representation, we need to question and to move through the 'disguise' of those emotions intentionally declared on the surface of discourse. Furthermore, the shift into the subjunctive mode can be linked to exactly such a moment of revealed artifice; to the moment in which a discourse inadvertently reveals its own fictionality and releases its 'third meaning' (Zelizer 2004). When approaching the continuous articulation of 'popular lupophobia' as a disguise in this Barthesian sense, one does notice that what really characterizes much of the DPP's grievances is not a deep-felt horror at the prospect of getting torn to pieces, but what might be described as a possessive annoyance at being inconvenienced in the free and unworried use of and access to nature. This kind of attitude is reminiscent of the poujadist populism Barthes also analysed, and at whose heart he located exactly a *petit-bourgeois* idea of 'moral-bookkeeping' in which a hatred of all things unquantifiable and therefore 'useless' (above all the 'idle intellectuals') was combined with the small shopkeeper's dread of being 'taken for a fool', of not getting what one is owed or what is rightfully one's property (Barthes 1997; Shields 2004). There is indeed no small measure of such possessive and self-righteous bellicosity when, for example, Karina Due declares that 'it cannot be right that children cannot play outside because there has to be room for wolves' (Vestergaard 2018). Or when Søren Espersen can readily admit that the only wolf attack in decades occurred when a rather incautious Canadian went running wearing earplugs in a known wolf habitat, because as Espersen simply exclaims 'that too [the behavior of the jogger *however careless*] is within our rights!' (Jyske Bank TV 2018).

Thus, the wolf-debate – and the DPP environmental imaginary more generally – is not simply the classical and fearful rejection of 'wild nature' in defence of the security of human societies. This would still grant to nature at least the privilege of being an entity to be feared, and to the wolf at least a measure of antagonistic reciprocity. But there is no real fear of getting eaten here. There is in reality rather a poujadist annoyance at somebody who infringes on our rights and our enjoyment of what is, after all, *our* countryside. The wolf is ultimately marked for eradication, not because it actually *is* dangerous, nor even because it is (mistakenly, but honestly) *believed* to be dangerous, but because – as a trespasser or a vagrant, rather than as an intruder or a threat – its mere presence might lessen our enjoyment, and thus 'cheat us' in the context to what thereby emerges as a possessive relationship between humans and their environment. Danish nature, or rather the Danish countryside is manifestly thought of as the property of the Danes, distributing the 'rights of use' solely to humans rather than to wolves. This points to an environmental imaginary configured around the anthropocentric hegemony of a human right not just to 'own' nature, but to on this basis rid it, not just of what is or is believed to be a threat, but of everything which in anyway lessen or fail to contribute to our unfettered enjoyment of it. As one critic of the DPP aptly put it, this entailed a 'design of nature' in which 'we only want the cute, the innocent and the useful – those with value (…) whereas those who are rough, ugly or slimy we don't like' (DR-TV 2018). Ultimately this is a hegemony which subordinates nature not only to human material needs and securities, but to human psychological preferences and phobias – ridding it of everything that might worry, annoy or inconvenience us. Søren Espersen ultimately unfolds the DPP's environmental imaginary in a single sentence, when he – in response to the proposition that the wolf was part of 'Danish nature' and thus should be accommodated – snapped: 'My fundamental claim is this: We don't have 'nature' in Denmark (…) and we cannot accommodate nature, at least not beyond the scale of frogs and the like' (Jyske Bank TV 2018).

Conclusion

The DPP's environmental imaginary can – through their positioning in, for example, the climate change debates of recent decades – be deemed both national in outlook and populist in its suspicion of political and scientific elites. During the climate change debate, the DPP proved adept at shifting between the factual ground of 'scientific' climate scepticism and a more ideological critique of the actors in the 'climate change lobby'. This however does not tell us very much about their 'environmental imaginary' in a wider sense. I have suggested that this might be more discernible by examining the party's discourse in the so-called wolf-debate in the spring of 2018.

In this debate the DPP also found itself confronted by various experts (zoologists, biologists), and again managed to largely move the discussion away from the 'facts' of wolves and towards a concern with 'popular lupophobia', calling for the recognition of – and political action against – the fact that people were afraid of wolves, groundless as that fear might be. I have argued that the fact that the DPP continuously advocate eliminating the wolf from the Danish countryside – not based on any provable danger, but solely in order to accommodate those who are nonetheless worried – bears witness to the extent of the anthropocentric hegemony that characterizes an environmental imaginary closer to the ideology of poujadist populism, than to that of fascism. Perhaps due to their ideological legacy and origins – which as noted is not one informed by fascist thinking or affiliations – the DPP has little in common with the organicist, ecological, and often more overtly racialist, imaginaries to be found towards the extreme end of the far-right spectrum. Their enviromental imaginary is by contrast almost radically anthropocentric. It is ultimately an idea of (Danish) nature in which not just human needs and security, but human enjoyment, ease and freedom from any worry or inconvenience, are the structuring laws. In this imaginary, the relationship between the Danes and their countryside is ultimately imagined in terms of property and the resultant rights of use.

The wolf is ultimately condemned not as a feared invader, but as an annoying trespasser in a countryside where the rights of occupation and enjoyment have long been passed wholly and totally to its human inhabitants.

Notes

1 Part of the corpus overlaps with that utilized in earlier work with Bernhard Forchtner (Forchtner and Kølvraa 2015). The corpus includes ca. 70 texts/media products. About a third of them – as well as the two one-hour long TV debates included – are associated with the 2018 wolf debate.
2 All quotes from the DPP and DPP members have been translated from Danish by the author.

References

Adelsteen, Pia (2017): Debat: Ulven hører ikke hjemme i Danmark, *BT*, 10 July.
Andersen, Henning K. (2018): Der er ikke plads til ulve i Danmark, *Effektivt Landbrug*, 21 March.
Anderson, Benedict (1983): *Imagined Communities: Reflections on the Origin and Spread of Nationalism*. London: Verso.
Arluke, Arnold and Sax, Boria (1992): Understanding Nazi animal protection and the Holocaust, *Anthrozoös*, 5(1): 6–31.
Arnds, Peter (2015): *Lycanthropy in German Literature*. Basingstoke: Palgrave.
Barthes, Roland (1977): *Image, Music, Text: Essays*. London: Fontana.
Barthes, Roland (1981): *Camera Lucida: Reflections on Photography*. New York: Hill and Wang.
Barthes, Roland (1997): *The Eiffel Tower, and other Mythologies*. Berkeley: University of California Press.
Barthes, Roland (2013): *Mythologies*. New York: Hill and Wang.
Bergmann, Brian (2018): Slut med ulve-frygt: Flere partier vil hegne ulven ind, *TV2Nord.dk*, 20 April.
Bettelheim, Bruno (1976): *The Uses of Enchantment: The Meaning and Importance of Fairy Tales*. Harmondsworth: Penguin.
Byrne, Tommy (2018): S og DF: Danske ulve skal hegnes inde, *Fyens stifttidende*, 19 April.
Caiani, Manuela and Parenti, Linda (2013): *European and American Extreme Right Groups and the Internet*. Burlington: Ashgate.
Chapoutot, Johann (2012): The Nazis and nature. Protectors or predators? *Vingtième Siècle. Revue d'histoire*, 113(1): 29–39.
Cosgrove, Denis (2004): Landscape and landschaft, *German Historical Institute Bulletin*, 35: 57–71.
Dohrmann, Jørn (2018): Ulve. Flyt ulvene væk fra mennesker, *JydskeVestkysten*, 10 May.
DPP (2018a): *Mærkesager*.https://danskfolkeparti.dk/politik/maerkesager/ [24 May 2018].
DPP (2018b): *Principprogram*.https://danskfolkeparti.dk/politik/principprogram/ [24 May 2018].
DR-Radio (2015): Natursyn, *P1*.
DR-TV (2018): *Debatten – Ulven Kommer*.
Esbjerg, Anette and Schleicher, Søren B. (2018): Andre dyr er langt farligere end ulven. Lille Rødhætte er ved at vinde over ulven, *Fyens Stiftstidende*, 5 May 2018.
Felton, Mark (2014): *Guarding Hitler: The Secret World of the Fuhrer*. London: Pen and Sword.

Forchtner, Bernhard (2019a): Far-right articulations of the natural environment: An introduction. In: Bernhard Forchtner (Ed). *The Far Right and the Environment: Politics, Discourse and Communication.* London: Routledge.

Forchtner, Bernhard (2019b): Nation, nature, purity: Extreme-right biodiversity in Germany, *Cultural Imaginaries of the Extreme Right. Special Issue of Patterns of Prejudice,* 53(3): 285–301, DOI: 10.1080/0031322X.2019.1592303.

Forchtner, Bernhard and Kølvraa, Christoffer (2015): The nature of nationalism. Populist radical right parties on countryside and climate, *Nature and Culture,* 10(2): 199–224.

Forchtner, Bernhard and Kølvraa, Christoffer (2017): Extreme right images of radical authenticity: Multimodal aesthetics of history, nature, and gender roles in social media, *European Journal of Cultural and Political Sociology,* 4(3): 252–281.

Forchtner, Bernhard, Kroneder, Andreas and Wetzel, David (2018): Being Skeptical? Exploring far-right climate-change communication in Germany, *Environmental Communication,* 12(5): 589–604.

Freud, Sigmund (1962): *Civilization and Its Discontents.* New York: Norton.

Fritz, Stephen G. (2004): *Endkampf: Soldiers, Civilians, and the Death of the Third Reich.* Lexington: University Press of Kentucky.

Fritzsche, Peter (1996): Nazi modern, *Modernism/modernity,* 3(1): 1–22.

Griffin, Roger (2007): *Modernism and Fascism: The Sense of a Beginning under Mussolini and Hitler.* Hampshire: Palgrave Macmillan.

Hervik, Peter (2011): *The Annoying Difference: The Emergence of Danish Neonationalism, Neoracism, and Populism in the post-1989 World.* New York: Berghahn Books.

Juul Christiansen, Flemming (2016): The Danish People's Party: Combining cooperation and radical positions. In: Tjitsken Akkerman, Sarah L. Lange and Matthijs Rooduijn (Eds). *Radical Right-Wing Populist Parties in Western Europe: Into the Mainstream?* London: Routledge.

Jyske Bank TV (2018): *Borgen Late Night.*

Kemp, Peter and Nielsen, Lisbeth (2009): *The Barriers to Climate Awareness.* Copenhagen, Denmark: Danish Ministry of Climate and Energy.

Kjærsgaard, Pia (2009a): Giv os propertionerne tilbage, *Pia Kjærsgaards Nyhedsbrev.*

Kjærsgaard, Pia (2009b): Klimafantaster og klimarealister, *Pia Kjærsgaards Nyhedsbrev.*

Kynde, Rikke and Hartung, Berit (2018a): Politikere toppes om ulven i Danmark, *BT,* 19 April.

Kynde, Rikke and Hartung, Berit (2018b): 'Vi skal ikke vente til ulven har angrebet et barn', *BT,* 18 April.

Langer, Walter C. (1972): *The Mind of Adolf Hitler: The Secret Wartime Report.* New York: Basic Books.

Lockwood, Matthew (2018): Right-wing populism and the climate change agenda: Exploring the linkages, *Environmental Politics,* 27(4): 712–732.

McKewon, Elaine (2012): Talking points ammo, *Journalism Studies,* 13(2): 277–297.

Miles, James Kristoffer (2017): Espersen sammenligner ulve-drab med slagtning af grise, *Ekstrabladet,* 26 December.

Mudde, Cas (2004): The populist zeitgeist, *Government and opposition,* 39(4): 541–563.

Olsen, Jonathan (1999): *Nature and Nationalism: Right-wing Ecology and the Politics of Identity in Contemporary Germany.* Basingstoke: Macmillan.

Olwig, Kenneth R. (2003): Natives and aliens in the national landscape, *Landscape Research,* 28(1): 61–74.

Owens, Craig (1980): The allegorical impulse: Toward a theory of postmodernism part 2, *October,* 13: 59–80.

Oxman, Elena (2010): Sensing the image: Roland Barthes and the affect of the visual, *SubStance*, 39(2): 71–90.
Palmer, Catherine (1998): From theory to practice: Experiencing the nation in everyday life, *Journal of Material Culture*, 3(2): 175–199.
Pedersen, Mette (2017): Efter heftig debat: Disse ulve er per definition livsfarlige, *Ekstrabladet*, 8 July.
Rahmstorf, Stefan (2004): *The Climate Sceptics*. Potsdam, Germany: Potsdam Institute for Climate Impact Research.
Ritzau (2009): Klima – den nye udlændingepolitik, *Børsen*, 21 September.
Ritzau (2017): Blodigt år for får: Danske ulve har dræbt 20 gange, *dr.dk*, 23 December.
Ritzau (2018): S og DF: Danske Ulve skal indhegnes, *Nordjyske.dk*, 19 April.
Roberts, Andrew (2009): *The Storm of War: A New History of the Second World War*. London: Allen Lane.
Shields, James G. (2004): An enigma still: Poujadism fifty years on, *French Politics, Culture & Society*, 22(1): 36–56.
Siim, Birte and Meret, Susi (2016): Right wing populism in Denmark – people, nation and welfare in the construction of the 'other'. In: Gabriella Lazaridis, Giovanna Campani and Annie Benveniste (Eds). *Rise of the Far Right in Europe: Populist Shifts and 'Othering'*. London: Palgrave Macmillan.
Stie, Hans-Henrik Busk (2018): Partier vil indhegne ulve i reservater – torskedumt, mener debattør. *TV2 Nyheder*, 19 April.
Thiis, Elisabeth (2017): DF vil dræbe ulvene efter nye angreb. *TV2 Nyheder*, 24 December.
Tønnessen, Morten (2016): The semiotics of predation and the umwelten of large predators. In: Timo Maran, Morten Tønnessen, Kristin Armstrong Oma, Laura Kiiroja, Riin Magnus, Nelly Mäekivi, Silver Rattasepp, Paul Thibault and Kadri Tüür (Eds). *Animal Umwelten in a Changing World. Zoosemiotic Perspectives*. Tartu, Estonia: University of Tartu Press, pp. 150–181.
TVMidtvest (2018): '*Flertal vil have indhegnet ulven*'.
Uekötter, Frank (2006): *The Green and the Brown: A History of Conservation in Nazi Germany*. Cambridge: Cambridge University Press.
Van Rensburg, Willem (2015): Climate change scepticism: A conceptual re-evaluation, *SAGE Open*, 5(2): 1–13.
Vestergaard, Morten (2018): Rapport: Der kan komme et nyt ulvekobbel hvert år i 5-10 år, *Jyllands-Posten*, 24 January.
Victor, George (1998): *Hitler: The Pathology of Evil*. Washington, DC: Brassey's.
White, Ed (2012): *How to Read Barthes' Image-Music-Text*. London: Pluto Press.
Wolfhagen, Rune (2011): Bagsiden: Klimaet, det er som det er, *Information*, 19 July.
Zelizer, Barbie (2004): The voice of the visual in memory. In: Kendall Phillips (Ed). *Framing Public Memory*. Tuscaloosa: University of Alabama, pp. 157–186.

8
THE FAR RIGHT AND CLIMATE CHANGE DENIAL

Denouncing environmental challenges via anti-establishment rhetoric, marketing of doubts, industrial/breadwinner masculinities enactments and ethno-nationalism

Martin Hultman, Anna Björk and Tamya Viinikka

Introduction

Sweden is a country with high credentials in both environmental politics as well as in welcoming refugees. However, over the past couple of years Sweden has seen how ideas from organised climate change denialist groups have merged with a far-right nationalist political party creating a new political landscape. This chapter that follows will analyse and explain this shift.

Denouncing climate change as a hoax and climate science results as a scam is not a new phenomenon (Dunlap and McCright 2015). It began in the United States in the late-1980s as a tactic by extractive industries such as ExxonMobile (Oreskes and Conway 2011). In the United States, and to some extent in Canada and Australia, the fossil fuel interests and conservative think tanks collaboration in marketing denial has been revealed in an array of studies covering 15 years (e.g. McCright and Dunlap 2003; Young and Couthinho 2013; Farrell 2016). Not many studies – up until this anthology – have thus studied how organised and lay climate change denial has spread in Europe as part of far-right nationalism (Hultman and Kall 2014; Forchtner and Kølvraa 2015; Jeffries 2017; Forchtner et al. 2018; Lockwood 2018). This tendency is just starting to get scholarly attention, despite its possibility of providing new and important knowledge, specifically of resistance towards effective climate politics as well as of broader political issues such as democracy, human rights and diversity. Only three empirical studies have been conducted so far on this specific topic. Regarding online communication in Germany, it is found that 'skeptic counterpublic is not restricted to voices pertaining to climate change, but forms an alliance of antagonism with other extreme fractions such as mysogynists, racists, and conspiracy theorists' (Kaiser and Puschmann 2017). In a more qualitative selected corpus of magazines and a blog ranging from anti-liberal to neo-Nazi, still on Germany, many of the familiar arguments and tropes regarding climate denialism have been found such as climate science being a religious cult, an elite money-making scam and invented by mainstream media (Forchtner et al. 2018). In a quantitative study of Norway based on Gallup data, a link between xenophobic values and climate denialism was found (Krange et al. 2018). Here we take a much broader perspective on far right and climate change denial when it is described in its historical and socio-political context, all exemplified with Sweden.

This chapter deals with the environmental communication in our current political landscape from 2013 with a specific focus on climate change denialism of *Sweden Democrats* (SD). After two consecutive elections (2010 and 2014), they have been firmly placed in an unneglectable swing role that Swedish governments of both liberal-conservative and social democratic-green shapes have needed to take into account. After the election 2018 they 'forced' the both centrist parties in Sweden (Liberal- and the Centre Party) to form a government with the Greens and Socialdemocrats. We follow the arguments from and texts written by SD politicians. Three complementing materials are analysed: First, texts originating from members of SD such as parliamentary speeches, debate articles in mass media as well internal member papers; second, texts originating from the organised climate change denial group *Stockholm Initiative* (SI); and third, text that overlaps both groups. We use a combination of qualitative actor-network and discourse methodology looking both at the actors and how they are connected in networks as well as what type of language they are using. The data set consists of one parliament debate from 2013 and 38 opinion pieces and proposals from 2014 to 2017; only printed articles written by SD have been chosen (Björk and Viinikka 2017). In addition to this, we also included tweets connected to those articles from leading SD politicians in our analysis. The search resulted in articles from some well recognised Swedish newspapers, but also Sweden Democrats own papers, *SD-Currier* (*SD-Kuriren*) and *Our Time* (*Samtiden*). Since 2003, the *SD-Currier* has been the official member magazine of the Swedish Democrats, but has existed since the formation of the party in 1988 (Axelsson and Borg 2014). *Our Time* is their web magazine. Regarding SI, we have analysed letters to editors and discussions on their own homepage called *Klimatupplysningen* (Climate Enlightenment) as well as secondary sources such as e-mails between SD and SI presented in mass media.

We start by looking at a wider context as well as the historical background of SD. Then we present their environmental politics overall and provide an in-depth analysis of the merge and rise of far-right nationalism and climate change denial actor-network-discourse, with focus on anti-establishment rhetoric, marketing of doubts, industrial/breadwinner masculinities and ethno-nationalism.

Historical and political context – climate change denial and the Sweden Democrats

Climate change denial became an increasingly prominent phenomenon in Sweden following the 2007 report from the Intergovernmental Panel on Climate Change (IPCC) (2007), the movie *An Inconvenient Truth* (2006) and the Stern report (2006), all of which is made clear in the book *Discourses of Global Climate Change* (Anshelm and Hultman 2014a). While causing such reactions globally, in Sweden three right-wing think tanks took up the rhetoric and two organised denial group was formed – though initially, Swedish parties showed no interest to deny climate science. Organized and party political climate change denial was marginal at this time, but the organised group SI, almost exclusively run by older influential men from large companies or professor emeritus from academia, made an significant impact. In the summer of 2008, the major Swedish newspapers and public radio granted space for articles and statements from SI, which happened directly after the network formed, demonstrating the influence of these actors in the debate (Anshelm and Hultman 2014a). During the period from 2008 to 2010 denialism still seemed to have limited influence on the Swedish party politics even though a couple of workshops were organised by liberal-conservative members of parliament in collaboration with SI (Anshelm and Hultman 2014b; Hultman and Kall 2014b).

The Sweden Democrats are a nationalist, social conservative party with roots in Nazism (Axelsson and Borg 2014; Ekman and Poohl 2010). The history of the party goes back to 1986 when the so called *Swedish Party* was formed from a merge between *Preserve Sweden Swedish* and *Progress Party*. Following the almost immediate split of the party, the Swedish Democrats were formed (Axelsson and Borg 2014; Wåg 2014). SD's first board consisted exclusively of men and were elected into parliament for the first time in 2010 (Ekman and Poohl 2010; Ekman 2014). With its history and present ideas regarding non-migration, anti-feminism and ethno-nationalism; researchers in Sweden have called them neo-Nazi (Peterson 2016), fascists (Arnstad 2015), right-wing populists (Widfeldt 2008), nationalists (Hellström et al. 2012), extreme right (Mulinari and Neergaard 2014), radical right (Erlingsson et al. 2014) or nationalist populist party (Oja and Mral 2013); none with an interest of analysing their climate change policy. The most recent classification made by one respected researcher on fascism in Sweden, judged them 'neo-fascists' and that is the classification we will hereby continue to use in this chapter on par with far-right nationalist (Berggren 2017). At the time of writing this chapter, SD got on average 18% of the votes in the national polls and received 15.5% in the EU-election 2019.

Environmental issues have been part of the Sweden Democrats politics since 1989. In the party programme from that year, the Swedish Democrats took up environmental issues which were on the Swedish political agenda at the time (Ekman and Poohl 2010; Peterson 2016). Back then, the focus was on protecting an etno-nationalist idea of 'Swedish' landscapes and 'Swedes' health and their animals, in line with their Nazi root ideology (Ekman and Poohl 2010). It was not until a couple of decades later that their environmental policies made it into the public debate, this time aligned with outspoken climate change denialism under the influence of old powerful men from the *Stockholm Initiative* (Baas 2016a). When they entered into parliament, Sweden Democrats got not only more visibility, but also gained support from the climate denialist network SI. The chairman of SI congratulated the leader of SD on the election (Baas 2016) and in comments about SD energy politics many supportive views are shown at SI's homepage (Klimatupplysningen 2012). The former SD Member of Parliament Thoralf Alfsson has been a very active part of discussions at the homepage of SI (Alfsson 2014). In 2018 the influential climate change denier Lars Bern did a series of talks in the right-wing nationalist television channel SwebbTV (Bern 2018b). How did this merge happen?

SD brings climate change denial into the parliament politics

The visibility and influence of organized climate change denial changed dramatically from the day when SD entered the Swedish parliament in 2010 and core members of SI started close collaboration with their energy and environmental spokesperson Josef Fransson. In 2013, Fransson then voiced climate denial ideas for the first time ever in the Swedish parliament when he described the Green Party as their main enemy, climate scientist as corrupt and government agencies as dominated by vegan extremists (Hultman and Kall 2014). Some years later, when the Paris Agreement generated even stronger climate change denial campaigns around the world aimed at undermining commitment to fulfil national goals (Dunlap et al. 2016). SD had firmly established themselves as the counter climate science political party. This resistance is similar to what happened in the United States since signing the Kyoto Protocol in 1998 (McCright and Dunlap 2003); today with a far-right nationalist twist to it (Forchtner and Kølvraa 2015; Krange et al. 2018; Jylhä et al. submitted). In Sweden, climate change denialism gained increased visibility during 2016–2017 (Vi-skogen 2017), a trend detectible also in the opinion pieces we have analysed from SD. SD voted no for the ratification of the Paris Agreement on 12 October 2016, in contrast to all other parties in the parliament. In October 2016, the Swedish Democrats presented a budget proposal for spring 2017 where the climate issue was given only half a page and using quotation marks around the word 'climate'. The Swedish Democrats also describe the climate debate as 'weird' in the same budget. They wanted to reduce funding for Swedish Meterological and Hydrological Institute (SMHI), with the argument that SMHI exaggerates the seriousness of climate change. In their next budget proposal, the Swedish Democrats wanted to reduce climate-related efforts by SEK 8 billion (Alestig 2017). Another very visible sign of the merge between climate change denialism and far-right nationalism was presented in autumn 2016 when a coalition of far-right movements in Europe awarded Václav Klaus their most prestigious prize European Freedom Award at a public ceremony in Sweden organised by SD. Since the start of the Stockholm Initiative organisation Klaus had been one of the few upheld politicians in the group because of his climate change denialism (Thauersköld Crusell 2009). This makes Klaus the perfect token of the merge between both groups as well as a possibility to understand the ethno-nationalists as connected by climate change denialism not least since most of the far right nationalist parties such as AfD, Ukip is part of the award committee. How does SD's climate change denial argumentation look like in forms of anti-establishment rhetoric, marketing of doubts, industrial/breadwinner masculinities and economic growth nationalism?

The forms of climate change denial

As noted in previous research, climate change denial comes in various forms and shapes. Which are the ones that are present in the Far Right?

The anti-establishment rhetoric

On 29 January 2013, in a debate on climate the Swedish Democrat Josef Fransson (the party's spokesperson on energy and environmental issues) said that 'the apocalyptic scenarios that many people believe in regarding the climate is false'. He described the Green Party as their biggest opponent and a communist nightmare (in accordance to the widespread watermelon metaphor) and spoke of climate researchers as only in it for the money (Hultman and Kall 2014). Presented in this way, the climate issue fitted a rhetorical pattern created by the SD, in which they present themselves as the party of the masses in contrast to the false elites (Hellström and Nilsson 2010). Fransson was in his speech to the parliament sarcastic towards all researchers, human rights organisations and politicians that have in his view 'built a lucrative career on global warming fear'. It has even gone so far, according to Fransson, that the vegan movement has 'taken over the Swedish Board of Agriculture', because the authority discusses the climate impact of meat. In his speech in front of the parliament, Fransson put forward many classical climate denial arguments such as science is not set, Swedish emissions are comparatively low (despite being around 10 tonnes per capita) and alluded to a conspiracy in which 'the elites' created the climate issue to be used as a dominating strategy, enforcing a communist economy on all humans (Fransson 2013).

There are very explicit similarities in the anti-establishment/anti-elite rhetoric employed by far-right and climate change deniers (e.g. Kaiser and Puschmann 2017). Those who dismiss climate science findings commonly claim that the mainstream scientists and politicians are distorting the scientific evidence due to their own personal interests, thereby intimidating and misleading the people (e.g. Anshelm and Hultman 2014b; Krange et al. 2018). Importantly, it has been put forward that the true motivation behind this far right idea in Sweden is a motivation to protect the existing power relations combined with xenophobic and anti-minority attitudes rather than anti-elite attitudes (Müller et al. 2014). It comes as no surprise that neo-fascists are trying to create an anti-establishment rhetoric with a similar perspective to climate change. SD markets themselves as a party for the man on the street who stands on the side of the people against the elite (Hellström and Nilsson 2010). The climate issue fits well into this pattern, not least in critique of the ecomodern hegemony in countries such as Sweden and Norway, as is also evident in the aforementioned parliamentary debate (Anshelm and Hultman 2014b; Krange et al. 2018). In countries such as Poland, and to some extent China, where climate change denial fit very well with the policies of the ruling party, the logic is more or less the opposite (Kundzewicz et al. 2017, Liu and Zhao 2017).

Marketing of doubts

One prime tactic of climate change denial since the late 1980s has been the marketing of doubts against climate science (Oreskes and Conway 2011). This has become a prime tactic of far-right movements as well; not least in the name of so called 'fake news', a concept far-right mass media machines induce on their opponents (Ott 2017). In 2014 Fransson wrote in the web magazine *News 24* that climate models are 'useless' and the UN IPCC presents conclusions that they do not have research to base their findings on (Fransson 2014). In the months leading up to the Paris meeting, Fransson was very active in social media as well as in mass media. On Twitter he exclaimed that the meeting was frivolous and suggested a YouTube clip featuring climate denialist Richard Lindzen, connected to the Cato Institute that is funded by Koch Industries. In one interview with the magazine *KIT*, Fransson claimed that there has been 'no significant warming the last 17–18 years' (Olsson and Agö 2015). In an autumn 2015 debate article, he claimed that higher levels of man-made carbon dioxide were a blessing as 'the soil become greener and crops grown larger due to the plants greater access to carbon dioxide. In a couple of decades, I think we look back at today's emergency alarm regarding the climate with amazement' (Fransson 2015). Fransson, being his party's energy and environment spokesperson during this period, is a significant voice, and his judgment is supported by other parliamentarians as well as municipality politicians of SD. They are using the whole palette of climate denialism arguments such as people against elites, corrupted research, the IPCC in communist overtake attempts as well as climate nationalism (e.g. Forchtner et al. 2018). In one official SD party report about EU energy and climate it was written that:

> Although Brussels feels better than at home on crazy climate policy, it is most important to seriously question many of the initiatives that the EU says it wants to implement. The prevailing climate and energy policy will eventually endanger the EU's states as it is huge, not to say astronomical resources that must be used.
>
> *(SD 2013)*

Before early 21st Century, far-right ideology has not explicitly referenced climate change issues (e.g. Forchtner et al. 2018). However, in line with the merge, explicit in Sweden from 2008 onwards, linking nationalist far-right ideology to climate change denial, dismissal of climate change has become one of the core issues among far-right politicians and their voters. Denying climate change was, for example, a major theme in Donald Trump's successful presidential campaign (Dunlap et al. 2016). It seems plausible the link between far-right nationalism and climate change denial reflects a continuation of the general correlation between right-wing ideology and denial from the early 1990s onwards, that has been identified in previous research (McCright and Dunlap 2011), but now with a much more aggressive, violent and explicit agenda.

Recently, core representatives of SD, such as Tomas Brandberg, continue to argue with climate change denial rhetoric, claiming it is 'the only party [which] discusses the great uncertainty in the climate models. It's sobering' (Brandberg 2017). Brandberg is a political secretary in the party, writing chronicles for two of their own papers. Previously, Brandberg had been employed by the Captus think tank, which is a conservative think tank formed in 2015, funded by the Foundation for Free Enterprise (Brandberg and Sanandaji 2008). Captus was one of the three conservative think tanks promoting climate change denials around 2008 in Sweden (Anshelm and Hultman 2014a). According to Brandberg (2017), a dismissive approach should be adopted against the climate models developed by SMHI and the UN Climate Panel. This is due to the difficulties in anticipating the climate of the future (Brandberg 2017) and the need for data from a collection period of at least 30 years (Brandberg 2015). Furthermore, Brandberg (2017) considers the SMHI is 'guessing' the Nordic climate for the next hundred years. Therefore, Brandberg believes that the SMHI should not receive government grants to speculate about the climate (Brandberg 2017). He summarises his standpoint saying: 'Many distinguished physicists believe that the cycles of the sun have a greater impact. Spontaneously, is it really unlikely that the earth's average temperature will be controlled by the sun!?' (Brandberg 2015). In a similar vein Martin Kinnunen, who is the SD's environmental policy new spokesperson since 2014 and member of the environment and agriculture committee, (Sveriges Riksdag odat.; Swedish Democrats 2014) claims: 'There is no certain knowledge today saying how increased carbon dioxide levels affect the climate' (Kinnunen 2016a). In bills sent into the parliament, Emilsson et al. (2014, 2015) and Filper et al. (2016b) write that: '(…) it is important that both the benefits and the disadvantages resulting from a warmer climate are included in the analysis'. SD spokespersons try to discredit the established authorities on climate change in Sweden by promoting uncertainties, thereby creating a gap between what they claim to be lay persons truer knowledge and the corrupt establishment. In their marketing of doubts SD uses arguments of warming trend denialism, if human beings stand behind this warming (attribution denialism), if it is good or bad (influential denialism) as well as the corruption of the science argument (Edvardsson Björnberg et al. 2017).

Industrial/breadwinner masculinities

Males are over represented in far-right nationalist movements (Mudde 2007) as well as holding climate change denial values in Sweden and many countries across the globe (for Sweden, see Wästberg and Lindvall 2017; Norway, see Krange et al. 2018; US, see McCright and Dunlap 2011; Brazil, see Jylhä et al. 2016). Since late 1980s climate change denial has been an activity for influential men with conservative values (Oreskes and Conway 2011). In research based on Gallup surveys in the United States, McCright and Dunlap (2011) found a correlation between self-reported understanding of global warming and climate change denial among conservative white men. In Sweden denial has historically, before the merge with SD, been articulated in the public realm by a small, homogeneous group of, almost exclusively, men and conservative think tanks. Research suggests that climate change denial is a form of identity and power position protective practise. These men have successful careers in academia or private industry, strong beliefs in a market society and a great mistrust of government regulation (Anshelm and Hultman 2014b).

To understand how far-right and climate change denialism merge, we can look at the group of climate denialists in Sweden, whose hub is the organisation Stockholm Initiative. A Swedish newspaper, *Expressen* (16 May 2016), revealed that SI Chairman Per Welander had congratulated Josef Fransson on his 2014 election victory and invited him to dinner. The Swedish Democrats' association with climate denialists from the Stockholm Initiative is done through think tanks such as Timbro and older male lobbyists like Lars Bern, Jonny Fagerström, Peter Stilbs and Per Welander (Baas 2016a, 2016b). These men have careers that would have been impossible without human exploitation of nature (as idea and in practice) and a constant fiery of fossil fuels. Acknowledging climate change would be to recognise how their own life project is unsustainable; and as individuals they today combine their climate change denialism with far-right nationalism when, for example, Lars Bern on his blog *Anthropocene* claim that SD is the only democratic party (Bern 2018a). The same applies to the neo-fascist position. To recognise the challenges and justice of climate change is to recognise that everybody in a country with general high consumption and emission have a responsibility. The problem (for these men) is that it is precisely the industrialised world dominated by white malestream norms and males in leading positions which bears the historical responsibility for climate change. That is an important aspect not recognised by neo-fascists; another aspect is that a climate policy which aims at limiting global warming to 1.5 or 2 degrees requires nation-states to give up some sovereignty. If the Paris Agreement is to have any real effect, national policy would be subjected to the global goal of limiting emissions. Hans Bergström, preface author of the climate denial Swedish book, *The Doomsday Clock*, is a key figure in this merge. He has been one of the influential persons to bring SD into the political discussions, and was the first one to use the description a Green Fatwá regarding climate science publically and has spoken about the real purpose of the EU's climate policy; that is, to transfer substantial amounts from European low-income groups to a United Nation fund that would pay out climate aid to African dictators (Bergström 2009). The merge between far-right nationalism and climate change denialism is based upon the ideological similarities in viewing the world from an industrial masculinities viewpoint, not wanting to let go of the colonial extractive logic that has served these men well, but violated the planet.

Nationalism – the climate friendly Swede and blaming all others

The nation plays a central role in the Sweden Democrats' texts and in their environmental policy in the way that global climate change is compartmentalised as a national issue. Of lately, SD has turned towards neo-liberalism supporting the marketisation of welfare systems (Wingborg 2016). A common argument the SD has for the emission reductions to take place abroad is Sweden's low national net emissions. Consequently, the Sweden Democrats argue that emission controls should take place in other parts of the world. The Swedish Democrats argue that climate policies could drive Swedish industry abroad. Therefore, Sweden should not adopt an ambitious climate policy as it will lead to increased emissions. Withholding the argument about Sweden's low net emissions as a nation, their Sweden Democrats national romanticism sometimes comes through: 'The natural love of the homeland is assumed to be a better and more natural way to protect nature than more abstract ideas about global climate agreements and the like' (Pettersson 2016). With that as a starting point, the nation is the prime subject and presented as being challenged by global issues: 'We in Sweden Democrats want greater focus on the environment in Sweden and our immediate area, while the other parties (including the Conservatives) often prefer to discuss global issues that Sweden has relatively little influence over' (Forsberg 2016).

In a similar vein, sociologists Mulinari and Neergaard (2014) discuss how members of SD have strong emotional connections to their dogs, cats and culturally produced ethno-Swedish landscapes. Such positive images of how we in Sweden care for our animals and the environment cannot be reconciled with Swedes high per capita environmental and climate impact. Despite Swedes on average contributing approximately 10 tons of carbon dioxide into the atmosphere every year and has on average an ecological footprint that needs 4.2 Earths to be sustained, SD argue that Swedes are not the problem (for similarities in Germany read Forchtner et al. 2018). For them the problem is everybody else; therefore, Sweden does not need to reduce emissions or lower its resources use: 'In view of the fact that carbon dioxide emissions increase significantly elsewhere in our world, we consider that more thorough impact assessments should be made that ambitious climate policy is unilaterally carried out in our part of the world with small emissions' (Emilsson et al. 2014; Fransson and Jansson 2014; read also 2015; Filper et al. 2016a, 2016b for similar statements). The nationalism manifests itself in the idea that ethno-Swedish jobs, living standards, rural areas and competitiveness do not need to be questioned (e.g. Bäckström Johansson and Fransson 2014; Forsberg 2016; Bäckström Johansson 2017). Kinnunen argues that Sweden Democrats feel a responsibility for Sweden's common assets in both the environment and economy issue (2016b). In a few very similar bills to the parliament, leading representatives of Sweden Democrats wrote that: 'The Swedish Democrats believe that if we are able to manage our industry and trade in the short and long term, and thus in the long run our jobs and welfare, we must consider what we actually release in relation to others when we decide on Sweden's emission reductions' (Emilsson et al. 2014, 2015; Fransson and Jansson 2014; Filper et al. 2016a).

The portrait of Sweden as a role model has not been a feature restricted to SD, but more or less dominates the Swedish climate debate overall. It has been a cornerstone of the liberal-conservative Swedish government's political proclamations on climate change between 2006 and 2014 and continued to be so with a Social democratic-Green party government 2014–2018 (Hultman and Anshelm 2017). First, it denotes that the country is part of an international competition, where it wears the leader's shirt. This metaphor implied that it is an advantage for Sweden to rapidly and forcefully take measures against climate change. Early changes in the national energy system and the development of 'carbon-free' technology, such as nuclear and transformations of production processes, were depicted as steps that could position the country favourably on the global economic market, since all countries sooner or later would have to go through the same process. Moreover, it would engender international respect and recognition for Sweden (Anshelm and Hultman 2014a). Here the climate change denialism of SD fit to well into the image of Sweden as a country that is already carrying its share.

Conclusion

With above reading of the Swedish example of far-right climate change denial we can see that this counter-science stance on this issue was not there from the beginning of these political parties, but actually created in a specific historical and socio-political context. The climate change denial of Sweden Democrats' is a combination of anti-establishment rhetoric, marketing of science doubts, industrial/breadwinner enactments as well as ethno-nationalism. They have been influenced to take this stance by organized climate change denial groups in a top-down process – later on spreading this to their voters through digital born far-right media such as SwebTV. Their climate policy is created through polarisation especially with the Green Party, but also with the hegemony of ecomodernism that takes climate science for real and say it will be solved via new technologies and market solutions. The in-depth case study of Sweden gives some clues to how the far right changed from focusing mostly on the ethno Swedishness of landscapes within the nation to rejecting climate change science as a global conspiration. The answer lies in the inability to see Swedes overall high per person climate impact due to the extensive use of resources and high level of consumption. Analyses of these trends draw our attention to far-right nationalism in the form of neo-fascist socio-political mobilisations in the United States and in Europe. For example, Poland in recent times has been seduced by the promise of making coal great again (Pulé and Hultman 2019). Workers and CEOs part of the extractive industries represents groups of society, which has changed towards far-right in many nations and vocally (and at times violently) supports far-right male leadership and agendas (Marcinkiewicz and Tosun 2015; Hochschild 2016; Jeffries 2017). Another clue to understanding the merge between far right and climate change denialism is to look beyond the nationalist curtain and engage in an analysis of trans-nationalistic authoritarianism (Lazaridis et al. 2016). Here, the European Freedom Award given to Václav Klaus by a coalition of far-right movements in Europe gives us insight into how transnational the far-right movement actually works. Their climate change denial is part of a global counter movement against action on global environmental issues that threaten our planet; ecocides with no boundaries. It might be possible to deeper understand this in terms of industrial/breadwinner masculinities identities, identities stuck in an industrial modern Fordist logic that is no longer materially present, who have now turned to authoritarianism (Pulé and Hultman 2019). The Earth faces a dire political situation in which climate denialism in many countries has merged with far-right actors, upheld, as we argue, by industrial/breadwinner masculinities and funded by global extractive industry fossil fuel companies. This intersection needs to be exposed and understood – not least the fact that men are much more a part of this movement than women – but also as not a given trajectory, but as a socio-political choice under influence of other actors in a political landscape.

References

Alestig, Peter (2017): Rockström om SD:s 'stora miss': Riskfylld strategi, *Svenska Dagbladet*, 15 January. www.svd.se/rockstrom-slar-tillbaka-mot-sd-inspireras-av-trump [23 April 2017].

Alfsson, Thoralf (2014): Grundstötning. http://thoralf.bloggplatsen.se/kategori/194333-klimathysteri/ [21 February 2015].

Anshelm, Jonas and Hultman, Martin (2014a): *Discourses of Global Climate Change: Apocalyptic Framing and Political Antagonisms*. London: Routledge

Anshelm, Jonas and Hultman, Martin (2014b): A green fatwā? Climate change as a threat to the masculinity of industrial modernity, *NORMA: International Journal for Masculinity Studies*, 9(2): 84–96.

Arnstad, Henrik (2015): Ikea fascism: Metapedia and the internationalization of Swedish generic fascism, *Fascism*, 4(2): 194–208.

Axelsson, Madelene and Borg, Kristian (2014): *Sverigedemokraternas svarta bok*. Stockholm, Verbal förlag.

Baas, David (2016a): Näringslivets hemliga uppvaktning av SD, *Expressen*, 16 May. www.expressen.se/nyheter/naringslivets-hemliga-uppvaktning-av-sd/ [10 September 2017].

Baas, David (2016b): SD-politik styrs dolt av klimatförnekare, *Expressen*, 19 October. www.expressen.se/nyheter/sd-politik-styrs-dolt-av-klimatfornekare/ [3 January 2017].

Bäckström Johansson, Mattias (2017): SD: Vi står inte bakom Energikommissionen, *Dagens samhälle*, 9 januari. www.dagenssamhalle.se/debatt/sd-vi-star-inte-bakom-energikommissionen-30687 [2 March 2017].

Bäckström Johansson, Mattias and Fransson, Josef (2014): Debatt: Så ser vi i SD på energipolitiken, *Dagens industri*, 17 October. www.di.se/artiklar/2014/10/17/debatt-sa-ser-vi-i-sd-pa-energipolitiken/ [2 March 2017].

Berggren, Lena (2017): *Fascismens återkomst*. www.blogg.umu.se/forskarbloggen/2017/02/fascismens-aterkomst/ [12 September 2017].

Bergström, Hans (2009): Global dimma, *Dagens Nyheter*. www.dn.se/ledare/kolumner/global-dimma/ [10 September 2017].

Bern, Lars (2018a): Kampen för att återinföra demokratin. https://anthropocene.live/2018/07/07/kampen-for-att-aterinfora-demokratin/ [9 July 2018].

Bern, Lars (2018b): Lördagsintervjun. www.swebbtv.se/blogg/137-lordagsintervju-20-del-4-lars-bern-om-globalisternas-fyra-stora-misstag [13 August 2018].

Björk, Anna and Viinikka, Tamya (2017): *'Världen går av all att döma inte under på grund av koldioxidutsläppen.' – Klimatförnekelse inom Sverigedemokraterna*. Candidate Essay, Institution of Tema, Linköpings universitet.

Brandberg, Tomas (2015): Nej, jordens resurser tar inte slut, *Samtiden*, 21 August. https://samtiden.nu/2015/08/nej-jordens-resurser-tar-inte-slut/ [9 March 2017].

Brandberg, Tomas (2017): Dags att kyla av klimatdebatten, *Samtiden*, 15 January. https://samtiden.nu/2017/01/dags-att-kyla-av-klimatdebatten/ [14 March 2017].

Brandberg, Tomas and Sanandaji, Nima (2008): Det orealistiska miljöpartiet, *Kvällsposten*, 4 March. www.expressen.se/kvp/ledare/det-orealistiska-miljopartiet/ [20 April 2017].

Dunlap, Riley, McCright, Aaron M. (2015): Challenging climate change – The denial countermovement. In: Riley E. Dunlap and Robert Brulle (Eds). *Climate Change and Society – Sociological Perspectives*. New York: Oxford University Press.

Dunlap, Riley, McCright, Aaron M. and Yarosh, Jerrod H. (2016): The political divide on climate change: Partisan polarization widens in the US, *Environment: Science and Policy for Sustainable Development*, 58(5): 4–23.

Edvardsson Björnberg, Karin, Karlsson, Mikael, Gilek, Michael, and Hansson, Sven Ove (2017): Climate and environmental science denial: A review of the scientific literature published in 1990–2015, *Journal of Cleaner Production*, 167: 229–241.

Ekman, Mattias (2014): Drömmen om ett rent Sverige. In: Madelene Axelsson and Kristian Borg (Eds). *Sverigedemokraternas svarta bok*. Stockholm: Verbal förlag, pp. 15–71.

Ekman, Mattias and Daniel Poohl (2010): *Ut ur skuggan – En kritisk granskning av Sverigedemokraterna*. Stockholm: Natur och Kultur.

Emilsson, Aron, Filper, Runar, Forsberg, Anders, Fransson, Josef and Kinnunen, Martin (2014): *En effektivare klimatpolitik*. Motion 2014/15:2834.

Emilsson, Aron, Filper, Runar, Forsberg, Anders, Fransson, Josef and Kinnunen, Martin (2015): *En effektivare klimatpolitik*. Motion 2015/16:2542.

Erlingsson, Gissur Ó., Vernby, Kåre and Öhrvall, Richard (2014): The single-issue party thesis and the Sweden Democrats, *Acta Politica*, 49(2): 196–216.

Farrell, Justin (2016): Network structure and influence of the climate change countermovement, *Nature Climate Change*, 6(4): 370–374.

Filper, Runar, Forsberg, Anders and Kinnunen, Martin (2016a): *En effektivare klimatpolitik*. Motion 2016/17:3301.

Filper, Runar, Forsberg, Anders and Kinnunen, Martin (2016b): *Med anledning av skr. 2015/16:87 Kontrollstation för de klimat- och energipolitiska målen till 2020 samt klimatanpassning*. Motion 2015/16:3321.

Forchtner, Bernhard and Kølvraa, Christoffer (2015): The nature of nationalism: Populist radical right parties on countryside and climate, *Nature and Culture*, 10(2): 199–224.

Forchtner, Bernard, Kroneder, Andreas and Wetzel, David (2018): Being Skeptical? Exploring far-right climate-change communication in Germany, *Environmental Communication*, 12(5): 589–604.

Forsberg, Anders (2016): Det finns ett parti med realistisk miljöpolitik, *Samtiden*, 11 March. samtiden.nu/2016/03/det-finns-ett-parti-med-realistisk-miljopolitik/ [9 March 2017].

Fransson, Josef (2013): Riksdagsdebatt 29 januari 2013, *Aktuell debatt om klimatförändringen*. www.riksdagen.se/sv/Debatter--beslut/Ovriga-debatter/Aktuella-debatter/Aktuell-debatt/?did=H0C120130129adanddoctype=ad [10 February 2014].

Fransson, Josef (2014): Klimatpolitiken har gått över styr, *24Nyheter*, 2 August 2014. Onlien at: https://nyheter24.se/debatt/774798-josef-fransson-sd-klimatpolitiken-har-gatt-over-styr [16 September 2017].

Fransson, Josef (2015): SD: Förhöjda halter av CO2 är en välsignelse, *Berslagsbladet Arboga tidning*, 16 December. www.bblat.se/opinion/insandare/sd-forhojda-halter-av-co2-aren-valsignelse [10 September 2017].

Fransson, Josef and Jansson, Mikael (2014) *Riksrevisionens rapport om klimat för pengarna*. Motion 2013/14:MJ23.

Hellström, Anders and Nilsson, Tom (2010): We are the good guys: Ideological positioning of the nationalist party Sverigedemokraterna in contemporary Swedish politics, *Ethnicities*, 10(1): 55–76.

Hellström, Anders, Nilsson, Tom and Stoltz, Pauline (2012): Nationalism vs. nationalism: The challenge of the Sweden Democrats in the Swedish public debate, *Government and Opposition*, 47(2): 186–205.

Hochschild, Arlie R. (2016): *Strangers in Their Own Land: Anger and Mourning on the American Right*. New York: The New Press.

Hultman, Martin and Anshelm, Jonas (2017): Masculinities of climate change. Exploring examples of industrial-, ecomodern-, and ecological masculinities in the age of Anthropocene. In: Marjorie Cohen (Ed). *Climate Change and Gender in Rich Countries*. London: Routledge, pp. 19–34.

Hultman, Martin and Käll, A.-S. (2014): Klimatskepticismen frodas på högerkanten. *Sydsvenskan*, 8 September. Onlien at: www.sydsvenskan.se/2014-09-08/klimatskepticismen-frodas-pa-hogerkanten [10 September 2017].

Jeffries, Elisabeth (2017): Nationalist advance, *Nature Climate Change*, 7(7): 469–471.

Jylhä, Kirsti M., Cantal, Clara, Akrami, Nazar and Milfont, Taciano (2016): Denial of anthropogenic climate change: Social dominance orientation helps explain the conservative male effect in Brazil and Sweden, *Personality and Individual Differences*, 98: 184–187.

Kaiser, Jonas and Puschmann, Cornelius (2017): Alliance of antagonism: Counterpublics and polarization in online climate change communication, *Communication and the Public*, 2(4): 371–387.

Kinnunen, Martin (2016a): SD: Svensk klimatpolitik blir dyrare och ineffektivare, *Altinget*, 23 June. www.altinget.se/miljo/artikel/sd-svensk-klimatpolitik-blir-dyrare-och-ineffektivare [2 March 2017].

Kinnunen, Martin (2016b): Repliker: 'Viktigt behålla och utveckla kärnkraften', *Dagens nyheter*, 22 November. www.dn.se/debatt/repliker/viktigt-behalla-och-utveckla-karnkraften/ [2 March 2017].

Klimatupplysningen (2012): Sverigedemokraterna – vettig men mycket tunn energipolitik. www.klimatupplysningen.se/2012/10/10/sverigedemokraterna-vettig-men-mycket-tunn-energipolitik/ [30 January 2014].

Krange, Olve, Kaltenborn, Bjorn P. and Hultman, Martin (2018): Cool dudes in Norway: Climate change denial among conservative Norwegian men, *Environmental Sociology*, vol. 5(1): 1–11.

Kundzewicz, Zbigniew W., Painter, James and Kundzewicz, Witold J. (2017): Climate change in the media: Poland's exceptionalism, *Environmental Communication*, 11: 1–15.

Lazaridis, Gabriella, Campani, Giovanna and Benveniste, Annie (2016): *The Rise of the Far Right in Europe*. London: Palgrave.

Liu, John Chung-En and Zhao, Bo (2017): Who speaks for climate change in China? Evidence from Weibo, *Climatic Change*, 140(3–4): 413–422.

Lockwood, Matthew (2018): Right-wing populism and the climate change agenda: Exploring the linkages, *Environmental Politics*, 27(4): 712–732.

Marcinkiewicz, Kamil and Tosun, Jale (2015): Contesting climate change: Mapping the political debate in Poland, *East European Politics*, 31(2): 187–207.

McCright, Aron M. and Dunlap, Riley E. (2003): Defeating Kyoto: The conservative movement's impact on US climate change policy, *Social Problems*, 50(3): 348–373.

McCright, Aron M. and Dunlap, Riley E. (2011): Cool dudes: The denial of climate change among conservative white males in the United States, *Global Environmental Change*, 21: 1163–1172.

Mudde, Cas (2007): *Populist Radical Right Parties in Europe*. Cambridge: Cambridge University Press.

Mulinari, Diana and Neergaard, Anders (2014): We are Sweden Democrats because we care for others: Exploring racisms in the Swedish extreme right, *European Journal of Women's Studies*, 21(1): 43–56.

Müller, Tim S., Hedström, Peter, Wennberg, Karl and Valdez, Sarah (2014): *Right-Wing Populism and Social Distance towards Muslims in Sweden: Results from a Nation-Wide Vignette Study*. Bachelor Essay, Institution of Social- and Welfare Studies, Linköpings universitet.

Oja, Simon and Mral, Brigitte (2013): The Sweden democrats came in from the cold: How the debate about allowing the SD into media arenas shifted between 2002 and 2010. In: Ruth Wodak, Majid Khosravinik and Brigitte Mral (Eds). *Right-Wing Populism in Europe: Politics and Discourse*. London: Bloomsbury Academic.

Olsson, Erik and Agö, Jenny (2015): Sverige har ett klimatskeptiskt riksdagsparti – gissa vilket? *KIT*. https://kit.se/2015/12/01/24969/sverige-har-ett-klimatskeptiskt-riksdagsparti-gissa-vilket/ [10 September 2017].

Oreskes, Naomi and Conway, Erik M. (2011): *Merchants of Doubt: How a Handful of Scientists Obscured the Truth on Issues from Tobacco Smoke to Global Warming*. New York: Bloomsbury.

Ott, Brian L. (2017): The age of Twitter: Donald J. Trump and the politics of debasement, *Critical Studies in Media Communication*, 34(1): 59–68.

Peterson, Abby (2016): The institutionalization processes of a neo-Nazi movement party: Securing social movement outcomes. In: Lorenzo Bosi, Marco Giugni and Katrin Uba (Eds). *The Consequences of Social Movements*. Cambridge: Cambridge University Press, pp. 314–337.

Pettersson, Simon O. (2016): Det behövs ett konservativt miljötänkande, *Samtiden*, 3 May. Onlien at: https://samtiden.nu/2016/05/det-behovs-ett-konservativt-miljotankande/ [9 March 2017].

Pulé, P., & Hultman, M. (2019). "Industrial/breadwinner masculinities. Understanding the complexities of climate change denial" in Kinnvall, C., & Rydstrom, H. (Eds.). (2019). *Climate Hazards, Disasters, and Gender Ramifications*. Routledge.

Samtiden (no date): *Samtiden*. http://samtiden.nu/ [12 March 2017].

SD-Kuriren (no date): *SD-Kuriren*. http://sdkuriren.se/ [12 March 2017].

Sveriges Riksdag (no date): *Ledamöter and partier – Martin Kinnunen*. www.riksdagen. se/sv/ledamoter-partier/ledamot/martin-kinnunen_57cbe134-829b-4fb8-9cc1-ce2f1cd02f3b [16 March 2017].

Swedish Democrats (2013): *Avvikande mening. Grönbok om klimat- och energipolitiken*. 2012/13:NU24.

Swedish Democrats (2014): *Sverigedemokraternas nya talespersoner*. www.mynewsdesk.com/se/sverigedemokraterna/pressreleases/sverigedemokraternas-talespersoner-1069208 [11 April 2017].

Thauersköld Crusell, Maggie (2009): *International Conference on Climate Change*. www. klimatupplysningen.se/2009/03/08/international-conference-on-climate-change/ [16 March 2018].

Vi-skogen (2017): *Varmare klimat – en iskall nyhet*. Botkyrka offset.

Wåg, Mathias (2014): Bruna rötter, blågul stam, bruna grenar – Sverigedemokraterna som plantskola för svensk extremhöger. In: Madelene Axelsson and Kristian Borg (Eds). *Sverigedemokraternas svarta bok*. Stockholm: Verbal förlag, pp. 95–115.

Wästberg, Olle and Lindvall, Daniel (2017): *Folkstyret i rädslans tid*. Stockholm: Fri Tanke.

Widfeldt, Anders (2008): Party change as a necessity–the case of the Sweden Democrats, *Representation*, 44(3): 265–276.

Wingborg, Mats (2016): *Den blåbruna röran: SD:s flirt med Alliansen och högerns vägval*. Leopard förlag. Stockholm.

Young, Nathan and Coutinho, Aline (2013): Government, anti-reflexivity, and the construction of public ignorance about climate change: Australia and Canada compared, *Global Environmental Politics*, 13(2): 89–108.

9

THE ALLURE OF EXPLODING BATS

The Finns Party's populist environmental communication and the media*

Niko Hatakka and Matti Välimäki

Introduction

In October 2016, political journalists attending a Finnish governmental party's energy policy news conference were astonished. Even though the Finns Party[1] had communicated its anti-environmental stances quite colourfully in the past, reporters kept asking themselves, did the party press officer just demand wind power plants to be disabled because they make bats explode? This chapter analyses the (anti-)environmental communication of the Finns Party (*Perussuomalaiset*, PS) – a centre-left populist anti-establishment party turned a populist radical right party. In this chapter, we provide both an overview of the Finns Party's environmental communication and a closer analysis of the performative aspects of Finnish populist anti-environmentalism in the media. We provide a historical overview of how the Finns Party's environmental communication has produced both inclusive and exclusive notions of 'the people'. In this stage, we focus our analysis on the Finns Party's direct communication with the public: official party platforms, the party paper, the party leaders' and Finns Party Members of Parliament's (MPs) blogs and Facebook posts. The aim is first to decode and explain the most salient programmatic features of the party's environmental communication, and second, to further explore the means of how these ideological contents are discursively disseminated by the party in the media. Focus is targeted especially on how the party utilises intentionally controversial elements in its anti-environmental communication to perform its populist core message in a hybrid media environment.

In liquid modernity (Bauman 2000), the political landscape has become more mediatised and geared towards the spectacular; politics has become more and more

---

* The authors of this chapter would like to thank research assistant Väinö Kuusinen, the Helsingin Sanomat Foundation and the Academy of Finland.

stylistic, like a performance (e.g. Mair 2013). In this chapter, we analyse anti-environmental populist communication both as thin-centred ideology (Mudde 2007: 23) and political communication style (Jagers and Walgrave 2007). In both aspects, we are interested in the discursive 'repertoires of performance that are used to create political relations' (Moffitt and Tormey 2014). Judith Butler (1990: 2) suggests that performativity has the power to 'inevitably produce what it claims merely to represent', and thus populism itself can be regarded as something that has to be performed and enacted (Jansen 2011: 82; Moffitt and Tormey 2014: 388). In the time of mediatised politics and the 24-hour news cycle (Cushion 2015), the production of populist representation of the people is performed by populist actors mainly in the media. Therefore we will discuss environmental communication as a discursive populist performance that consists of various repertoires of constructing 'the people', 'the elites' and 'the elites' allies' (Jagers and Walgrave 2007) and how this performance is disseminated and circulated in the public sphere.

According to Moffitt and Tormey (2014: 389) 'performance ... is not merely a one-sided relationship in which a politician "performs" for a passive audience, but rather a feedback loop whereby the performance can actually change or create the audience's subjectivity, and this in turn can change the context and efficacy of the performance'. Therefore it is vital to try and understand not only the discursive contents of populist environmental communication but also its performance in relation to the media landscape and various publics it manifests in. Additionally to unpacking the party's programmatic environmental communication (total of 30 programmes from 1995 to 2017), the chapter provides a discourse analysis on a particular Finns Party campaign against wind power in 2016. The campaign focused on the alleged health and environmental hazards of wind turbines – and became highly salient in the Finnish public sphere partially because of the campaign's controversial elements (e.g. claiming that turbines make bats explode and make people sick). We analysed 142 news articles and Finns Party politicians' 41 blog posts and 92 Facebook posts related to the campaign in the contexts of the hybrid media system (Chadwick 2013) and Ruth Wodak's (2013, 2015) 'right-wing populist perpetuum mobile'. Thus, we are not only interested in how the controversial campaign message was communicated by the party, but also how this message was circulated in professional and online media – and how this 'feedback loop' of the populist performance (Moffitt and Tormey 2014: 389) was to be re-appropriated by the party. Before moving on to the Finns Party's programmatic and mediated environmental communication, we first set the context by providing background on the far-right and environmental politics in Finland.

Far-right and environmental politics in Finland: An overview

The far-right and right-wing radicalism have enjoyed little support in Finland after the mid-1930s. Like most of Europe, Finland experienced far-right political restlessness and even political violence between the World Wars. In addition to the

Finnish civil war of 1918, radical militant nationalism first manifested in voluntary and unofficial military campaigns to establish a so-called Greater Finland by invading northwestern areas of Soviet Russia that were populated to a substantial degree by Finnish speakers. Ultra-nationalist radicals participated in several armed rebellions and revolts between 1918 and 1922, until the military campaigns ended in the chaotic withdrawal of all Finnish armed forces from Soviet soil. After this, the most significant organisation to advance the idea of a Greater Finland and the unity of the Finnic peoples was The Academic Karelia Society (AKS) founded in 1922. The ideological spirit of the AKS was present in the only substantial attempt for a far-right coup in Finland, which took place in 1932 when the so-called *Lapua* movement tried to mobilise an offensive against the centre-right government in Mäntsälä. The *Lapua* movement was a combination of peasant populism, upper-class nationalism, and fierce anti-communism. Without proper legitimacy the coup-attempt was unsuccessful, the *Lapua* forces stepped down and the movement's organizational structure was dismantled (Ahti 1990; Jussila et al. 1999: 139–166). A party called Patriotic People's Movement (*Isänmaallinen kansanliike*, IKL) was formed to build on the *Lapua* movement's heritage. IKL's goals were ultra-nationalist, anti-democratic, anti-Soviet and anti-communist, and the party was able to gain 14 seats in both 1933 and 1936 parliamentary elections and 8 seats in 1939. After Finland's defeat to the Soviet Union in the Continuation War in 1944, both IKL and AKS were banned on the insistence of the Soviets (Nygård 1982).

Since the 1940s, far-right actors have not had significant parliamentary or municipal representation in Finland. During the Cold War, the Finnish far-right had no support or feasible means of organising, and the few vocal activists were tightly monitored by the Finnish Security Police (Jokinen 2015). After the collapse of the Soviet Union, Finnish politics and culture experienced a reinvigoration of patriotism in the mid-1990s (Kinnunen and Jokisipilä 2012). Finnish neo-patriotism manifested in its most extreme form in relatively small but violent neo-Nazi groups that formed in larger Finnish cities throughout the 1990s, originally as responses to increasing numbers of refugees fleeing to Finland. In the past 15 years, ultra-nationalist organisations such as *Suomen Sisu* and outright far-right organisations such as the Finnish Resistance Movement (*Suomen Vastarintaliike*) have reappeared in the public sphere, and they mostly focus on organising rallies and social media campaigns in disseminating their message. In the 2010s, several members of *Suomen Sisu* became Finns Party MPs, and the party has been repeatedly challenged for its connections, especially with online far-right action (Hatakka et al. 2017, 2018). As with most radical right parties (Olsen 1999; Gemenis et al. 2012), environmental issues have not been especially salient in the agendas of the Finnish far-right. Instead, the various mostly unconnected groups have for the most part concentrated on opposing their self-proclaimed enemies in Finnish society, i.e. especially communist and Russian/Soviet influence from the 1910s to the end of the Cold War, and the left-wing liberals and minorities in the post-Cold War period (Mickelsson 2011; Jokinen 2011: 226–238; Kotonen 2018).

From the 1950s to the 2010s, Finnish environmental politics followed a common trend, shifting away from local and national environmental protection towards global climate questions. The Finnish Green movement began to emerge in the 1970s, when the post-war generation became interested in post-material ideas of global consciousness and shared environmental responsibility (see e.g. Inglehart 1977; Dalton 2009). During the political organisation of the Finnish Green movement in the 1980s, several distinct yet organisationally fuzzy factions formed within it. The one with a moderate program of environmental protection and progressive social politics became the most popular. Bearing resemblance to many other West European countries, this ideological line manifested in the forming of the Green League (*Vihreä liitto*) in 1988 (Järvikoski 1991; Jokinen and Järvikoski 1997). Some of the more radical environmentalist factions, like the one exemplified by the extremist green ideologist Pentti Linkola, went as far as to facilitate eco-fascism and anti-humanism. His most extreme texts suggest mandatory population control, forced sterilisations, ending foreign aid and immigration, and founding of a 'green police' to suppress all opposition (Linkola 2004). Whereas the Green League's moderate agenda has gained substantial parliamentary representation and governmental power in the 1990s, 2000s and 2010s, more radical and fringe ideas of the original Green movement have practically disappeared from the Finnish public sphere (Mickelsson 2015: 245–252).

Since the early 1980s, all Finnish parliamentary parties have incorporated environmental policy in their programmes, although some have had environmental protection in their agenda for much longer (see e.g. Agrarian League 1914[2]). Often the central points of departure in the parties' programmes in the 2000s and 2010s have been related to international environmental cooperation, perceived causes of climate change, and levels of support for preferring renewable energy sources over coal and nuclear energy (e.g. National Coalition 2011; SDP 2009). In the 2010s, Finnish governments' programmes have emphasised international cooperation to tackle climate change and environmental hazards, to reduce greenhouse gas emissions, and to increase the use of bio-fuels and renewable energy (e.g. Finnish government 2011, 2017).

The Finns Party's environmental policy

In this volume, 'the far-right' refers to parties that are right of the centre-right – ranging from the radical right to neo-Nazism. Though currently the Finns Party can be described as a populist radical right party, the party's roots are not in the far-right. Its predecessor, the Finnish Rural Party (*Suomen Maaseudun Puolue*, SMP), was a centre-leftist agrarian-populist party that aimed to voice the concerns of the 'forgotten people', referring to Finnish rural citizens disillusioned with the structural change that came with the industrialisation and urbanisation of Finland in the 1960s (Palonen 2017: 232). The Finns Party was founded in 1995 as a direct successor of SMP, and in the beginning the party had a similar yet modernised and more urban, centre-leftist and anti-elitist populist agenda (Ruostetsaari 2011). The party began to gain more nativist sway later in the 2000s, when the party started cooperating with Finnish anti-immigration movements and individuals who had ideological resonance and personal connections to the far-right (Hatakka 2017: 2026–2028).

Throughout the 2010s, the party had two distinct factions: one came from the Finnish Rural Party's centre-leftist tradition whereas the other resembled other European radical right populist parties with anti-immigration agendas (Jungar 2016). In the summer of 2017, the anti-immigration faction led by Member of the European Parliament (MEP) Jussi Halla-aho took over the party leadership when he and three of his supporter MPs were elected as chairmen by the party congress. As a result, the party was split in two, as 20 MPs left to form their own parliamentary group. The post-split Finns Party can be categorised as a 'populist radical right party' (Mudde 2007) with a nativist agenda, but still the party cannot be regarded as anti-democratic or extremist (on the concepts, see e.g. Mudde 1995; Carter 2018). Yet, the party and especially some of its individual politicians enjoy solid support among the more extreme Finnish far-right (Palonen 2017: 306–309; Hatakka et al. 2018). The party's shift from rural populism towards the populist radical right is visible also in the party's environmental communication.

Even though appreciation of Finnish nature is present in the party's communication, environmental policy has not been a strong focal point in the Finns Party's programmes, until the late 2010s. The party's earlier environmental discourses that derive from the rural-populist tradition have focused especially on preserving and protecting the Finnish countryside. A significant share of SMP's support came from Finnish farmers (Ruostetsaari 2011), and the theme of protecting the traditional small-scale, family-style farming runs throughout the programmes until the early 2010s (PS 1995, 2001, 2011). The traditional landscape of the Finnish countryside plays a fundamental role in the Finns Party's ideological heartland, an imaginary of uncorrupted time and place where everything is well (e.g. Taggart 2006). In this heartland, Finnish nature provides sustenance, work, means for industry and individual rejuvenation and relaxation (PS 2003, 2007, 2011). Especially in the earlier programmatic work of the 1990s and 2000s, 'the environment' is used as a very concrete term referring to the local environment – mainly cultivated countryside – where people are autonomous and free to utilise nature and to reign over it in a harmonious yet authoritative manner. The party portrays nature as part of Finnish national identity and as something that should be treasured as an asset, not only for honest hardworking Finns but also for the competitiveness of the Finnish economy in the international market (PS 2003, 2011). Yet, forms of environmental protection that do not visibly affect the immediate environment have appeared on the party's programmes mainly when the party has challenged them.

Sovereignty of the people is vital to understand certain consistent inconsistencies in the party's stances towards environmental communication (see also Forchtner and Kølvraa 2015; Forchtner et al. 2018). The party has repeatedly communicated its support for local toxin-free agriculture and keeping Finnish nature clean (PS 2003, 2011), but at the same time the party has avidly opposed environmental protection policies. The party has argued that various forms of environmental regulation breach the people's autonomy over their land and that they are economically detrimental to the ordinary tax-payer. Regulation of sewage systems in remote areas (PS 2011), seal hunting quotas (PS 2008), fur farming restrictions (PS 2003, 2011), 'green taxation' on petrol (PS 2001, 2017), and the zoning of vast natural conservation areas (PS 2008, 2009) are all framed as restrictions to rural people's fundamental rights such as freedom to choose their trade and place of residence. Therefore in the earlier programmes, opposing environmental protection or posing a challenge to 'green values' can be regarded as defending of the interests of the people the party claims to represent. Also in the later programmes environmental protection is supported as long as it is 'reasonable' (PS 2017) or, in other words, as long as it does not endanger the sovereignty of the people.

Towards the late 2000s and early 2010s – as the party transformed towards the radical right – the party programmes became more detached from the original rural populist tradition of focusing on the Finnish periphery and its inhabitants. The turn to more general anti-green national protectionism or economic nationalism is in line with nationalist-populist and far-right movements' general suspicion towards international cooperation and supranational decisions. Until the late 2000s, especially the EU and the global market were framed as the main enemies stripping the Finnish people of their agricultural and environmental autonomy (PS 2007, 2008). In its official platforms, the Finns Party has never questioned the validity of climate change as a natural phenomenon, but it has severely questioned whether international cooperative means to curb climate change are efficient and reasonable from the point of view of the Finnish people (PS 2017; see also Forchtner and Kølvraa 2015). The party has regarded supranational environmental protection, especially, as a hindrance for the competitiveness of the Finnish economy. In the programmes, environmental protection is framed as a trade off to economic growth (PS 2011, 2015a, 2017; see also Gemenis et al. 2012). The relatively small Finnish industry sector is portrayed by the party as a victim of the dictates of industrial giants like China and the United States (PS 2011). Thus the party has argued that relatively clean economies should not have to bear as much responsibility for emission regulation (PS 2014, 2015a, 2019a, 2019b). The Finns Party has also stressed that the Finnish energy policy should be aimed to boost employment and to cater to the demands of Finnish industry (PS 2017, 2019b). Their suggested solution has been to increase the utilisation of peat and forestry by-products for energy production instead of investing in low-emission renewable energy-sources, especially wind power (PS 2017, 2019b). The main argument here is that forestry- and peat-based bio-energy would provide more jobs locally in the long run.

The turn in the Finns Party's environmental communication – and also its increasing opposition to wind power – is not solely economic or structural; it has strong undercurrents of purely discursive anti-environmentalism that is used in performing the populist idea of the party's enemies – namely the green elites. Especially in the Finns Party MPs' online communication, post-modern environmentalism is framed to derive from the same idealist origins as multiculturalism, the current main target and perceived threat communicated by the party (see e.g. PS 2015b). Green policies and environmentalist views have been framed as 'hysteria' (*vihervouhotus*, Elo 2014; PS Facebook 2010), 'frivolousness' (*viherhumppa*, Hakkarainen 2015; Lohela 2011) or as 'elite dictates' (PS 2010). The party's communication consistently regards environmentalism as ideologically illogical, unpatriotic, economically unviable, elitist, and negligent of the average people's troubles. Green values and environmental protection policies are therefore presented as a polar juxtaposition to the Finns Party's 'reasonable' (PS 2017, 2019b) environmental agenda, which is framed to be based on common sense, patriotism, supporting the national economy, and heeding to the people and their day-to-day problems.

Campaign against wind power as populist performance

In the fall of 2016, just a few months preceding the Finnish communal election, the Finns Party office launched a campaign that attacked wind power on a previously unprecedented scale. The party organised a press conference claiming that wind turbines are health hazards, and that Finnish elites do not recognise the acute danger that turbines pose to people who live in their vicinity. The press conference featured a tearful man telling the story of his family's misfortune of living close to a wind power plant. In this narrative, his children and wife had been getting various symptoms from the 'undetectable ultrasound' emitted by the nearby turbines (Waris 2016). The Finns Party's press officer – a former steel industry lobbyist – built on the narrative claiming that more than 10% of all Finns were in imminent danger and that wind turbines, or at least further investments in wind power, should be halted until the proclaimed health hazards had been properly investigated (Suomen Uutiset 2016a). The idea has originated from the UK, where especially United Kingdom Independence Party (UKIP) has addressed dangers of wind power (see e.g. Batel and Devine-Wright 2018).

The campaign contained most ideological aspects of populist political communication, especially people-centrism, anti-elitism and expressing intent to restore the people's sovereignty (see e.g. Ernst et al. 2017; Forchtner et al. 2018). The narrative portrayed the dismay of an average hardworking taxpayer, whose heartfelt concerns for his family's well-being had been arrogantly sidestepped by the corrupt political and scientific establishment. The perceived hazards of wind power were framed by the campaign as a 'direct result of blue-red-and-green climate politics' – referring to an alleged alliance between foreign energy companies and centre-right and leftist pro-green political elites – which 'costs billions of euros to tax-payers' (Suomen Uutiset 2016a). The campaign heavily discredited the elites by suggesting that they, in cooperation with the National Institute for

Health and Welfare (*Terveyden ja hyvinvoinnin laitos*, THL), were actively trying to conceal the 'truth' of wind power being not only economically unviable but also hazardous to the health of hundreds of thousands of Finns (Suomen Uutiset 2016a). The party demanded that the people's concerns would be taken seriously and that their right to express their sincere opinions would not be silenced. This aforementioned 'people' did not only consist of the individuals directly afflicted by the dangers of wind power, but broadly to all who are willing to question elite policies that did not cater to the well-being of ordinary working citizens. The party's press officer was clear in expressing that the Finns Party was 'the only party without linkages to the wind power industry' and that they were 'the sole political force that fights injustice and defends the rights of the people' (Suomen Uutiset 2016a).

The campaign narrative thus contained clear explications of the people, of the elites, and of the party as means for restoring the jeopardised sovereignty of the people. Additionally to the campaign's populist ideological core, the press conference's message contained several stylistic elements of populist communication (e.g. Moffitt and Tormey 2014; Ernst et al. 2017) – especially emotionalisation, dramatisation, heeding to 'common sense' over elite forms of knowledge, colloquial language, and polarising black-and-white rhetoric that clearly defined 'us' and 'them'. What topped off the already controversial and intriguing human-interest story was that the press officer additionally claimed that wind turbines were a danger to wildlife – namely that 'they make bats' insides explode' (Suomen Uutiset 2016a). The already resonant populist narrative combined with a seemingly outrageous idea of aerial mammals in peril took both journalistic and social media by storm; and for a couple of weeks the Finnish public sphere was frenzied with discussion about the Finns Party's opposition to 'the wind power lobby'. To close the conference, the party press officer suggested that their supporters should prepare for a 'massive defaming campaign' that was to ensue after the conference, due to 'the huge political and economic linkages of the parliamentary parties and the wind power industry' (Suomen Uutiset 2016a). He was not wrong about the campaign causing a backlash.

The human-interest side of the afflicted-family-narrative was especially resonant with the tabloids (e.g. Waris 2016) that have been claimed to be stylistically prone to adhere to populist communication (e.g. Mazzoleni 2008). The daily newspapers on the other hand, were outright critical in their coverage of the conference, and featured a number of experts denying that wind turbines would neither cause health issues in humans nor make bats blow up (e.g. Hartikainen 2016; Huovinen 2016). Online, the party's semi-absurd message was heavily ridiculed especially by the party's political rivals and ideological opponents: for example, a left-wing MP wondered on Twitter, 'whether the Finns Party office's tap water has been contaminated with psychedelics' (Sarkkinen 2016). The party's campaign was thus severely challenged by journalists, experts and online commentators for dismissing science and for juxtaposing physics with individuals' personal experiences. Nevertheless, as a result of the Finns Party's insistence, after the news conference the Finnish government declared to carry out an impartial investigation on the health hazards of wind turbines. The Green League leader Ville Niinistö blamed the government for buying into the Finns Party's 'mumbo jumbo' (*huuhaa*, Myllymäki 2016). After the investigation, THL's researchers declared that symptoms experienced by people living close to wind power plants are most likely a result of a nocebo effect, meaning that negative expectations of being afflicted by turbines would be the cause of the negative experiences. The researchers further added that perceived health hazards cannot be scientifically proven nor entirely discredited (Lanki et al. 2017).

The party reacted aggressively to the initial backlash to frame the understatement and denial of the perceptions and illness of the individuals (i.e. 'the people') as proof that the political and scientific establishment ('the elites') are not paying attention to the concerns of 'the people' (Suomen Uutiset 2016b). As an accusation reversal, the party's press officer questioned the reliability of THL's researchers (Hartikainen 2016). The press officer especially blamed the selection of experts to have been based on political preferences to manipulate the results and to get desired interpretations of the issue (Kantomaa 2017; Waris 2017) – a common strategy used by environmental sceptics (see e.g. Jacques et al. 2008). The campaign was strongly personified to the press officer Matti Putkonen, but also many Finns Party MPs commented on it, especially on their blogs, framing criticism as politically motivated smearing, belittling and making a mockery of people's experiences (Suomen Uutiset 2016b): even stating that 'the fact that there is plenty of research on the topic, does not remove the experiences of the people' (Mattila 2016). Other parties' and especially the Greens League's criticism was framed as 'proof of their real colors' of not being worried about environmental hazards and that the Finns party is 'the only party that takes people's concerns of health seriously' (Meri 2016).

Not everyone in the party was entirely happy with the anti-intellectual communicative approach chosen by the party office, but they were clear not to express their dissatisfaction in public. For example the party leader Timo Soini refused to comment on the health hazards of wind turbines, excusing himself from commenting by stating that 'the issue belongs to the party office that chooses its preferred means of waking the people up' (Ijäs 2016). Thus, when asked to comment on the mini-spectacle revolving around exploding bats, notable party figures like the party leader and the parliamentary group leader sidestepped the queries, but eagerly utilised the free airtime to explain, for example, how the party supports investments in peat- and forestry-based bio energy that 'would benefit both the regional and national economies' (e.g. Ahtokivi 2016). It is also worth noting, that while the mediated criticism focused on the most outrageous anti-scientific claims made by the party's press officer, other nearly as controversial claims – like the one suggesting that foreign wind energy companies were secretly allied with Finnish political elites and researchers to manipulate the truth – remained entirely unchallenged.

Conclusion

Matthew Lockwood (2018) suggests that right-wing populist parties' anti-environmental stances can be explained either by a structural or ideological approach. The Finns Party's environmental programmatic communication is partially structurally explainable; for example, the party's focus on lowering 'green taxation' of petrol and advocating for investing in high-emission peat energy plants to create jobs in rural areas are direct responses to the increasing precariousness of the party's constituencies in the Finnish periphery. But in some aspects – illuminated here by the 'exploding bats campaign' – right-wing populist environmental communication can also consist of mere discursive performance that is nearly void of all structurally motivated ideological content except for the thin centre or core of populism. This performance portrays a reasonable, down-to-earth and virtuous people being patronised by out-of-touch elites who are willing to endanger the people for unrealistic bleeding-heart-liberal environmental objectives. The fact that anti-environmental communication by right-wing populist parties can take forms that are more robustly ideological or just 'empty' populist performance supports the stance that it can be useful to regard populism both as thin-centred ideology and political style (e.g. Jagers and Walgrave 2007: 337–338).

A third approach for understanding populist anti-environmental communication would be to consider it as an effort to pursue short-term strategic goals. The Finns Party's campaign against wind power was without a doubt designed to maximise attention and to mobilise potential supporters. The party press officer stated later that especially his reference to bats was deliberately provocative: 'The bat was fashioned into a lure and cast in to the sea of journalists. And jackpot!' (Poutanen 2016) Ruth Wodak has used the term 'right-wing populist perpetuum mobile' to portray the phenomenon of right-wing populist parties utilising scandalous ambivalent rhetoric to unleash media events that can last for weeks (Wodak 2015: 19–20). These intentionally provocative communication strategies – that previously have been studied in the context of racism accusations (e.g. Wodak 2013; Hatakka et al. 2017) – provide populist parties with free media publicity and allow them to reach wider audiences than what their resources would otherwise allow. In other words, the strategic creation of controversy facilitates the acting out of mediated populist performances in the hybrid media system. Whereas immigration-related provocative communication can be beneficial for right-wing populist parties, building populist performances around such a flammable topic still poses the actual risk of the parties becoming affiliated with racism (Hatakka 2017; Hatakka et al. 2017). Therefore, using anti-environmental communication to trigger the 'perpetuum mobile' could prove to be less risky for the parties – though it might make them seem anti-scientific or prone to 'mumbo jumbo'.

The Finns Party's 'exploding bats' campaign utilised several performative features that made it resonant for journalists, political opponents and supporters alike. The performance was in many ways compatible with news values: it was unexpected, it utilised untraditional communication, and it created a provocative conflict with the elites' and people's perceived experiences. The party was able to use this resonance to gain extremely high salience of the party's stances both in journalistic and social media – and even if the publicity was negative and condescending, the bad press was actively appropriated by the party for the construction of the people and the elites. Before the Finns Party's breakthrough election of 2011, the party conducted a similarly controversial campaign that targeted state-supported 'faux and pretentious modern art' (PS 2011), causing a similar response that allowed the acting out of a nearly identical populist performance (Hatakka 2012: 383–384).

For the transformed Finns Party that has abandoned its rural populist tradition, environmental communication appears secondary – unless it is being used for triggering a populist performance. This is where the party differs from many far-right movements, whose environmental communication is more organically-motivated and thus closer to ecological thinking (Olsen 1999; Forchtner and Kølvraa 2015; contributions by Boggs (2019), Forchtner and Özvatan (2019) in this volume). Environmental politics – even when only addressed strategically as a topic among others – provides fertile ground for populist radical right and identity struggles, and for creating conflict with pro-environment actors. In the 2019 Finnish parliamentary elections, the increasingly salient and consensual outcry for combating climate change was actively challenged by the party to successfully establish a discursive boundary of difference between the 'people' and the 'elites'. It is most likely that the Finns Party will continue to circulate provocative anti-environmental narratives to unleash mediated populist performances.

Notes

1 The party changed its English translation to 'the Finns Party' in 2012. Some researchers still use the old unofficial translation 'the True Finns'.
2 One of the first party platforms from the still extant Finnish parties to address environmental issues came from the Agrarian league. The programme addressed the problem of deforestation from the overexploitation of forests by sawmills.

References

Agrarian League (1914): *Maalaisliiton ohjelma*. Programme. Helsinki: The Agrarian League.
Ahti, Martti (1990): *Kaappaus? Suojeluskuntaselkkaus 1921, Fascismin aave 1927, Mäntsälän kapina 1932*. Helsinki: Otava.
Ahtokivi, Ilkka (2016): Perussuomalaiset: EU:n taakanjakoehdotus kohtuuton Suomelle. *Verkkouutiset*. www.verkkouutiset.fi/perussuomalaiset-eun-taakanjakoehdotus-kohtuuton-suomelle-58368/ [19 April 2018].
Batel, Susana and Devine-Wight, Patrick (2018): Populism, identities and responses to energy infrastructures at different scales, in the United Kingdom: A post-Brexit reflection, *Energy Research & Social Science*, 43: 41–47.
Bauman, Zygmunt (2000): *Liquid Modernity*. Cambridge: Polity Press.
Butler, Judith (1990): *Gender Trouble*. New York: Routledge.

Boggs, Kyle (2019): The rhetorical landscapes of the 'alt right' and the Patriot Movements: Settler entitlement to native land. In: Bernhard Forchtner (Ed). *The Far Right and the Environment: Politics, Discourse and Communication*. London: Routledge.

Carter, Elisabeth (2018): Right-wing extremism/radicalism: Reconstructing the concept, *Journal of Political Ideologies*, 23(2): 157–182.

Chadwick, Andrew (2013): *The Hybrid Media System. Politics and Power*. Oxford: Oxford University Press.

Cushion, Stephen (2015): *News and Politics: The Rise of Live and Interpretive Journalism*. Abingdon: Routledge.

Dalton, Russell (2009): Economics, environmentalism and party alignments: A note on partisan change in advanced industrial democracies, *European Journal of Political Research*, 48(2): 161–175.

Elo, Simon (2014): Vihervouhotus on lopetettava! 17 June 2014. www.facebook.com/408001925602/posts/10154264074230603 [19 April 2018].

Ernst, Nicole, Engesser, Sven, Büchel, Florin, Blassnig, Sina and Esser, Frank (2017): Extreme parties and populism: And analysis of Facebook and Twitter across six countries, *Information, Communication & Society*, 20(9): 1347–1364.

Finnish government (2011): *Hallitusohjelman strateginen toimeenpanosuunnitelma – kärkihankkeet ja vastuut*. Programme. Helsinki: The Finnish government.

Finnish government (2017): *Valtioneuvoston selonteko kansallisesta energia- ja ilmastostrategiasta vuoteen 2030*. Report. Helsinki: Ministry of Economic Affairs and Employment.

Forchtner, Bernhard and Kølvraa, Christoffer (2015): The nature of nationalism: Populist radical right parties on countryside and climate, *Nature and Culture*, 13(2): 199–224.

Forchtner, Bernhard, Kroneder, Andreas and Wetzel, David (2018): Being skeptical? Exploring far-right climate change communication in Germany, *Environmental Communication*, 12(5): 589–604.

Forchtner, Bernhard and Özvatan, Özgür (2019): Beyond the 'German forest': Environmental communication by the far right in Germany. In: Bernhard Forchtner (Ed). *The Far Right and the Environment: Politics, Discourse and Communication*. London: Routledge.

Gemenis, Kostas, Katsanidou, Alexia and Vasilopoulou, Sofia (2012): The politics of anti-environmentalism: Positional issue framing by the European radical right. *Paper presented at the PSA Annual Conference*, 3–5 April 2012, Belfast.

Hakkarainen, Teuvo (2015): 'Mistä voimme säästää?', 1 April 2015. www.facebook.com/198979100169761/posts/813196978747967 [19 April 2018].

Hartikainen, Jarno (2016): Linjataan ensin, tutkitaan sitten. *Helsingin Sanomat*, 7 October 2016.

Hatakka, Niko (2012): Journalismin perussuomalainen uudelleentulkinta sosiaalisessa mediassa. In: Ville Pernaa and Erkka Railo (Eds). *Jytky. Eduskuntavaalien 2011 mediajulkisuus*. Turku: Kirja-Aurora, pp. 350–394.

Hatakka, Niko (2017): When logics of party politics and online activism collide: The populist Finns Party's identity under negotiation, *New Media and Society*, 19(12): 2022–2038.

Hatakka, Niko, Niemi, Mari and Välimäki, Matti (2017): Confrontational yet submissive: Calculated ambivalence and populist parties' strategies of responding to racism accusations in the media, *Discourse & Society*, 28(3): 262–280.

Hatakka, Niko, Niemi, Mari and Välimäki, Matti (2018): "Perussuomalaiset ei ole rasistinen puolue". Näin populistipuolue vastaa median rasismisyytöksiin. In: Mari Niemi and Topi Houni (Eds). *Media ja populismi. Työkaluja kriittiseen journalismiin*. Tampere: Vastapaino.

Huovinen, Jorma (2016): Tutkija ampuu alas perussuomalaisten väitteet infraäänen terveyshaitoista. *Aamulehti.* www.aamulehti.fi/kotimaa/tutkija-ampuu-alas-perussuomalaisten-vaitteet-infraaanen-terveyshaitoista-23980390/ [19 April 2018].

Ijäs, Johannes (2016): Timo Soini tunnustaa avoimesti: "Joo. Virhe tuli." – ja nostaa esiin myös meriuposkuoriaiset. *Demokraatti* [Finnish SDP's party paper, online]. https://demokraatti.fi/timo-soini-tunnustaa-avoimesti-joo-virhe-tuli-ja-nostaa-esiin-myos-meriuposkuoriaiset/ [19 April 2018].

Inglehart, Ronald (1977): *The Silent Revolution: Changing Values and Political Styles among Western Publics.* Princeton: Princeton University Press.

Jacques, Peter, Dunlap, Riley and Freeman, Mark (2008): The organisation of denial: Conservative think tanks and environmental scepticism, *Environmental Politics,* 17(3): 349–385.

Jagers, Jan and Walgrave, Stefaan (2007): Populism as political communication style: An empirical study of political parties' discourse in Belgium, *European Journal of Political Research,* 46(3): 319–345.

Jansen, Robert (2011): Populist mobilization: A new theoretical approach to populism, *Sociological Theory,* 29(2): 223–244.

Jokinen, Christian (2015): *Terrorismista ja sen torjunnasta: Suojelupoliisi ja kansainvälinen terrorismi 1958–2004.* Turku: University of Turku.

Jokinen, Pekka and Järvikoski, Timo (1997): Ympäristösosiologian ja ympäristöpolitiikan perusteita. In: Juhani Pietarinen, Jokinen Pakka, Järvikoski Timo, Hoffren Jukka and Gustafsson Jaana (Eds). *Ympäristönsuojelu ja yhteiskunta.* Turku: Turun yliopiston täydennyskoulutuskeskus, pp. 45–98.

Jokinen, Thomas (2011): Ääriliikkeet Suomessa 1990- ja 2000-luvuilla. In: Anssi Kullberg (Ed). *Suomi, terrorismi ja SUPO.* Helsinki: WSOY, pp. 225–249.

Jungar, Ann-Cathrine (2016): From the mainstream to the margins? The radicalization of the True Finns. In: Tjitske Akkerman, Sarah de Lange and Matthijs Rooduijn (Eds). *Radical Right-Wing Populist Parties in Western Europe: Into the Mainstream?* Abingdon: Routledge, pp. 113–143.

Jussila, Osmo, Hentilä, Seppo and Nevakivi, Jussi (1999): *From Grand Duchy to a Modern State: A Political History of Finland since 1809.* London: Hurst & Company.

Järvikoski, Timo (1991): Ympäristöliike Suomessa. In: Vuokko Aromaa et al. (Eds). *Ympäristö ja aika,* Jyväskylä: Gummerus, pp. 150–177.

Kantomaa, Raija (2017): Räjähtävistä lepakoista saarnannut perussuomalaisten Putkonen lyttää ministeriön tuulivoimatutkimuksen: "Tuloksia manipuloitu". *MTV news.* www.mtv.fi/uutiset/kotimaa/artikkeli/rajahtavista-lepakoista-saarnannut-perussuomalaisten-putkonen-lyttaa-ministerion-tuulivoimatutkimuksen-tuloksia-manipuloitu/6473922#gs.ho=XJAE [19 April 2018].

Kinnunen, Tiina and Jokisipilä, Markku (2012): Shifting images of 'Our wars': Finnish Memory Culture of World War II. In: Tiina Kinnunen and Ville Kivimäki (Eds). *Finland in World War II History, Memory, Interpretations.* Leiden: Brill, pp. 435–482.

Kotonen, Tommi (2018): *Politiikan juoksuhaudat.* Helsinki: Atena.

Lanki, Timo, Turunen, Anu, Maijala, Panu, Heinonen-Guzenev, Marja, Känkälä, Sami, Toivo, Tim, Toivonen, Tommi, Ylikoski, Jukka and Yli-Tuomi, Tarja (2017). *Tuulivoimaloiden tuottaman äänen vaikutukset terveyteen.* Helsinki: Ministry of Economic Affairs and Employment.

Linkola, Pentti (2004): *Voisiko elämä voittaa?* Pieksämäki: Tammi.

Lockwood, Matthew (2018): Right-wing populism and the climate change agenda: Exploring the linkages, *Environmental Politics,* 27(4): 712–732.

Lohela, Maria (2011): Perussuomalaiset ei tanssi viherhumppaa, 4 February 2011. Online at: www.facebook.com/108427315890237/posts/140042646057909 [19 April 2018].
Mair, Peter (2013): *Ruling the Void: The Hollowing of Western Democracy*. London: Verso.
Mattila, Pirkko (2016): *On lupa puhua tuulivoiman terveysvaikutuksista*. Kaleva.fi. https://blogit.kaleva.fi/suomalainen-blogi/on-lupa-puhua-tuulivoimaloiden-terveysvaikutuksist [19 April 2018].
Mazzoleni, Gianpetro (2008): Populism and the media. In: Daniele Albertazzi and Duncan McDonnell (Eds). *Twenty-first Century Populism. The Spectre of Western European Democracy*. Basingstroke: Palgrave Macmillan, pp. 49–64.
Meri, Leena (2016): *Perussuomalaisten ajankohtaiskatsaus*. leenameri.fi. www.leenameri.fi/blogi/2016/11/27/14190 [19 April 2018].
Mickelsson, Rauli (2011): Suomalaisten nationalistipopulistien ideologiat. In: Matti Wiberg (Ed). *Populismi: Kriittinen arvio*. Helsinki: Edita, pp. 147–174.
Mickelsson, Rauli (2015): *Suomen puolueet: Vapauden ajasta maailmantuskaan*. Tampere: Vastapaino.
Moffitt, Benjamin and Tormey, Simon (2014): Rethinking populism: Politics, mediatisation and political style, *Political Studies*, 62(2): 381–397.
Mudde, Cas (1995): Right-wing extremism analyzed: A comparative analysis of the ideologies of three alleged right-wing extremist parties (NPD, NDP, CP'86), *European Journal of Political Research*, 27(2): 203–224.
Mudde, Cas (2007): *Populist Radical Right Parties in Europe*. Cambridge: Cambridge University Press.
Myllymäki, Laura (2016): Vihreiden Niinistö ilmastostrategiasta: "Hallitus kuuntelee työmies Putkosen huuhaata". *Aamulehti*. www.aamulehti.fi/kotimaa/ville-niinisto-tuulivoiman-terveysvaikutusten-arviointi-huuhaata-matti-putkosen-politiikan-seurausta-24090198/ [19 April 2018].
National Coalition (2011): *Kokoomuksen ympäristöpoliittinen ohjelma*. Programme. Helsinki: The National Coalition Party.
Nygård, Toivo (1982): *Suomalainen äärioikeisto maailmansotien välillä: ideologiset juuret, järjestöllinen perusta ja toimintamuodot*. Jyväskylä: University of Jyväskylä.
Olsen, Jonathan (1999): *Nature and Nationalism: Right-Wing Ecology and the Politics of Identity in Contemporary Germany*. New York: St. Martin's Press.
Palonen, Emilia (2017): Timo Soinin populismin perusta 1988–2017. In: Emilia Palonen and Tuija Saresma (Eds). *Jätkät & jytkyt Perussuomalaiset ja populismin retoriikka*. Tampere: Vastapaino, pp. 221–248.
Poutanen, Pauli (2016): Putkonen selitti lepakkopuheitaan: "Lein lepakosta vieheen, jonka heitin toimittajamereen". *Iltalehti*. www.iltalehti.fi/uutiset/2016101322454096_uu.shtml [19 April 2018].
PS (*Perussuomalaiset*) (1995): *Oikeutta kansalle – Perussuomalaisen puolueen yleisohjelma*. Programme. Helsinki: The Finns Party.
PS (2001): *Lähiajan tavoiteohjelma*. Programme. Helsinki: The Finns Party.
PS (2003): *Eduskuntavaaliohjelma*. Programme. Helsinki: The Finns Party.
PS (2007): *Eduskuntavaaliohjelma 2007: Oikeudenmukaisuuden, hyvinvoinnin ja kansanvallan puolesta!* Programme. Helsinki: The Finns Party.
PS (2008): *Perussuomalaisten kunnallisvaaliohjelma 2008 – Äänestäjän asialla*. Programme. Helsinki: The Finns Party.
PS (2009): *Perussuomalaisten EU-vaaliohjelma 2009*. Programme. Helsinki: The Finns Party.

PS (2010): *Perussuomalaiset tulevat haastamaan vihreän sanelupolitiikan*, 23 February 2010. www.facebook.com/181106108341/posts/10150096401305328 [19 April 2018].
PS (2011): *Suomalaiselle sopivin. Perussuomalaiset r.p:n eduskuntavaaliohjelma*. Programme. Helsinki: The Finns Party.
PS (2014): *Perussuomalaisten EU-vaaliohjelma*. Programme. Helsinki: The Finns Party.
PS (2015a): *Perussuomalaisten eduskuntavaaliohjelma – pääteemat*. Programme. Helsinki: The Finns Party.
PS (2015b): *Perussuomalaisten maahanmuuttopoliittinen ohjelma 2015*. Programme. Helsinki: The Finns Party.
PS (2017): *Arjesta se alkaa – Perussuomalaisten kuntavaaliohjelma 2017*. Programme. Helsinki: The Finns Party.
PS (2019a): *Perussuomalaisten eduskuntavaaliohjelma 2019 – Äänestä Suomi takaisin*. Programme. Helsinki: The Finns Party.
PS (2019b): *Perussuomalainen ympäristö- ja energiapolitiikka*. Programme. Helsinki: The Finns Party.
Ruostetsaari, Ilkka. (2011): Populistiset piirteet vennamolais-soinilaisen puolueen ohjelmissa. In: Matti Wiberg (Ed). *Populismi. Kriittinen arvio*, Helsinki: Edita, pp. 94–146.
Sarkkinen, Hanna (2016): Onkohan PS:n puoluetoimiston hanaveteen liuennut psykedeelejä, 6 October 2016. https://twitter.com/HSarkkinen/status/783994117709758465 [19 April 2018].
SDP (2009): *Euroopan parlamentin vaalien vaaliohjelma 2009*. Programme. Helsinki: The Social Democratic Party of Finland.
Suomen Uutiset (2016a): Putkonen tuulivoimaloiden terveysuhista: Kymmenet perheet lähteneet evakkoon – vakavalla altistusalueella 650 000 henkeä. www.suomenuutiset.fi/putkonen-tuulivoimaloiden-terveysuhista/ [19 April 2018].
Suomen Uutiset (2016b): Putkonen: Miksi media ja THL pelkäävät avoimuutta ja läpinäkyvyyttä? www.suomenuutiset.fi/putkonen-media-thl-pelkaavat-avoimuutta-lapinakyvyytta/ [19 April 2018].
Taggart, Paul (2006): Populism and representative politics in contemporary Europe, *Journal of Political Ideologies*, 18(1): 269–288.
Waris, Olli (2016): Biologi Putkosen lepakkopuheista: "Tässä on kyse väärinkäsityksestä". *Iltalehti*. www.iltalehti.fi/uutiset/2016100622422920_uu.shtml [19 April 2018].
Waris, Olli (2017): Zyskowicz haastoi jälleen professori Lavapuroa: "Missä tällainen perustuslaki on säädetty? Venezuelassa?" *Iltalehti*. www.iltalehti.fi/politiikka/201703022200079177_pi.shtml [19 April 2018].
Wodak, Ruth (2013): 'Anything goes!' The Haiderization of Europe. In: Ruth Wodak, Majid KhosraviNik and Brigitte Mral (Eds). *Right-wing Populism in Europe: Politics and Discourse*. London: Bloomsbury, pp. 23–37.
Wodak, Ruth (2015): *The Politics of Fear: What Right-wing Populist Discourses Mean*. London: Sage.

PART IV
Central Europe

10
THE ECOLOGICAL COMPONENT OF THE IDEOLOGY AND LEGISLATIVE ACTIVITY OF THE FREEDOM PARTY OF AUSTRIA

Kristian Voss

Introduction

Throughout its history as one of the oldest and most successful contemporary far-right parties in Western Europe, the *Freiheitliche Partei Österreichs* (FPÖ) has established a significant record of advancing nature protection. Neither a political ploy nor a recent development, the FPÖ has been pioneering with nature protection in Austria. In 1968, the FPÖ was the first party to include a policy statement advocating nature protection (Riedlsperger 1998: 33) and was the party most associated with nature protection during the 1975 national election (Dolezal 2008: 124–125). Since then, nature protection continues to be a salient, fundamental, and a comprehensive concern for the FPÖ, similar as other far-right parties across Western Europe (Voss 2014). This ecological component of ideology and legislative activity of the FPÖ, a party with origins tied to pan-Germanism, national liberalism, and National Socialism, is related to the ethnic nationalist embrace of anti-anthropocentrism and organicism. As an especially relevant party of the contemporary far-right party family and in Austria, the FPÖ is positioned to likewise be influential regarding nature protection. The FPÖ is one of the oldest (official foundation on 7 April 1956), most successful (almost always garners the third highest vote share at national legislative elections, frequently on par with the *Sozialdemokratische Partei Österreichs* [SPÖ] and *Österreichische Volkspartei* [ÖVP] since the 1990s, and 26.91% of votes in 1999 is the second highest success for the party family), most experienced (member of one coalition national government with the SPÖ and three with the ÖVP, including the recent government under Sebastian Kurz after a result of 26% at the 2017 national legislative election), only far-right party represented in the Austrian Parliament, and a modal party concerning overall ideology for the far right.

Through a mixed methods analysis of party documents and legislative motions, this chapter attempts to understand the ideology and legislative activity of the FPÖ regarding nature protection, or to answer the following questions: What is the nature protection program of the FPÖ, and why does the FPÖ believe in and advance nature protection in a legislature? Specifically, this chapter will focus on the analysis of recent party documents and motions from the last three legislative periods from 2006 to 2017; the integration of party documents from as far back as 1959 will provide further evidence of the long-term ecological component of FPÖ ideology. All the positions and motions devoted to nature protection have been identified and analysed for a general position, justification and policy preference/s or proposal/s. Such positions and motions cover 15 issue areas and almost all of the 78 sub-issues previously identified as constituting far-right nature protection in Western Europe (see Voss 2014). While a qualitative analysis of party documents and motions is centrally necessary to understand the intricacies of ideology and legislative activity, a quantitative analysis also provides supplementary information to compare the salience of nature protection over time and with other Austrian parties.

Analysed party documents from the FPÖ include more traditional documents, or manifestos concisely outlining principles (*Parteiprogramm*) and election priorities (*Wahlprogramm*), and very in-depth documents detailing ideology and policy (*Handbuch*). Both kinds of party documents are appropriate data indicating the policy positions and emphases of a party (Klingemann et al. 1994: 56). Manifestos are additionally relevant and authoritative as documents ratified by and thus representative of the party as a whole, and no sentence appears 'without a purpose' (Budge 2001: 51–56). One way to attempt to legislatively fulfil promises in Austria is through independent motions (*Selbständige Anträge von Abgeordneten*); motions submitted in a relevant committee by at least five parliamentarians, with the ultimate goal of passage in a plenary session of the *Nationalrat* to force the government to implement its demands. Regardless of the outcome, motions are still important signalling devices of positions and demands (Jenny and Müller 2001: 299). Largely initiated by individual opposition parliamentarians with higher seniority and expertise, motions provide evidence of specific nature protection positions and demands endorsed by the party. Debates and voting results provide another detailed layer of valuable information. Legislative proposals, government bills and parliamentary questions are alternative sources of data less appropriate than motions for this chapter (see Voss 2014: 186–187).

Following this introduction, a mixed methods analysis of party documents and legislative motions will show that nature protection is a salient, comprehensive and fundamental concern for the FPÖ, covering an array of issues and sub-issues. Second, I will argue that nature protection for the FPÖ is fundamentally tied to the ethnic nationalist ecological framework of anti-anthropocentrism and organicism, and has continuity with National Socialism. Third, a comparison of the voting behaviour and debates on legislative motions further indicates the intensity of the FPÖ regarding nature protection, and how and why the nature protection of the FPÖ differs from the Greens and less extreme right *Bündnis Zukunft Österreich* (BZÖ), when both were represented in the *Nationalrat*. Finally, some concluding thoughts focus on the nature protection priorities of the FPÖ in its recent coalition government with the ÖVP.

Salient, comprehensive and fundamental nature protection

Nature protection is a salient, comprehensive and fundamental concern for the FPÖ regarding both ideology and legislative activity. During the three legislative periods from 2006 to 2017, the FPÖ has taken a leading role promoting nature protection in the *Nationalrat*. Although responsible for almost the same percentage of nature protection motions from 2006 to 2008 as the Greens, as shown in Table 10.1, the FPÖ clearly outpaces the Greens during the next two legislative periods with roughly 37% and 34% of the total nature protection motions for all parties. Additionally, the FPÖ devotes a higher percentage of their motions to nature protection than the Greens during the latest two full legislative periods. As the most salient issue for legislative activity, the FPÖ devotes more motions to nature protection than, for example, law and order or immigration. From 2008 to 2013 and 2013 to 2017, the 28.7% and 25% of FPÖ motions devoted to nature protection significantly outpace the 5.7% and 6.2% or 3.4% and 6.3% of motions devoted respectively to law and order or immigration.

From ideology to legislative activity, the FPÖ comprehensively focuses on almost all of the issues and sub-issues of far-right nature protection. Practicing what they preach, the FPÖ covers 96.1% and 88.5% of the 78 nature protection sub-issues respectively in their party documents and legislative motions. Such a comprehensive development is unsurprising, since the FPÖ has long devoted significant attention to nature protection. Calling in 1979 for a 'fundamental reorientation' of the troubled relationship between nature and humans, the FPÖ (2013: 165) has continued to advocate for a 'fundamental reassessment' of modern developments, especially the negative effects of globalization and Europeanization on nature and nation. Whether communicated as an 'ecological generational contract' in 1997 or a 'special responsibility' and a 'question of survival' in 1990, the FPÖ has long partly justified the protection of nature as a duty to ancestors and descendants of the nation. Accordingly, the FPÖ has proposed, as also shown in Table 10.2, the following comprehensive program of nature protection.

Agriculture

Industrial agriculture and its corresponding reliance on artificial substances and monoculture, according to the FPÖ, are destructive to nature and nation. Out of concern for the 'massive loss of biodiversity' and human health, the FPÖ (2013: 65–66) vehemently rejects genetically modified organisms (GMOs). Linking the Common Agricultural Program of the EU with 'irreparable damage' to nature

TABLE 10.1 Salience of nature protection for activity in a legislature (motions) in Austria from 2006 to 2017

		Grünen	SPÖ	N	ÖVP	TS	EZÖ	FPÖ	Total	Average
2013–2017	# of nature motions by party	139	16	54	16	85	–	162	472	79
	% of total nature motions	29.4%	3.4%	11.4%	3.4%	18.0%	–	34.3%	100%	–
	% of nature of party motions	19.2%	8.5%	8.5%	9.2%	26.9%	–	25.0%	–	16.2%
	% of nature of total motions	5.2%	0.6%	2.0%	0.6%	3.2%	–	6.0%	17.6%	2.9%
2008–2013	# of nature motions by party	208	64	–	64	–	172	296	804	161
	% of total nature motions	25.9%	8.0%	–	8.0%	–	21.4%	36.8%	100%	–
	% of nature of party motions	28.8%	27.5%	–	27.4%	–	28.7%	29.0%	–	28.3%
	% of nature of total motions	7.4%	2.3%	–	2.3%	–	6.1%	10.5%	28.6%	5.7%
2006–2008	# of nature motions by party	103	33	–	28	–	38	101	303	61
	% of total nature motions	34.0%	10.9%	–	9.2%	–	12.5%	33.3%	100%	–
	% of nature of party motions	34.7%	30.6%	–	32.9%	–	19.9%	27.4%	–	29.1%
	% of nature of total motions	9.8%	3.1%	–	2.7%	–	3.6%	9.6%	28.9%	5.8%
Total	# of nature motions by party	450	113	54	108	85	210	559	1,579	226
	% of total nature motions	28.5%	7.2%	3.4%	6.8%	5.4%	13.3%	35.4%	100%	–
	% of nature of party motions	25.8%	21.4%	8.5%	22.0%	26.9%	26.6%	27.5%	–	22.7%
	% of nature of total motions	6.9%	1.7%	0.8%	1.7%	1.3%	3.2%	8.5%	24.1%	3.4%

Abbreviations: SPÖ, Sozialdemokratische Partei Österreichs; ÖVP, Österreichische Volkspartei; TS, Team Stronach; BZÖ, Bündnis Zukunft Österreich; FPÖ, Freiheitliche Partei Österreichs.

TABLE 10.2 Nature protection in FPÖ party documents from 1975 to 2013 and legislative motions from 2006 to 2017

Sub-issue	Motions from 2006 to 2008	Motions from 2008 to 2013	Motions from 2013 to 2017	Party Documents
Agriculture:				
Industrial Agriculture (−)	–	583/A(E), 706/A(E), 1349/A(E)	185/A(E), 2250/A(E)	1983: 17, 1986, 1990, 2008a: 36, 40, 2008b: 15, 2009: 60, 2011a: 72, 2013: 68–69
Common Agriculture Policy (−)	756/A(E)	1349/A(E)	439/A(E)	1997: 31, 1996, 2008a: 36–38, 2009: 57–58, 2011a: 73–74, 2013: 68–71
Genetically Modified Organisms (−)	135/A(E), 779/A(E)	504/A(E), 586/A(E), 583/A(E), 909/A(E), 1102/A(E), 1268/A(E), 1281/A(E), 1457/A(E), 505/A(E), 667/A(E)	191/A(E), 257/A(E), 1807/A(E)	1990, 1996, 2002, 2006: 10–11, 2008a: 38–39, 2008b: 15, 2009: 53–59, 2011a: 68–75, 2011b: 8, 2013: 64–72, 174–175
Self-Sufficiency (+)	–	583/A(E)	–	1983: 17–18, 1990, 1997: 31, 2002, 2006: 11, 2008a: 36–40, 2008b: 14–15, 2009: 58–60, 2011a: 72–76, 2011b: 7, 2013: 68–69, 74, 283
Natural Agriculture (+)	779/A(E)	583/A(E), 796/A(E), 1879/A 795/A(E)	520/A(E), 186/A(E), 749/A(E), 2158/A(E)	1979, 1983: 18, 30, 1986, 1990, 1994, 1996, 1997: 29–31, 2002, 2006: 11, 2008b: 36–40, 2009: 56–60, 2011a: 72–78, 2011b: 4, 2013: 68–69
Artificial Substances (−)	789/A(E)	1255/A(E), 1551/A(E), 1113/A(E), 2282/A(E)	–	1990, 1995, 1997: 31, 2011a: 69–70, 2013: 66

(*Continued*)

TABLE 10.2 (*Continued*) Nature protection in FPÖ party documents from 1975 to 2013 and legislative motions from 2006 to 2017

Sub-issue	Motions from 2006 to 2008	Motions from 2008 to 2013	Motions from 2013 to 2017	Party Documents
Agricultural Biodiversity (+)	754/A(E), 155/A(E)	1457/A(E), 2266/A(E)	—	1997: 31, 2005: 27
Natural Food (+)	—	613/A(E), 614/A(E), 706/A(E), 1367/A(E), 573/A(E), 623/A(E), 1255/A(E), 2170/A(E)	2267/A(E)	2009: 63, 2011a: 81
National and Local Products (+)	320/A(E), 169/A(E), 330/A(E)	214/A(E), 305/A(E), 41/A(E), 231/A(E), 1367/A(E), 2152/A(E), 1422/A(E)	408/A(E), 417/A(E), 479/A(E), 1416/A(E). 600/A and 600/A 1393/A(E), 873/A(E), 2123/A(E), 1664/A(E), 2135/A(E), 438/A(E), 1807/A(E)	2002, 2011a: 73–74, 2013: 70
Assistance and Promotion (+)	637/A(E), 756/A(E), 910/A	1460/A(E), 655/A(E), 1818/A(E)	185/A(E), 2212/A72/A(E), 192/A 195/A(E), 611/A(E), 650/A(E), 193/A(E), 1047/A(E), 1341/A(E), 185/A(E)	2008b: 37–41, 2009: 57–61, 2011a: 73–78, 2013: 72–73
Small, Family, and Youth (+)	756/A(E)	583/A(E)	—	1983: 17, 1986, 1990, 1995, 1997: 30–31, 2002, 2006: 11, 2008b: 40, 2009: 60, 2011a: 76, 2011b: 7, 2013: 72–73
Multifunctionality in Agriculture (+)	591/A(E)	220/A(E)	—	1990, 1994, 1997: 31, 2008b: 36, 2009: 56, 2011a: 72–73, 2013: 64

(*Continued*)

TABLE 10.2 (*Continued*) Nature protection in FPÖ party documents from 1975 to 2013 and legislative motions from 2006 to 2017

Sub-issue	Motions from 2006 to 2008	Motions from 2008 to 2013	Motions from 2013 to 2017	Party Documents
Animal Protection, Rights, and Welfare				
Protection…: General (+)	775/A(E)	340/A(E)	2302/A(E)	1994, 2002, 2006: 11, 2008: 40, 2009: 60, 2011a 77–78, 2011b: 4, 2013: 74–76
Cruelty (–)	–	1470/A(E), 1471/A(E), 1774/A(E)	1093/A(E)	2002, 2008b: 41, 2009: 60–61, 2011a: 78–79
Factory Farming (–)	–	1471/A(E)	–	1994, 2006: 11–12, 2008b: 40, 2009: 60, 2011a: 77–78, 2013: 74
Ritual Slaughter (–)	–	180/A	–	2006: 12, 2008b: 41, 2009: 62, 2011a: 79–80, 2013: 76
Transportation Restrictions (+)	–	889/A(E), 1183/A(E), 1422/A(E)	408/A(E)	1999, 2002, 2006: 11, 2008b: 40, 2009: 61–62, 2011a: 78–79, 2013: 75
Pet Ownership and Trade (+/–).	655/A(E)	834/A(E), 694/A(E), 1091/A(E), 1106/A(E), 1670/A(E), 1215/A(E), 1216/A(E), 1274/A(E), 603/ A(E), 719/A(E), 2058/A(E)	1940/A(E)	2002, 2009: 61, 2011a: 78–79, 2013: 74–76
Animal Experimentation (–)	–	–	–	2002, 2006: 11, 2008b: 41, 2009: 62, 2011a: 75, 2013: 76
Ecological Hunting Ethic (+)	–	–	1095/A(E), 1181/A	1994, 2009: 62–63, 2011a: 80–81, 2013: 77–78

(*Continued*)

TABLE 10.2 (*Continued*) Nature protection in FPÖ party documents from 1975 to 2013 and legislative motions from 2006 to 2017

Sub-issue	Motions from 2006 to 2008	Motions from 2008 to 2013	Motions from 2013 to 2017	Party Documents
Conservation, Preservation, Protection, and Restoration of Nature				
Nature Parks and Reserves (+)	463/A(E)	–	510/A(E), 1232/A(E), 1285/A(E), 1714/A(E), 2137/A(E)	1983: 30–31, 1986, 1990, 2009: 63–64, 2011a: 81–82, 2013: 78–79
Biodiversity (+)	577/A(E)	583/A(E)	1232/A(E), 1285/A(E), 1714/A(E)	2008b: 39, 2009: 64, 2011a: 82, 2011b: 4; 2013: 78–79
Sustainable Forestry/ Silviculture (+)	432/A(E)	39/A(E), 2298/A(E)	1181/A	1983: 30, 1986, 1990, 1994, 1997: 31, 2002, 2011a: 77, 2013: 72–74
Water Conservation (+)	267/A(E)	683/A(E), 913/A(E), 808/A(E), 969/A(E)	1355/A(E), 1713/A(E)	1975, 1983: 30, 1990, 1995, 2002, 2006: 6–7, 2008a: 14, 2008b: 35, 2009: 50, 2011a: 66, 2013: 63, 213–214
Restoration (+) Eco- Leisure and Tourism (+)	258/A(E), 463/A(E), 591/A(E), 772/A(E)	497/A(E), 908/A(E), 1019/A(E), 1550/A(E), 1578/A(E), 1755/A(E), 2352/A(E), 2353/A(E)	1241/A(E), 1579/A(E), 1943/A(E), 2137/A(E)	1994, 2006: 7, 2009: 51, 2011a: 67 1990, 2008b: 117–118, 2009: 187, 2011a: 191–195, 2013: 197, 201
Economics				
Anthropogenic Pollution (–)	–	–	–	1975, 1979, 1990, 2008b: 134, 2009: 209, 2011a: 223
Harmony Economic Model (+)	–	2316/A	–	1975, 1979, 1983: 29, 1986, 1990, 1994, 1997: 21, 28–29, 2013: 171–173

(*Continued*)

TABLE 10.2 (*Continued*) Nature protection in FPÖ party documents from 1975 to 2013 and legislative motions from 2006 to 2017

Sub-issue	Motions from 2006 to 2008	Motions from 2008 to 2013	Motions from 2013 to 2017	Party Documents
Nature Protection Standards (+)	–	996/A(E), 808/A(E), 1937/A(E), 1879/A	282/A(E), 2030/A(E)	1975, 1979, 1983: 30, 1990, 1997: 29, 1996, 2002, 2013: 63
'Polluter Pays' Principle (+)	–	520/A, 169/A	–	1975, 1983: 30, 1990, 1995
Environmental Remediation (+)	–	169/A, 1329/A(E), 1936/A(E)	–	1990
Ecological Taxation (+/–)	–	–	–	1990, 1994, 1997: 21, 29–30
Energy				
Fossil Fuels (–)	1074/A 198/A(E), 153/A(E), 412/A(E), 477/A(E), 608/A(E)	206/A(E), 211/A(E), 212/A(E), 1329/A(E), 1343/A(E), 1779/A223/A(E)	566/A(E)	1999, 2008: 14, 2009: 49–50, 2011a: 63–64, 2013: 60, 214–215
Renewable Energy (+):	–	254/A(E), 153/A(E), 444/A(E), 795/A(E), 575/A(E)	199/A(E), 211/A(E), 215/A(E), 217/A(E), 341/A(E), 959/A(E), 2101/A(E), 1954/A(E), 2180/A(E), 412/A(E), 2320/A, 223/A(E)	1979, 1983: 32, 2002, 2008a: 14, 2008b: 30–31, 2009: 40–41, 2011a: 57–59, 2013: 54–56, 214–215
Biomass (and Bio-Fuels), (+)	–	226/A(E), 323/A(E), 430/A(E), 1252/A(E), 841/A(E), 1669/A(E), 2320/A	–	1983: 18, 1990, 1997: 31, 1983: 18, 33, 2002, 2006: 11, 2008b: 43, 2009: 66, 2011a: 84, 2013: 55, 81
Hydropower (+)	–	507/A(E), 218/A(E), 810/A(E)	–	1983: 33, 2006: 6, 2011a: 62, 2013: 55
Hydrogen (+)	–	223/A(E)	–	2012:03
Geothermal (+)	–	–	–	1983: 33, 2013: 55

(*Continued*)

TABLE 10.2 (*Continued*) Nature protection in FPÖ party documents from 1975 to 2013 and legislative motions from 2006 to 2017

Sub-issue	Motions from 2006 to 2008	Motions from 2008 to 2013	Motions from 2013 to 2017	Party Documents
Solar (+)	153/A(E), 156/A(E)	211/A(E), 280/A(E), 405/A(E), 791/A(E), 225/A(E), 1137/A(E), 209/A(E), 1291/A(E)	176/A(E), 177/A(E), 250/A(E)	1979, 1983: 33, 2002, 2008a: 14, 2008b: 33–34, 2009: 43–47, 2011a: 60–62, 2013: 55, 58
Wind (+)	—	223/A(E)	—	1979, 1983: 33, 2002, 2008a: 14, 2008b: 31, 2009: 41, 2011a: 59, 63, 2013: 59
Nuclear Energy (−)	24/A(E), 576/A(E), 613/A(E), 44/A(E), 724/A(E), 683/A(E)	683/A(E), 837/A(E), 753/A(E), 979/A(E), 344/A(E), 510/A(E), 654/A(E), 44/A(E), 49/A(E), 50/A(E), 1043/A(E), 1280/A(E), 208/A(E), 1687/A(E), 181/A(E), 2012/A(E), 811/A(E), 1146/A(E), 1317/A(E), 1318/A(E), 1518/A(E), 1837/A(E), 1861/A(E), 2033/A(E), 2059/A(E), 2097/A(E), 2180/A(E), 2354/A(E)	587/A(E), 461/A(E), 256/A(E), 1192/A(E), 1725/A(E), 1726/A(E), 1727/A(E), 1920/A(E), 521/A(E)	1975, 1979, 1983: 29–32, 1986, 1990, 2002, 2006: 10, 2008b: 31–33, 2009: 40–45, 2011a: 65–66, 2011b: 4, 2013: 55, 62
Energy Conservation/Efficiency (+)	194/A(E), 39/A(E), 153/A(E), 614/A(E), 727/A(E)	197/A(E), 211/A(E), 233/A(E), 1092/A(E), 210/A(E), 1933/A(E), 1545/A(E), 493/A(E)	300/A(E), 292/A(E), 407/A(E), 1133/A(E), 1134/A(E), 2095/A(E), 2281/A(E)	1975, 1979, 1983: 32–33, 1990, 2002, 2008a: 14, 2008b: 32–35, 2009: 41, 2011a: 59, 2013: 57–58

(*Continued*)

TABLE 10.2 (*Continued*) Nature protection in FPÖ party documents from 1975 to 2013 and legislative motions from 2006 to 2017

Sub-issue	Motions from 2006 to 2008	Motions from 2008 to 2013	Motions from 2013 to 2017	Party Documents
Energy Independence (+)	153/A(E), 265/A(E)	211/A(E), 2180/A(E), 223/A(E)	633/A(E)	1983: 32, 2008a: 14, 2008b: 30–35, 2009: 40, 49, 2011a: 57, 63, 208, 2013: 55, 59
Fish and Marine Policy				
CFP (–) and Re-Nationalization (+)	—	—	—	2017:43
Sustainable Fishing (+)	—	—	—	—
Marine Protection (+)	—	537/A(E)	1443/A(E), 1442/A(E)	2013:22
Aqua/Mariculture (–/+)	—	—	1443/A(E), 1442/A(E)	—
Human Health and Bio-Ethics				
Preventive Medicine (+)	388/A(E), 321/A(E), 471/A(E), 505/A(E), 644/A(E), 539/A(E), 574/A(E), 801/A(E), 836/A(E)	285/A(E), 490/A(E), 529/A491/A(E), 492/A(E), 503/A(E), 361/A(E), 723/A(E), 84/A(E), 897/A(E), 178/A(E), 872/A(E), 604/A(E), 965/A(E), 992/A(E), 1061/A(E), 1075/A(E), 1145/A(E), 804/A(E), 1151/A(E), 1378/A(E), 1241/A(E), 1552/A(E), 573/A(E), 619/A(E), 620/A(E), 185/A(E), 621/A(E), 1158/A(E), 1535/A(E), 403/A(E), 1651/A(E), 1575/A1740/A(E), 624/A(E), 668/A(E), 1757/A(E), 809/A(E), 192/A(E), 973/A(E), 1090/A(E), 2346/A(E), 2347/A(E), 2356/A(E)	157/A(E), 284/A(E), 387/A(E), 376/A(E), 973/A(E), 519/A(E), 899/A(E), 375/A(E), 1342/A(E), 1344/A(E), 1173/A(E), 1266/A(E), 1172/A(E), 1395/A(E), 1464/A(E), 1339/A(E), 1529/A(E), 1744/A(E), 1804/A(E), 1805/A(E), 1806/A(E), 1571/A(E), 1990/A(E), 2097/A(E), 2303/A(E)	2002, 2008a: 8, 2008b: 94, 110, 131, 2009: 178, 205–206, 2011a: 179–180, 220, 2011b: 8, 2013: 187, 226–227, 229–232

(*Continued*)

TABLE 10.2 (*Continued*) Nature protection in FPÖ party documents from 1975 to 2013 and legislative motions from 2006 to 2017

Sub-issue	Motions from 2006 to 2008	Motions from 2008 to 2013	Motions from 2013 to 2017	Party Documents
Natural Treatments (+)	721/A1091/A(E), 1106/A(E), 968/A(E)	173/A(E), 175/A(E), 176/A(E), 1511/A(E), 1599/A(E), 1961/A(E)	481/A(E), 483/A(E), 653/A(E)	2006: 10, 2008a: 8, 2008b: 133, 2009: 207, 2011a: 223, 2013: 229
Abortion (−)	648/A(E), 649/A(E), 647/A(E)	1364/A(E), 1421/A(E), 167/A(E), 168/A(E), 1907/A(E)	918/A(E), 2124/A(E), 2125/A(E)	2006: 10, 2008a: 8, 2008b: 81, 133, 2009: 133, 208, 2011a: 127, 223, 2013: 134, 143, 160–161, 229–230
Bioethics (+)	594/A(E)	421/A(E), 1462/A(E), 1534/A(E), 1179/A(E), 624/A(E)		1997: 2, 2002, 2008b: 39, 136, 2009: 214, 2011a: 226, 2011b: 8–10, 2013: 232–233
Immigration and Nationalism				
Immigration (−)	—	—	799/A(E), 670/A(E), 1339/A(E), 2051/A(E), 1683/A 2129/A(E)	1990, 1994, 1995, 1997: 6, 2002, 2006: 3, 2008b: 21, 2011a: 4, 33, 2013: 31, 284
Nature Degradation (−)	—	—	1093/A(E)	1990, 1997: 6, 2008a: 2, 2011a: 37–38, 2013: 35
National Diversity (+)	240/A(E), 256/A(E), 431/A(E), 445/A(E), 515/A646/A(E), 686/A(E), 766/A(E), 785/A(E), 860/A(E), 885/A	45/A(E), 237/A(E), 346/A(E), 238/A(E), 1565/A(E), 1903/A(E), 404/A, 1801/A(E), 48/A(E), 1512/A(E), 2274/A(E), 2312/A(E)	151/A(E), 386/A(E), 389/A(E), 820/A(E), 985/A(E), 1280/A(E), 1871/A(E), 1774/A(E)	1990, 1994, 1997: 1, 5–6, 12, 2002, 2008b: 18, 21, 165–166, 2009: 24, 27, 238, 245, 2011a: 4, 29, 33, 256–257, 263–265, 2011b: 11, 2013: 48, 267–269, 275–276, 283

(*Continued*)

TABLE 10.2 (*Continued*) Nature protection in FPÖ party documents from 1975 to 2013 and legislative motions from 2006 to 2017

Sub-issue	Motions from 2006 to 2008	Motions from 2008 to 2013	Motions from 2013 to 2017	Party Documents
Individual, Consumer, and Community Education, Protection, and Participation				
Personal Responsibility (+)	–	–	–	1979, 2008b: 17, 2009: 23, 2011a: 27
Public Participation and Input (+)	–	1320/A(E), 1954/A(E)	–	1983: 30, 1990, 2013: 79–81
Ecological Consciousness (+)	93/A(E), 893/A	1877/A(E)	292/A(E), 1199/A(E), 1133/A(E), 1464/A(E), 1200/A(E), 1134/A(E), 1198/A(E), 1988/A(E), 1989/A(E), 1987/A(E), 2158/A(E), 1579/A(E), 2001/A(E)	1979, 1983: 29, 1986, 1994, 2006: 12, 2008b: 158, 2009: 230, 2011a: 78, 248
Product Information and Safety (+)	508/A(E), 571/A(E), 330/A(E), 513/A(E),	202/A(E), 214/A(E), 41/A(E), 231/A(E), 613/A(E), 614/A(E), 653/A(E), 706/A(E), 839/A(E), 178/A(E), 832/A(E), 872/A(E), 968/A(E), 969/A(E), 232/A(E), 964/A(E), 806/A(E), 1060/A(E), 1537/A(E), 1570/A(E), 1278/A(E), 2175/A(E), 1506/A(E), 1422/A(E)	415/A(E), 410/A(E), 1419/A(E), 1607/A(E), 1609/A(E), 1892/A(E), 1893/A(E)	1986, 1990, 2002, 2006: 11–12, 2008b: 37–41, 131–134, 2009: 57–61, 206–209, 2011a: 73–79, 220–224, 2013: 55, 70–72, 75–76, 230

(*Continued*)

TABLE 10.2 (Continued) Nature protection in FPÖ party documents from 1975 to 2013 and legislative motions from 2006 to 2017

Sub-issue	Motions from 2006 to 2008	Motions from 2008 to 2013	Motions from 2013 to 2017	Party Documents
International Relations				
Global Ecological Concerns (+)	–	215/A(E), 347/A(E), 802/A(E), 909/A(E), 1268/A(E), 1281/A(E), 181/A(E)	2250/A(E)	1990, 2002, 2008b: 41, 2009: 61, 2013: 77),
Ecological Development Aid (+)	–	2066/A(E)	–	1990, 2011b: 11?
Import Restrictions (+)	–	841/A(E), 1420/A(E), 1669/A(E)	–	1990, 1997: 21, 2002, 2008b: 100, 2009: 166, 2011a: 166
Anthropogenic Climate Change (−)	93/A(E), 152/A(E), 153/A(E), 328/A(E), 772/A(E)	206/A(E), 211/A(E), 796/A(E), 221/A(E)	795/A(E), 223/A(E)	2008b: 31, 2009: 40–41, 49–50, 2011a: 57–63, 2013: 56–60
Science and Technology				
Ecological Science and Tech. (+)	227/A(E)	763/A(E), 433/A(E), 609/A(E), 853/A(E), 1535/A(E)	–	1979, 1983: 30, 1986, 1990, 2002, 2008b: 31, 2009: 40, 2011a: 57, 206
Spatial Planning				
Urbanization (−)	260/A(E)	–	–	1983: 29, 1990, 2008b: 42–43, 2009: 66, 2011a: 84
Traditional Harmony Style (+)	693/A(E), 717/A(E)	737/A213/A(E), 1145/A(E), 1723/A(E), 2213/A(E), 2021/A(E), 2345/A(E)	301/A(E), 2105/A(E), 1876/A(E)	1983: 18, 1994, 1997: 32, 2002, 2008b: 161, 2009: 238, 2011a: 256, 2013: 260

(*Continued*)

TABLE 10.2 (*Continued*) Nature protection in FPÖ party documents from 1975 to 2013 and legislative motions from 2006 to 2017

Sub-issue	Motions from 2006 to 2008	Motions from 2008 to 2013	Motions from 2013 to 2017	Party Documents
Rural Revitalization (+)	–	–	246/A(E), 414/A(E)	1983: 18, 1990, 1997: 30–31, 2008b: 36, 2009: 56–57, 2011a: 73, 2013: 133
Traditional National Culture and Nature				
Traditional Culture (+)	–	316/A(E), 1138/A(E), 1176/A(E), 1059/A(E), 1147/A(E), 1148/A(E), 1517/A(E), 1567/A(E), 1319/A2272/A(E)	433/A(E), 1308/A(E), 1864/A(E), 2311/A(E)	1983: 18–20, 1997: 32, 2009: 236–237, 2011a: 255–256, 2011b: 10, 2013: 207–208, 258–264, 267
Rural and Peasant Life (+)	–	756/A(E)	–	1983: 20, 1986, 1997: 30, 2008a: 14, 2008b: 36, 2009: 56–57, 2011a: 73, 2013: 68–69
Subnational Diversity (+)	–	–	–	1983: 20, 2002, 2009: 238, 2011a: 256, 2013: 260
Transportation				
Harmonized Infrastructure (+)	40/A(E), 43/A(E), 180/A(E), 87/A(E), 286/A(E), 287/A(E), 64/A, 426/A, 502/A(E), 282/A(E)	329/A(E), 335/A(E), 338/A(E), 319/A, 791/A(E), 172/A(E), 562/A(E), 1329/A(E), 1348/A(E), 334/A(E), 1353/A(E), 342/A(E), 1423/A, 1777/A(E), 2275/A(E), 2301/A(E), 321/A	510/A(E), 229/A(E), 1191/A, 224/A(E), 2106/A(E), 174/A	1983: 34, 2013: 211–213

(*Continued*)

TABLE 10.2 (*Continued*) Nature protection in FPÖ party documents from 1975 to 2013 and legislative motions from 2006 to 2017

Sub-issue	Motions from 2006 to 2008	Motions from 2008 to 2013	Motions from 2013 to 2017	Party Documents
Public Transportation (+)	598/A(E)	318/A(E), 336/A(E), 345/A(E), 1329/A(E), 833/A(E), 1269/A(E), 1595/A(E), 1974/A(E), 2020/A(E), 2174/A(E), 171/A(E), 495/A(E)	187/A(E), 265/A(E), 155/A(E), 152/A(E), 1944/A(E)	1990, 2008b: 34, 126, 2009: 190, 2011a: 210, 2013: 211–213, 217
Multimodal Transportation (+)	503/A(E)	794/A(E), 658/A(E), 2160/A(E)	848/A(E)	1983: 34, 1990, 1997, 2002, 2008b: 32–33, 119, 2009: 43, 189, 194, 2011a: 65, 189, 206–207, 2013: 61, 211–213
Eco-Friendly Individual Trans. (+)	195/A(E), 197/A(E), 504/A(E), 531/A(E), 325/A(E), 729/A(E)	190/A(E), 356/A(E), 347/A(E), 203/A(E), 704/A(E), 763/A(E), 807/A(E), 899/A(E), 1330/A(E), 1331/A(E), 334/A(E), 1440/A(E), 1254/A(E), 2181/A(E)	570/A(E), 808/A(E), 585/A(E), 228/A(E)	1990, 2006: 10, 2008b: 33, 119, 2009: 43–44, 189, 2011a: 64–65, 189, 206, 2013: 60–61
Foreign Transit and Traffic (−)	–	–	578/A(E)	1990, 1997, 1999, 2002
Cycling and Pedestrian Zones (+)	–	1055/A(E), 966/A(E)	–	2009: 191–192, 2011a: 211, 2013: 218

(*Continued*)

TABLE 10.2 (*Continued*) Nature protection in FPÖ party documents from 1975 to 2013 and legislative motions from 2006 to 2017

Sub-issue	Motions from 2006 to 2008	Motions from 2008 to 2013	Motions from 2013 to 2017	Party Documents
Waste Management				
Recycling and Disposal (+)	199/A(E), 608/A(E), 326/A(E), 728/A(E)	212/A(E), 207/A(E), 216/A(E), 1045/A(E), 220/A(E)	219/A(E), 1615/A(E)	1979, 1983: 30, 1990, 2008b: 35, 2009: 51, 2011a: 67, 2013: 63–64, 196–197
Littering (−)	–	–	–	2009: 51–52, 2011a: 67–68, 2013: 64
Sewage and Waste Treatment (+)	480/A(E)	–	–	–

Note: All independent motions referenced can be located at www.parlament.gv.at/PAKT/VHG/.

and nation, the FPÖ (2013: 65–69) concomitantly condemns multinational corporations for seeking to establish 'irreversible food dictatorship[s],' especially through free trade agreements. In response, specific FPÖ legislative efforts against industrial agriculture include bans on the cultivation and importation of GMOs, animal products from GMO feed, seed chemically treated with neonicotinoids, chlorpyrifos, and patenting life forms, the last condemned as a 'food weapon' wielded by corporations. Whether 'friendly farming methods' in 1979 or 'holistic agriculture policy' in 1990, FPÖ discourse has consistently focused on the promotion of small to medium-sized Austrian family farms utilizing more traditional and natural methods. Along with polyculture of rare, old and valuable seed varieties more resistant to pests and disease, the FPÖ (214/A(E)) promotes local agriculture as essential to avoid higher pollution and energy consumption levels associated with the 'non-sensical' transportation of imported food. Soil carbon sequestration, green manure, crop rotation and the conversion of equipment to vegetable oil are among the other nature friendlier ideas advanced through FPÖ legislative activity. The cherished 'natural balance' between nature and nation, the FPÖ (2013: 68) also believes, depends on a 'free and powerful peasant class.' Direct and indirect support demanded in FPÖ motions thus include subsidies for only 'real farmers,' not industrial farms; tax breaks; promotion of multi-functionality; specialized educational programs in sustainable agriculture and tourism; milk quotas; authorizing the production and sale of organic hay food products; and a product labelling system to promote Austrian products. Additionally, the FPÖ (2011a: 73–74) touts the importance of 'consumer patriotism' to encourage the purchase of Austrian products.

Animal protection, rights and welfare

Whether regarding companionship, sale, livestock, work, science or wildlife, the FPÖ strives to improve animal protection. Tied to demands for the inclusion of animal protection in the constitution, the FPÖ (2013: 75) is particularly adamant that a penal provision is 'most urgent and paramount' and 'long overdue' to combat animal cruelty, including in a motion to establish a stricter imprisonment term of 3 years. Dog licenses, pet databases, training courses and free veterinary care for abandoned animals are all FPÖ proposed efforts to improve humane pet ownership. When this cannot be widely guaranteed for pets requiring special care, the FPÖ favours strict restrictions, such as a ban on certain snakes. Likewise, the FPÖ also opposes the sale of pets at fairs and events, and demands compliance audits and mandatory basic training for retailers and breeders. Smuggling of animals, according to the FPÖ, remains a problem requiring improved border controls. Relatedly, the FPÖ (2013: 75) argues that the illegal transportation of livestock is particularly problematic, an 'outgrowth of greed' aided by insufficient EU regulations; therefore, the FPÖ proposes motions to discourage products associated with the long transportation of animals, and to reduce stress during necessary transportation, such as by covering and softening metal floors. Associating some of the most severe neglect and abuse of animals with factory farms, the FPÖ (2013: 74) instead touts small-scale and natural farming as integrating a 'natural, respectful treatment of ... animals.' As with economics, the FPÖ (749/A(E): 53; 180/A) rejects the toleration of 'pure animal suffering' for religion or multiculturalism, and thus motions to ban the 'barbaric method' of ritual slaughter without anesthetization. Finally, the FPÖ promotes animal protection regarding animal experimentation and sustainable hunting ethics.

Conservation, preservation, protection and restoration of nature

Conservation positions and legislative efforts of the FPÖ focus on minimizing or reversing the negative effects of human activity on individual species of organisms, ecosystems and landscapes. Cherishing the 'ecological balance' between nation and nature, the FPÖ (2013: 77–79) promotes ecological sensitive spatial planning and the creation of 'green bridges' as a way for organisms to traverse human-made infrastructure. Like with conserving hedgerows, shelterbelts, and copses through smaller-scale agriculture, the FPÖ calls for the maintenance of natural habitat on personal properties and use of nesting aids to protect wild bees and plants. Bees have a 'hard life' today without crucial habitat, the FPÖ (1714/A(E)) laments, because properties are too 'well-kept, uniform' and meadows hayed too early. The FPÖ also specifically motions to protect biodiversity from destruction, such as native from non-native organisms, the Isel River from hydropower, Lake Skadar from nuclear power, or the Danube River from EU channelization and canalization plans, the last an 'ecologically unjustified' project that the FPÖ (913/A(E)) warns will destroy retention areas and riparian vegetation that provide habitat, ensure water quality and limit flooding. Concomitantly, the FPÖ has promoted reforestation with mixed species and restoration of regulated rivers to their natural state, respectively since at least 1959 and 1994. Warning in 1986 that protecting the forest was the 'need of the hour,' the FPÖ prioritizes restoring the 'vitality of the forest' to safeguard biodiversity and water quality, and prevent landslides. Although 'nature conservation has absolute priority,' the FPÖ (2137/A(E); 2013: 195) is still a big proponent of funding and allowing access to national parks, and ecotourism.

Economics

The harmony economic model advanced by the FPÖ at least dates back to their 1975 recommendations that nature protection 'takes precedence over profit and personal gain' and economic development should be 'in harmony' with nature. Ecological quality assessments of water, improvement of water quality and remediation of polluted areas are among the nature protection standards proposed by the FPÖ. While shifting away from ecological taxes as ineffective and unfair, the

FPÖ instead favours tax cuts and investments to incentivize an assortment of nature-friendlier technologies and behaviours across nature protection issues and sub-issues. Additionally, the FPÖ optimistically proposes innovative solutions to combat pollution, such as photocatalytic coatings to purify air and reduce greenhouse gas emissions.

Energy

Viewing nature protection and energy as 'two sides of the same coin,' the FPÖ (2013: 59, 64) promotes domestic renewable energy utilization as necessary to break from the 'fatal dependence' on nature-destructive fossil fuels. Opposed to the Nabucco and Nord Stream pipelines and the installation of fossil fuel-based heating systems in new constructions, the FPÖ (252/A(E): 99) also 'fundamentally' rejects hydraulic fracturing for natural gas as a 'great danger to humans and the environment with unforeseeable consequences.' Of the opinion that coal 'cannot be the way into the future,' the FPÖ (566/A(E)) supports removing tax exemptions and then redirecting funds to the expansion of renewable energy. Referring to the promotion of solar-, wind- and hydro-power as 'our way,' the FPÖ (1725/A(E): 36) has also for decades endorsed biofuels, hydrogen and geothermal as viable renewable sources of energy. In particular, the FPÖ (2013: 60) favours solar energy for its 'largest remaining exploitable potential,' and thus has submitted motions for photovoltaic systems on public buildings, storage systems, investing in the domestic solar energy industry, promotional campaigns and the creation of a nationwide cadastre of optimal sites. Nuclear energy is one alternative source of energy absolutely rejected by the FPÖ as far back as 1975 as an existential threat to the 'natural balance.' Accordingly, the FPÖ submits many motions explicitly against current or future nuclear power and/or waste storage in Albania, Czechia, Germany, Italy, Slovakia and Slovenia. Furthermore, the FPÖ also opposes EU funding of nuclear energy, and to withdraw from or reorient Euratom (The European Atomic Energy Community) to focus on the phase-out of nuclear energy. By assuming a 'leadership role' against nuclear energy, the FPÖ (256/A(E)) believes that Austria could help usher Europe into a 'more positive, sustainable future.' Finally, the FPÖ aims to reduce energy consumption and inefficiency through state guidance and energy-saving technology, including heat pumps, heat recovery ventilation, thermostatic mixing valves, external shading, solar air-conditioning, light-emitting diodes, energy-efficient devices, devices with actual power switches, and discontinuing daytime running lights on motor vehicles.

Fish and marine policy

Despite the geography of Austria, the FPÖ still supports nature-friendlier fish and marine policies. Like with agriculture, the FPÖ (1443/A(E): 370) vehemently opposes genetically modified (GM) salmon as 'greed for profit' that threatens native organisms via escapes and human health. In the past, the FPÖ also strove to minimize the impact of hydropower on river flow continuum for native fish through the construction of fish ladders.

Human health and bio-ethics

Endorsing a 'rethinking' of the supremacy of curation, the FPÖ (2013: 226–227) instead believes that prevention should be the 'primary focus' of health policy. Three related pillars of this FPÖ health policy include preventing exposure, preventive medicine and promoting a healthy lifestyle. Bisphenol in thermal paper and baby bottles, acrylamide in food, microplastics in cosmetics, and lead in water are all examples of carcinogenic, toxic, and mutagenic substances that the FPÖ strives to ban or limit. Additionally, the FPÖ warns of the threat of electromagnetic pollution, especially salivary gland cancer in children from cellular phones, and antibiotic resistant bacteria from nano-silver applications for consumer goods. Preventive medicine efforts endorsed by the FPÖ in motions include improved access to cancer detection examinations, and free immunizations to prevent several different diseases. Motions from the FPÖ to promote the final pillar of prevention policy include reintroducing more rigorous physical exercise in school, facilitating sports, motivating people through famous athletes, proper nutrition, natural food and combating substance abuse. Finally, the FPÖ supports alternative treatments as more natural, and considers combatting abortion as the same as protecting other life forms in nature.

Immigration and nationalism

Immigration, the FPÖ contends, is a major threat to nature and nations. Austria is 'not a country of immigration,' the FPÖ (2013: 31) declares, partially because it is a densely populated country at carrying capacity. In addition to banning ritual slaughter without anesthetization, the FPÖ (1093/A(E)) insists that stricter punishments for animal cruelty are more necessary because of 'cultural and religious reasons' or that 'young people with a migration background' are often the perpetrators. Likewise, the FPÖ justifies restrictions on immigration to prevent the spread of disease. Most importantly, the FPÖ conceptualizes the preservation of nations the same as the conservation of biodiversity, thus immigrants are implicitly analogous to invasive, non-native species. Multiculturalism and 'cultural levelling' are ultimate threats, according to FPÖ (2013: 260, 269) discourse, with the Balkans an example of the failure of the 'the musings of multicultural dreamers.' This ethnic nationalism is also on display in FPÖ (389/A(E); 386/A(E)) motions promoting the protection and self-determination of fellow ethnic Germans outside the country 'fighting for their survival' and dependent on Austria for 'preservation,' or in South Tyrol, Sappada and Slovenia, and the interests of Sudeten and Carpathian German expellees.

Individual, consumer, and community education, protection and participation

In addition to a strong state, the FPÖ also emphasizes the importance of individuals for nature protection. Not only through referendums, such as on harmonized infrastructure or Euratom, the FPÖ also envisions citizen participation in renewable energy projects, environmental impact assessments, and through tax-deductible donations to nature protection organizations. First communicating the importance of cultivating a 'new environmental awareness' in 1979, the FPÖ has continued to support using the state to educate the population in general about nature protection, and specifically children about anthropogenic climate change and animal welfare. To facilitate nature-friendlier decisions, the FPÖ has since 1986 promoted a clear and effective 'eco-label' for a wide array of products, including or associated with GMOs, GM micro-organisms, foreign origins, non-anesthetized slaughter, industrial agriculture, long transportation times, unhealthy ingredients, specific absorption rate level of mobile phones, energy efficiency, and water quality.

International relations

For decades, the FPÖ has explicitly touted the intensification of European and other international cooperation to institute and ensure nature protection. Preventing GMOs, conserving migratory bird habitat, protecting the Arctic, combating illegal logging, and promoting renewable energy and electric mobility are all international goals raised in FPÖ motions. Additionally, the FPÖ aims to combat anthropogenic climate change by limiting fossil fuels and industrial agriculture; the FPÖ (2013: 60) though communicates that carbon credit trading is a 'rip off' and not conducive for nature protection. Unfortunately international organizations, the FPÖ (181/A(E)) claims, often do not sufficiently address nature and nation protection, including the International Atomic Energy Agency 'paralysed' by an 'incomparably more powerful nuclear lobby.' Worse, international organizations, the FPÖ also believes, are responsible for nature destruction, especially in the developing world. For example, the FPÖ (1669/A(E)) chastises the EU for turning people into 'kind of modern cotton and palm oil slaves' for biofuels, and thus demands an EU ban on biofuel imports from countries associated with nature destruction and human rights violations. Additionally, the FPÖ promotes fair trade, and preventing or reducing genetic engineering, expulsion of indigenous people and farmers, and deforestation. Concomitantly, the FPÖ criticizes multinational corporations for manufacturing in countries with lower nature protection standards.

Science and technology

A complicated relationship, the FPÖ embraces or rejects science and technology depending on its compatibility with nature and nation protection. While welcoming new technological developments regarding renewable energy, energy efficiency, and nature-friendlier transportation, the FPÖ warns against GMOs, unnatural food, nano-silver, and electromagnetic pollution from new electronics. For the former, the FPÖ does not hesitate to call for research, development and distribution, but for the latter, the FPÖ calls for bans or restrictions.

Spatial planning

The traditional style of spatial planning advanced by the FPÖ involves the harmonization of nature and nation. In tandem with already discussed conservation motions, the FPÖ (895/A(E): 196) also endorses green roofs as a 'reconciliation of man and nature.' Additionally, the FPÖ calls for the improved protection of oil tanks from damage to prevent contamination. Next, the FPÖ strives to ensure that urban sprawl, particularly large retail shopping centres, does not encroach on nature and destroy the vibrancy of city centres. Furthermore, the FPÖ criticizes high-voltage power lines as a threat to human health, nature and landscapes, and instead favours partial underground cabling. The FPÖ similarly calls for protecting unique cultural landscapes, such as the Wachau region from power plants and dams or generally from Romani encampments. Finally, the FPÖ favours the rehabilitation of brownfield sites over the development of greenfield areas, creating green spaces, and improving or maintaining rural communities, presented in their discourse as most in harmony with nature and essential for national survival.

Traditional national culture and nature

Warning of the 'identity destruction and alienation of people from their roots' tied to globalization, multiculturalism and other modern developments, the FPÖ (2013: 258–259) instead promotes traditional culture, language, historical sites and museums as crucial components of preserving nations, and implicitly their connection to nature. As a result, the FPÖ (2013: 260) demands that the protection, preservation and promotion of Austrian culture, including the 'richness and diversity' of local culture, should be enshrined in the constitution. Specifically, FPÖ motions on culture include retaining pre-Christian Germanic traditions, mandatory state broadcasts of folk music, protecting festivals, and educating children in fairy tales, national myths and folk songs. The preservation of such national culture, the FPÖ (2013: 68) communicates, depends on a 'free and powerful peasant class.' Finally, the FPÖ has motioned to preserve and enhance museums dedicated to Austrian folklore and worldwide cultural diversity.

Transportation

The transportation policy advanced by the FPÖ follows a similar harmonization strategy as other components of its nature protection program. As with agriculture or spatial planning, the FPÖ believes that the necessary construction, modernization and expansion of transportation infrastructure should avoid or minimize the destruction and disturbance of nature and nation. For example, the FPÖ opposes the extension of certain truck parking spaces to protect nature, and supports finding an 'environmentally friendly location' thus avoiding substantial tree removal. Likewise, the FPÖ explains that small scale detours rather than large roadways are preferable to protect mountain environments, or that a third runway plan for the international airport in Vienna would greatly increase noise pollution and other health hazards. When infrastructure is necessary, the FPÖ endorses minimizing the negative effects through noise-reduction measures and air pollution monitoring. To encourage the use of public transportation, the FPÖ supports electronic ticketing, nationwide valid tickets, exchange of commuter tax for rail cards, discounts for senior citizens, and ending distance-related tolls for buses. As a part of their 'holistic, multi-modal transportation strategy' and discourse, the FPÖ (2160/A(E)) strives to make the transportation of freight by rail, road, water and air more nature and nation friendlier, including through motions to improve airplane technology, electrify rail lines, tax vehicles depending on engine class and emissions and to raise night-time truck speed limits to a level when engines emit less pollution.

Nature-friendlier individual transportation is the other central focus of FPÖ legislative efforts. By integrating new infrastructure technology and incentivizing purchases, the FPÖ hopes to encourage the use of electric-, hybrid-, and biodiesel-powered vehicles. For example, the FPÖ endorses support for the purchase of pure electric-powered vehicles; tax relief for the construction of private charging stations for electric vehicles; the implementation of 'ubriticity,' including the use of street lights as electric service stations; establishment of electric car mechanic training courses; and acoustic warning signals to safely accommodate the transition to electric vehicles. Additional park and ride areas and permitting right turns on some red lights are two more FPÖ efforts to reduce traffic and corresponding pollution. Furthermore, the FPÖ aims to reduce pollution by mandating brake and other filter systems for new motor vehicles, and increase control of motor vehicles from Eastern Europe, which it has referred to in 1999 as 'rolling bombs [or] particularly desolate vehicles.' Considering cycling and pedestrian zones as even more nature and nation friendlier forms of transportation, the FPÖ (966/A(E)) touts adult-supervised groups of children walking to school as an example of a 'smart and ecological way.'

Waste management

Reduction and recycling are the two major components of FPÖ waste management motions. Concerned about the 'nuisance of food destruction,' the FPÖ (1198/A(E)) endeavours to reduce food waste in society and business. Explicitly promoting recycling since at least 1979, the FPÖ has continued to submit nature-friendlier motions, such as the use of re-usable packaging and the eventual transition to bioplastic bags, from materials provided by local farmers, and the re-purposing of building materials. Other FPÖ efforts include the proximate disposal of waste, converting waste into fuel, and an accessible deposit and collection system for batteries. Especially concerned with carcinogenic waste from abroad, the FPÖ calls for stricter oversight and mandatory notification of citizens of any local handling, storage and disposal.

Anti-anthropocentrism and organicism of FPÖ nature protection

Fundamentally, nature protection in party documents and legislative motions of the FPÖ is tied to the ethnic nationalist framework of anti-anthropocentrism and organicism and the goal of re-harmonizing nature and nation. Such an ecological framework and discourse, most widely embraced by ethnic nationalists associated with at least Romanticism to National Socialism, is based on the connected core ideas that humans are not above, but a part of nature, and that this nature is a dynamic system of interconnected wholes (see Voss 2014: 41–74). The position and legislative activity on animal protection most explicitly showcases the FPÖ connection to anti-anthropocentrism. During debates on motions, the FPÖ expresses the anti-anthropocentric views that 'animals are not commodities, but sentient beings' (1480/A(E): 363), 'compassionate creatures,' or 'living beings with feelings' (642/A(E): 80–82). Furthermore, the FPÖ boasts that it is the 'voice for the animals' (2002/A(E): 104) striving to protect animals based on the 'respect for all life' (642/A(E): 82), including by legally codifying anti-anthropocentrism. As a result, the FPÖ harshly criticizes the anthropocentric pursuit of profit maximization as especially responsible for the poor treatment of animals, whether associated with industrial agriculture, smuggling, puppy-mills, or abusive owners. Similarly, the FPÖ is adamant that the well-being of animals is more important than religion and often science, transportation, spatial planning or energy.

Anthropocentric behaviour and its corresponding habitat destruction and water, soil, air, noise, electromagnetic or light pollution, the FPÖ laments, are connected to the disruption of the cherished organic harmony between nature and humans. The utilization and/or expansion of GMOs, artificial agriculture substances and techniques, coal, hydraulic fracturing, fossil fuel-powered vehicles, nuclear energy, waterway engineering, urban sprawl, high tension voltage power lines and immigration are a number of some of the specific anthropogenic activities that the FPÖ blames for simultaneously threatening nature and nation. Instead, the FPÖ believes in and focuses legislative activity on re-establishing harmony between nature and nation, or banning and discouraging perceived unnatural or unharmonious practices and substances and mandating and encouraging perceived natural or harmonious practices and substances. Traditional and more natural agriculture, conservation, renewable energy, energy efficiency, harmonized spatial planning, importation restrictions, harmonized transportation infrastructure, public transportation, nature-friendlier individual vehicles, reducing waste, and recycling are a number of ways promoted by the FPÖ to limit the anthropogenic disruption of nature. Additionally, the promotion of a healthy lifestyle through natural food, immersion in nature,

nature protection education, preventive medicine, alternative medicine, physical fitness, limiting exposure to dangerous substances, and support of rural and farming communities and traditional culture are all ways that the FPÖ endeavours to keep the nation in harmony with itself and nature.

This embrace of the ethnic nationalist framework on nature protection indicates a level of continuity of the ideology and legislative activity of the FPÖ with the past, especially National Socialism, as well as a common bond with other contemporary far-right parties across Western Europe. The FPÖ shares a number of similarities with the nature protection program of the National Socialists, including regarding animal protection and rights (Giese and Kahler 1939; Sax 2000); conservation (Lekan 2004); more natural forms of agriculture (Riodran 1997: 25) and silviculture (Imort 2005: 43–46); glorification of rural communities (Bramwell 1985); historic preservation (Hagen 2009); more natural lifestyle for both ecological and human health reasons, including through a mixed natural diet (Treitel 2009) and preventive medicine and homeopathy (Proctor 1999); redemption of technology (Rohrkrämer 1999: 30, 44–49); and harmonizing human-made structures with nature and nation (Shand 1984; Lekan 2005: 80–85). Nature protection differences between past and present exist, but are unsurprisingly more the result of social, political, and economic conditions and scientific knowledge than core ideology changes. Today, the FPÖ shares a similar ecological component of ideology and legislative activity with other parties in the far-right family, as research has indicated for most significant parties or actors across Western Europe (Voss 2014) or Germany (Forchtner et al. 2018; Hurd and Werther 2016; Olsen 1999), Denmark, and United Kingdom (Forchtner and Kølvraa 2015).

Comparing FPÖ nature protection legislative activity with other Austrian parties

Votes and parliamentary debates on nature protection motions provide another layer of evidence shedding light on the ecological component of FPÖ ideology and legislative activity. While voting results for motions individually submitted by the FPÖ, BZÖ, and Greens, as shown in Table 10.3, indicate high unity between the far right and Greens, most motions end up tabled or rejected. This development, the FPÖ contends, is evidence of the ÖVP and SPÖ not holding genuine beliefs in meaningful nature protection. For example, the FPÖ has criticized the ÖVP and SPÖ for the following: inaction, insufficient action, and negative action regarding animal protection (2286/A(E): 63; 2002/A(E): 104; 2302/A(E)); dishonesty and use of 'slogans,' 'word play,' and 'empty promises' regarding nuclear energy (1725/A(E): 36–37); incompetence regarding GMO salmon and free trade (1443/A(E): 371); partisanship for rejecting FPÖ motions (1714/A(E): 163); selling Austria 'down the river' and making it a 'hostage' to the nuclear power industry (1280/A(E): 263–264); or lacking knowledge and consistency regarding illegal transportation (889/A(E)). While the ÖVP and SPÖ often

TABLE 10.3 Vote results for individual opposition party sponsored nature protection motions from 2008 to 2017

	Grünen		BZÖ/TS		FPÖ	
2013–2017						
	GRÜ: GRÜ	12.9%	TS:TS	0.0%	FPÖ: FPÖ	5.3%
	GRÜ: GRÜ, FPÖ	11.4%	TS:TS, FPÖ	26.9%	FPÖ: FPÖ, TS	26.3%
	GRÜ: GRÜ, TS	14.3%	TS:TS, GRÜ	7.7%	FPÖ: FPÖ, GRÜ	2.7%
	GRÜ: GRÜ, N	19.4%	TS:TS, N	0.0%	FPÖ: FPÖ, N	0.0%
	GRÜ: GRÜ, FPÖ, TS	9.7%	TS:TS, FPÖ, GRÜ	38.5%	FPÖ: FPÖ, TS, GRÜ	26.3%
	GRÜ: GRÜ, TS, N	9.7%	TS:TS, GRÜ, N	7.7%	FPÖ: FPÖ, GRÜ, N	10.5%
	GRÜ: GRÜ, FPÖ, TS, N	12.9%	TS:TS, FPÖ, GRÜ, N	19.2%	FPÖ: FPÖ, TS, GRÜ, N	10.5%
	GRÜ: All Parties	9.7%	TS: All Parties	0.0%	FPÖ: All Parties	10.5%
	Other	0.0%	Other	0.0%	Other	7.9%
total N		31		26		38
2008–2013						
	GRÜ: GRÜ	12.9%	BZÖ: BZÖ	10.5%	FPÖ: FPÖ	8.8%
	GRÜ: GRÜ, FPÖ	11.4%	BZÖ: BZÖ, FPÖ	13.2%	FPÖ: FPÖ, BZÖ	19.8%
	GRÜ: GRÜ, BZÖ	14.3%	BZÖ: BZÖ, GRÜ	10.5%	FPÖ: FPÖ, GRÜ	7.7%
	GRÜ: GRÜ, FPÖ, BZÖ	47.1%	BZÖ: BZÖ, FPÖ, GRÜ	61.8%	FPÖ: FPÖ, BZÖ, GRÜ	55.0%
	GRÜ: All Parties	10.0%	BZÖ: All Parties	2.6%	FPÖ: All Parties	6.6%
	Other	4.3%	Other	1.3%	Other	2.2%
total N		70		76		91

reject such claims and FPÖ motions as superfluous, too ecological, or unfeasible, it is clear that the FPÖ is comparably more pro-nature protection.

Generally agreeing on nature protection, the FPÖ instead differs with the BZÖ and Greens more on fundamentals and specific policy. While the FPÖ draws more from Green Romanticism or political ecology, the Greens are more partial to Green rationalism or environmentalism and support globalism and multiculturalism and oppose the connection between nature and nations. Differences between the FPÖ and BZÖ regarding nature protection corroborate one of the major findings of Voss (2014), or that the more extreme or ethnic nationalist a far-right party, the more salient, comprehensive and fundamental the ecological component of ideology and legislative activity. The less extreme and more liberal BZÖ covers 58.6% and 70.0% of the 78 nature protection sub-issues respectively in party documents and legislative motions, figures significantly lower than the 96.1% and 88.5% for the FPÖ. The issue of so-called 'cheat ham' highlights the differences: the more far-right FPÖ favours a ban on artificial ham as unnatural food, while the less far-right BZÖ favours the more liberal approach of accurate product labelling and allowing consumers to opt out of purchasing. While principally supportive, the FPÖ (2140/A(E): 163; 895/A(E): 211; 805/A(E): 187) also occasionally differs with other parties concerning the specifics of nature protection efforts in the *Nationalrat*, rejecting motions on a solar register and green roofs as redundant and improper or food information and allergen regulations as counterproductive and bolstering industrial agriculture. Conversely, the Greens (642/A(E): 82) rejected FPÖ and Team Stronach (TS) efforts on cat castration to protect native nature as too excessive, while liberal NEOS (1725/A(E): 42; 2250/A(E): 62) unsurprisingly objects to most of the FPÖ nature protection legislative activity, calling FPÖ motions against nuclear energy as too extreme or motions against free trade agreements and industrial agriculture as 'anti-American resentment.'

Conclusion

On 18 December 2017, the FPÖ formed a coalition government with the ÖVP, which eventually dissolved on 22 May 2019 following the so-called Ibiza affair. The FPÖ did not head what is now the Ministry of Sustainability and Tourism, although this had less to do with their commitment to nature protection than the ÖVP and its central agriculture faction, which has controlled the ministry for 12 consecutive governments since 1987. However, with Norbert Hofer, the FPÖ's energy and environmental spokesman from 2006 to 2015 responsible for many nature protection motions, heading the Ministry for Transport, Innovation and Technology, and Beate Hartinger-Klein, the Animal Protection Minister and Minister of Labor, Social Affairs, Health and Consumer Protection, the FPÖ had the means to enact nature protection. While in government, Minister Hofer favoured investments in railway expansion, shifting more freight from road to rail, hydrogen vehicles, charging stations for electric vehicles on motorways, reducing traffic congestions to curb pollution, park and ride facilities, noise abatement and making Austria energy 'self-sufficient'.

Unveiled as #mission2030, the FPÖ (2018c) committed to a 'sustainable and ecological future' with all electricity generated from renewable energy by 2030 and CO_2-neutral transportation by 2050. Furthermore, the FPÖ (2018a) heralded this 'end of the fossil age' as also including bans on the installation of oil heating systems in 2020 and through the '100,000 Roofing Program' of photovoltaic systems. Other FPÖ prioritisations in government included the protection of bees, European-wide ban on fur farming, air-conditioned transport of animals, agriculture apprenticeships, combatting plastic pollution, protecting workers from carcinogens and nationwide daily physical education classes in schools. Overall, the FPÖ (2018b) pursuit in government of what it referred to as an 'environmental policy with heart and mind' directly integrates past motions. While front-stage moderation (i.e. the exclusion or rebuke of more extreme politicians like Susanne Winter and Barbara Rosenkranz or Andreas Mölzer and Martin Graf) under Heinz-Christian Strache, party leader from 2005 to 2019, possibly indicates an ongoing genuine backstage ideological moderation, a process that similarly occurred under Jörg Haider, Hofer (2019) vows that nature protection will be a central 'responsibility and duty' of his leadership of the FPÖ; as the self-proclaimed 'true Greens', the FPÖ (2019) has accordingly already teamed up with the SPÖ against the ÖVP to implement its long-demanded ban on the herbicide glyphosate. Cultivated over decades, nature protection has been and remains a salient, comprehensive and fundamental concern for the FPÖ.

References

Bramwell, Anna C. (1985): *Blood and Soil: Richard Walther Darré and Hitler's "Green Party."* Buckinghamshire: Kensal Press.

Budge, Ian (2001): Validating the Manifesto Research Group approach: Theoretical assumptions and empirical confirmations. In: Michael Laver (Ed). *Estimating Policy Position of Political Actors*. New York: Routledge, pp. 50–65.

Dolezal, Martin (2008): Austria: Transformation drive by an established party. In: Hanspeter Kriesi, Edgar Grande, Romain Lachat, Martin Dolezal, Simon Bornschier and Timotheos Frey (Eds). *West European Politics in the Age of Globalization*. Cambridge: Cambridge University Press, pp. 105–129.

Forchtner, Bernhard and Kølvraa, Christoffer (2015): The Nature of nationalism: 'Populist radical right parties' on countryside and climate, *Nature and Culture*, 10(2): 199–224.

Forchtner, Bernhard, Kroneder, Andreas and Wetzel, David (2018): Being skeptical? Exploring far-right climate change communication in Germany, *Environmental Communication*, 12(5): 589–604.

Freiheitliche Partei Österreichs (1975): *Alternative 75. Wahlplattform der Freiheitlichen Partei Österreichs*. The Manifesto Document Collection. Manifesto Project.

Freiheitliche Partei Österreichs (1979): *Wahlplattform der Freiheitlichen Partei Österreichs für die Nationalsratswahl am 6. Mai 1979*. The Manifesto Document Collection. Manifesto Project.

Freiheitliche Partei Österreichs (1983): *Wahlprogramm der Freiheitlichen Partei Österreichs (FPÖ) für die Nationalratswahl 1983*. The Manifesto Document Collection. Manifesto Project.

Freiheitliche Partei Österreichs (1986): *Für eine Politik ohne Privilegien*. The Manifesto Document Collection. Manifesto Project.

Freiheitliche Partei Österreichs (1990): *Für Österreichs Zukunft - An der Erneuerung führt kein Weg mehr vorbei!* The Manifesto Document Collection. Manifesto Project.

Freiheitliche Partei Österreichs (1994): *Österreich-Erklärung.* The Manifesto Document Collection. Manifesto Project.

Freiheitliche Partei Österreichs (1995): *20 Punkte für den „Vertrag mit Österreich.* The Manifesto Document Collection. Manifesto Project.

Freiheitliche Partei Österreichs (1996): *Österreichische Bundesregierung: Versprochen - Gebrochen.* www.FPÖ.or.at/eu/versprochen1.htm [2 August 1997].

Freiheitliche Partei Österreichs (1997): *Programm der Freiheitlichen Partei Österreichs.* www.FPÖ.or.at/programm/einleitung.html [14 July 1997].

Freiheitliche Partei Österreichs (2002): *Sozial & gerecht. Lebenswert & leistbar. Zukunftsorientiert & modern. Wir gestalten Österreich mit Sicherheit.* The Manifesto Document Collection. Manifesto Project.

Freiheitliche Partei Österreichs (2005): *Das Parteiprogramm der Freiheitlichen Partei Österreichs.* www.FPÖ.at/fileadmin/Contentpool/Portal/PDFs/Parteiprogramme/Parteiprogramm_dt.pdf [3 December 2012].

Freiheitliche Partei Österreichs (2006): *Wahlprogramm der Freiheitlichen Partei Österreichs FPÖ. Nationalratswahl 2006.* www.FPÖ.at/fileadmin/Contentpool/Portal/PDFs/Dokumente/FP_-Wahlprogramm_NR-Wahl_2006.pdf [3 December 2012].

Freiheitliche Partei Österreichs (2008a): *Österreich im Wort.* www.FPÖ.at/fileadmin/Contentpool/Portal/wahl08/FP_-Wahlprogramm_NRW08.pdf [3 December 2012].

Freiheitliche Partei Österreichs (2008b): *Handbuch freiheitlicher Politik. Ein Leitfaden für Führungsfunktionäre und Man- datsträger der Freiheitlichen Partei Österreichs.* www.erhoert.at/Politinfos/Parteiprogramme/Handbuch_freiheitlicher_Politik_web.pdf [3 December 2012].

Freiheitliche Partei Österreichs (2009): *Handbuch freiheitlicher Politik. 2. Auflage. Ein Leitfaden für Führungsfunktionäre und Mandatsträger der Freiheitlichen Partei Österreichs.* www.orts-FPÖ.At/elements/files/hb_FPÖ2.pdf [3 December 2012].

Freiheitliche Partei Österreichs (2011a): *Handbuch freiheitlicher Politik. 3. Auflage. Ein Leitfaden für Führungsfunktionäre und Mandatsträger der Freiheitlichen Partei Österreichs.* www.FPÖ.at/fileadmin/Content/portal/PDFs/_dokumente/2011_handbuch_gesamt_web.pdf [3 December 2012].

Freiheitliche Partei Österreichs (2011b): *Parteiprogramm der Freiheitlichen Partei Österreichs (FPÖ).* www.FPÖ.at/fileadmin/Content/portal/PDFs/2012/2012_parteiprogramm_web.pdf [3 December 2012].

Freiheitliche Partei Österreichs (2013): *Handbuch freiheitlicher Politik. Ein Leitfaden für Führungsfunktionäre und Man-datsträger der Freiheitlichen Partei Österreichs 4. Auflage/2013.* www.FPÖ.at/fileadmin/Content/portal/PDFs/_dokumente/2013_handbuch_freiheitlicher_politik_web.pdf [14 June 2013].

Freiheitliche Partei Österreichs (2018a): #mission 2030: Regierung setzt Maßnahmen zum Klimaschutz. www.fpoe.at/artikel/mission-2030-regierung-setzt-massnahmen-zum-klimaschutz/ [20 April 2018].

Freiheitliche Partei Österreichs (2018b): Rauch: "Ambitionierte Umweltpolitik spiegelt sich im Budget wider!". www.fpoe.at/artikel/rauch-ambitionierte-umweltpolitik-spiegelt-sich-im-budget-wider/ [20 April 2018].

Freiheitliche Partei Österreichs (2018c): Rauch: "Klima- und Energiestrategie Wegweiser für nachhaltiges und ökologisches Österreich!" www.fpoe.at/artikel/rauch-klima-und-energiestrategie-wegweiser-fuer-nachhaltiges-und-oekologisches-oesterreich/ [20 April 2018].

Freiheitliche Partei Österreichs (2019): Nationalrat: FPÖ gibt grünes Licht für Glyphosat-Verbot. www.fpoe.at/artikel/nationalrat-fpoe-gibt-gruenes-licht-fuer-glyphosat-verbot/ [12 June 2019].

Giese, Clemens and Kahler, Waldemar (1939): *Das deutsche Tierschutzrecht: Bestimmungen zum Schutz der Tiere*. Berlin: Duncker & Humblot.

Hagen, Joshua (2009): Historic preservation in Nazi Germany: Place, memory, and nationalism, *Journal of Historical Geography*, 35(4): 690–715.

Hofer, Norbert (2019): Klimaschutz geht uns alle an! www.facebook.com/story.php?story_fbid=2390493477894606&id=1650580648552563 [7 June 2019].

Hurd, Madeline and Werther, Steffen (2016): The militant media of neo-Nazi environmentalism. In: Heike Graf (Ed). *The Environment in the Age of the Internet: Activists, Communication and the Digital Landscape*. Cambridge: Open Book Publishers, pp. 137–170.

Imort, Michael (2005): "Eternal forest-eternal Volk". The rhetoric and reality of National Socialist forest policy. In: Franz-Josef Brueggemeier, Mark Cioc and Thomas Zeller (Eds). *How Green Were the Nazis? Nature, Environment, and Nation in the Third Reich*. Athens: Ohio University Press, pp. 43–72.

Jenny, Marcelo and Müller, Wolfgang C. (2001): Die Arbeit im Parlament. In: Wolfgang C. Müller, Marcelo Jenny, Barbara Steininger, Martin Dolezal, Wilfried Philipp and Sabine Preisl-Westphal (Eds). *Die österreichischen Abgeordneten: Individuelle Präferenzen und politisches Verhalten*. Wien: WUV-Universitätsverlag, pp. 261–370.

Klingemann, Hans-Dieter; Richard L. Hofferbert and Ian Budge (1994): *Parties, Policies, and Democracy*. Boulder: Westview Press.

Lekan, Thomas M. (2004): *Imagining the Nation in Nature: Landscape Preservation and German Identity, 1885–1945*. Cambridge: Harvard University Press.

Lekan, Thomas M. (2005): "It shall be the whole landscape!": The Reich nature protection law and regional planning in the Third Reich. In: Franz-Josef Brueggemeier, Mark Cioc and Thomas Zeller (Eds). *How Green Were the Nazis? Nature, Environment, and Nation in the Third Reich*. Athens: Ohio University Press, pp. 73–100.

Olsen, Jonathan (1999): *Nature and Nationalism: Right-Wing Ecology and the Politics of Identity in Contemporary Germany*. New York: St. Martin's Press.

"Pelztierhaltung: Hartinger-Klein für EU-weites Verbot." *ORF*, 18 Mar 2019. https://orf.at/stories/3115613/

Proctor, Robert N. (1999): *Nazi War on Cancer*. Princeton: Princeton University Press.

Riedlsperger, Max (1998): The Freedom Party of Austria: From protest to radical right populism. In: Hans-Georg Betz and Stefan Immerfall (Eds). *The New Politics of the Right: Neo-Populist Parties and Movements in Established Democracies*. London: Palgrave Macmillan, pp. 27–44.

Riordan, Colin (1997): Green ideas in Germany: A historical survey. In: Colin Riordan (Ed). *Green Thought in German Culture*. Cardiff: University of Wales Press, pp. 3–41.

Rohkrämer, Thomas (1999): Cultural criticism in Germany 1880–1933, a typology, *History of European Ideas*, 25(6): 321–339.

Sax, Boria (2000): *Animals in the Third Reich: Pets, Scapegoats, and the Holocaust*. New York: Continuum International Publishing Group.

Shand, James D. (1984): The Reichsautobahn: Symbol for the Third Reich, *Journal of Contemporary History*, 19(2): 189–200.

Treitel, Corinna (2009): Nature and the Nazi diet, *Food and Foodways*, 17(3): 1–20.

Voss, Kristian (2014): *Nature and Nation in Harmony: The Ecological Component of Far Right Ideology*. Unpublished PhD thesis. Florence.

11

THE ENVIRONMENTAL COMMUNICATION OF JOBBIK

Between strategy and ideology

Anna Kyriazi

Introduction

This chapter investigates how environmentalism fits into the political program and rhetoric of the Movement for a Better Hungary (*Jobbik Magyarországért Mozgalom*), commonly known as Jobbik. In line with the core argument of this volume, I see environmentalism as perfectly compatible with the agenda of the far right, the obvious link being nationalism, including, among others, notions of a natural and spiritual connection between a 'people' and their 'homeland'. But is nationalist ideology the only link between the far right and the environment? This chapter sets out to fully investigate what kind of ideological and strategic associations underlie the neglected relationship between the Hungarian far right and the environment.

In recent years Hungary has undergone a nationalist revival, from party competition (Gessler and Kyriazi forthcoming), to civil society (Molnár 2016) to popular culture (Feischmidt and Pulay 2017). Jobbik occupies a central position in this neo-nationalist universe, which is why it is the focus of this chapter. I begin by examining Jobbik's environmental positions and ideology as they develop in three consecutive manifestos (2010, 2014, 2018). I then scrutinize how these are communicated to the public through social and conventional media. I base my interpretations on analytical methods that focus on both text and discourse (Guest et al. 2012; Wodak and Meyer 2009). In order to grasp changing dynamics, I adopt a longitudinal perspective starting with Jobbik's appearance in the Hungarian political space and ending with the most recent national elections held in April 2018.

Before turning to the empirical analysis of Jobbik's environmental communication, I briefly outline the context in which this unfolds, including the recent history of the Hungarian green movement and the emergence of Jobbik in the first years of the new millennium.

Environmentalism in Hungary

In Hungary, like elsewhere in the region (Carmin and Fagan 2010), environmentalism was central to the anti-Communist opposition. Given the lack of political freedoms, environmentalism represented a semi-legal channel through which people could get involved in social and political activism without risking full-out confrontation with the government, in part due to the universal appeal that the protection of nature and the environment have, and in part because the communist leadership tactically permitted some dissent in this area (Berg 2000). Environmentalists were indeed successful in forcing policy change. Most famously, in 1989 Hungary unilaterally cancelled the Gabčikovo-Nagymaros hydroelectric dam project on the Danube River, long opposed by experts and public opinion alike (Botcheva 1996). The eventual fall of state-socialism delinked the green movement from anti-communism, which accounted for its widespread popularity in the first place. Moreover, while all major post-communist parties incorporated green proposals into their programs, in the context of post-transition economic hardship, environmental concerns were crowded out by material considerations (Berg 2000).

Nonetheless, the green movement grew further in the 1990s, becoming one of the most important actors in Hungarian civil society. Around the turn of the millennium and in the context of widespread disillusionment with the political establishment, activists organized into a new party: Politics Can Be Different (*Lehet Más a Politika*, heretofore LMP). LMP wanted to go beyond the traditional left-right divide, blending conservative, left and liberal ideologies. It pledged to combat political corruption, and championed social justice, ecological sustainability and the rule of law (Tóth and Grajczjár 2015). LMP entered the Hungarian parliament in 2010 and, despite a period of major infighting, it still plays a role in Hungarian politics. Like Jobbik, LMP is highly critical of the previous political elites and globalism, but it rejects far-right intolerance and extremism (Bíro-Nagy et al. 2013: 250).

At this point, it is worthwhile to briefly examine the demand for green politics in Hungary, which appears to be modest. Surveys show that though respondents support environmental protection and are aware of the severity of environmental degradation, they are far less willing to make sacrifices for the sake of the environment (Table 11.1). Moreover, in the aftermath of the worldwide economic crisis people's interest in and sympathy for nature and the environment has somewhat decreased (WWF 2016).

Jobbik in the Hungarian political space

Jobbik emerged in the context of this economic downturn, itself also compounded by a political crisis. Founded in 2003, it rose to prominence in 2006, when a speech by the freshly re-elected Prime Minister Ferenc Gyurcsány, in which he admitted to lying to citizens about the economy, was leaked, leading

TABLE 11.1 Opinions related to the environment in Hungary (in percentage)

Important to this person looking after environment: *very much like me/like me*	73
Global warming or the greenhouse effect: *very serious/somewhat serious problem*	93
Pollution of rivers, lakes and oceans: *very serious/somewhat serious problem*	96
Membership of environmental organization: *active member*	1
The Government should reduce environmental pollution, but it should not cost me any money: *strongly agree/agree*	78
I would agree to an increase in taxes if the extra money were used to prevent environmental pollution: *strongly agree/agree*	40
I would agree to an increase in taxes if the extra money were used to prevent environmental pollution: *strongly disagree/disagree*	58

Source: World Values Survey Wave 5: 2005–2009.

TABLE 11.2 Jobbik electoral results, 2006–2019 (percentage and rank)

	2006	*2009*	*2010*	*2014*	*2018*	*2019*
National Assembly	2 (5th)[a]		17 (3rd)	20 (3rd)	19 (2nd)	
European Parliament		15 (3rd)		15 (2nd)		6 (5th)

Source: National Election Office.

Note
a Allied with the Hungarian Justice and Life Party.

to massive and protracted protests. Since its electoral break-through in the 2009 European Parliament elections, Jobbik has played an important role in Hungarian politics (see Table 11.2). However, more recently Jobbik's position was weakened. The resignation of long-time, charismatic leader, Gábor Vona, in the aftermath of the 2018 April elections was followed by turmoil and infighting among leadership. Eventually Tamás Sneider was elected new party leader, while László Toroczkai, the candidate representing a more radical position, was excluded. A number of Jobbik members went on to form a new party, "Mi Hazánk Mozgalom" ("Our Homeland Movement"). The party ran in the 2019 European Parliament elections winning 3% of the vote and, therefore, zero seats. At the same time, Jobbik's vote share decreased sharply.

Jobbik is a far right populist party. It is populist because it espouses a dichotomous view of society, one that is split between two homogenous and antagonistic groups – 'the pure people' and 'the corrupt elite' – and because it argues that politics should express the *volonté générale* (general will) of the people (Mudde 2016: 68). Jobbik criticizes capitalism and globalization (Varga 2014), seen as threatening the political unity, economic strength, and cultural authenticity of the 'nation'. It is also an ardent proponent of the fight against corruption, promising a crackdown on 'politician crime' as well as 'Gypsy-crime'. Nationalism, anti-establishment attitudes, anti-Semitism and especially anti-Roma sentiment are the best predictors of support for Jobbik (Karácsony and Róna 2011).

Jobbik's paramilitary connections, radical statist ideology and racist discourse (Kyriazi 2016) situate it on the extreme end of the European far-right party landscape, along with the Greek Golden Dawn, the British National Party and the National Democratic Party of Germany (Mudde 2014). However, while initially Jobbik championed rebellious radicalism (Tóth and Grajczjár 2015), in the run-up to the 2014 elections, it attempted to moderate its appeal (Kovarek and Farkas 2017). It was in this context that Jobbik's environmentalism became more visible. While this shift has not been extensively analyzed as yet, Bíro-Nagy et al. (2013) attribute it to LMP's weakness, on the one hand, and to the ideological affinity of environmentalism and extreme nationalism, on the other hand, laconically remarking that 'the dramatization of environmental conflict meshes well with Jobbik's xenophobia and criticism of multinational corporations' (242). I concur with this assessment, even though, as I shall show, there are other factors too – strategic and principled – that underlie Jobbik's environmentalism.

More specifically, in this paper I outline three distinct uses of Jobbik's environmentalism. First, it serves to emphasise the party's nationalist ideology, e.g. by equating nature with the homeland that should be revered and protected. Second, it is a strategic tool in that it enables Jobbik to catalogue and criticise the party's "enemies," including the "establishment," the government, and left/liberal parties and activists. Third, environmentalism also contributes to the effort to project a more moderate image by tapping an issue that is not directly related to identity-politics with which Jobbik has been most closely associated.

Jobbik's environmental agenda

Programmatic elements

The eco-social national economy

As with other far right actors, Jobbik's concern for nature links to the threat of globalism and modernity (see, e.g. Forchtner et al. 2018). Since Jobbik's early days, former party leader Gábor Vona has argued that 'the economy has to be constrained in the interest of a humane environment (eco), a dignified life (social), and Hungariandom (nation)' (Vona 2009), coining the term 'eco-social national economy' to denote this ideal. In an article discussing the concept in details, he claims that 'the greed of the economy has destroyed nature'. He further writes:

> Global capitalism, based on multinational capital's freedom of movement – more precisely, freedom to pillage – has set alarm bells ringing. Humanity and the globe cannot continue the forced course of economic development. (…) this large global upswing has indeed produced noteworthy economic figures, but its beneficiary was almost exclusively the international financial sector, while the people have been physically and mentally crippled,

the social chasm in the world has continued to grow, and our planet has been giving out all the more frequent and alarming immune responses to humanity's new 'Babylonian revolt'.[1]

(Vona 2009)

This text presents two diametrically opposed antagonistic forces. The one currently dominating the world is an artificial, unnatural and inhumane set of influences, which measures success only in numbers. The other is the nation: a natural unit of social organization, 'the largest natural human community that is capable of historical progress'. However powerful, global capitalism nonetheless faces imminent defeat: it is only a matter of time before the natural world will rise to subdue the artificial world, bringing an end to humanity's modern-time hubris, just as God punished the Babylonians for their arrogance in biblical times. The current decay will be replaced by rejuvenation: the natural will defeat the artificial; divine creation will once again take precedence over the man-made. These are the fundamentals of palingenetic ultranationalism (Griffin 2000), Jobbik's core ideology. The tropes described above have occupied centre stage in Jobbik's successive manifestos, though nationalist discourse has been gradually scaled back. It is to the discussion of these programmatic documents that I now turn.

Manifestos

Jobbik has had an environmental program since its inception. The party's first manifesto, the 'Bethlen Gábor Program', was published in 2007 and contained a section entitled 'Stop the Destruction of the Environment'. It called for harsher penalties for environmental crime, the expansion of domestic energy production, and stronger consumer protection. Jobbik's 2010 manifesto, 'Radical Change', expanded the 13-page-long Bethlen Gábor Program into an exhaustive, 88-page collection of extensive proposals for Hungary's 'national sovereignty and social justice'. Environmental protection came up in several parts, but the main program was laid out in a section entitled 'In Harmony with Nature', introduced as follows:

> With the collapse of the global market economy emphasis has been gradually shifting towards local interests, strengthening the role communities that are living in harmony with their surroundings, that are more and more self-sufficient in terms of energy and food, and that are strong, and rich, both materially and spiritually. The profit-centered approach gives way to a human and nature-centered approach, in which people are not mere instruments, but parts and important members of a community. This community is the nation whose basic building block is the family.
>
> (Jobbik 2010: 23)

Echoing Vona's above-cited article, this passage also casts the global market economy as destructive and inhumane and makes a call to shift from the global to the local. According to this vision, people can only lead meaningful and prosperous lives in natural habitat groups, i.e. communities based on close-knit familial relations. The emphasis on self-sufficiency, which comes up across the document, is an unmistakable reference to the nationalist ideal of sovereignty and autarchy. The threat of globalization to national freedom is a central preoccupation of the far right in Hungary and beyond (see: Forchtner et al. 2018). Yet, as we shall see, while Jobbik criticizes environmental degradation as a result of global capitalism, it readily subordinates the protection of nature to the economic preponderance of the nation, which it sees as vital for the exercise of self-determination.

Possessive forms: 'our air', 'our cities', 'our water resources', convey the deep interconnection between people and their surroundings. The manifesto harshly criticizes the 'establishment', i.e. all post-Communist governments in Hungary (left and right), 'speculators' and multinational corporations. An overarching theme in the party's detailed proposals is the need for stronger regulation and harsher penalties. The linkage between environmentalism, quality of governance (most commonly, the fight against corruption and/or inefficiency) and a strong law and order agenda has remained the cornerstone of the party's environmental program over the years.

The subsequent 2014 manifesto, entitled 'We Say It, We Solve It' (a reference to the party's battle against 'political correctness'), builds on and expands this agenda. It devotes a separate section to Jobbik's environmental program, though environmental issues are included also in other parts (agriculture, energy, etc.). Compared to the previous one, this manifesto shows further ideological crystallization: it makes clear the ideological link between environmentalism and nationalism while also scaling up criticism of political opponents. Accordingly, the language is also stronger. Lexical choices ('corruption', 'mafia'), typesetting (bold and italic) and sarcasm ('decorative ministry', 'cash cow') convey outspoken, righteous indignation. This manifesto dismisses the apparent incompatibility of nationalism and environmentalism, claiming that the nationalists are the only 'true' environmentalists, as the far right typically does, (see: Olsen 1999, Chapter 1 in this volume by Forchtner 2019).

> Jobbik is a national party and (...) we believe that a green ideology and a robust national politics are not antithetical, but, conversely, presuppose one another. Only those who love, appreciate and respect the land on which they live, along with its history, culture and society are capable of devising a truly harmonious, authentic program for environmental protection. (...) It is therefore time to break with the Western lie that environmental protection is a left-liberal privilege. (...) We, the Jobbik, worship the Carpathian basin, our homeland, and we worship our nation,

the Hungarians. It follows directly from this that we are most determined to fight for the protection of our environment (…)

(Jobbik 2014: 68)

The concept of harmony is again central to this program, which posits, in the spirit of organic nationalism, the deeper metaphysical interconnection of man and nature, people and homeland, from which the party's 'genuine' concern for the environment arises. Further, the circle of the nation's 'enemies' widens to include not only the ruling political class, but also the 'West' as well as environmentalists, typically associated with the left. The latter are mocked as 'clowns' and 'bored young upper middle class people looking for a superficial adventure and a purpose in life' (Jobbik 2014: 68). Thus, through its environmental agenda, Jobbik affirms its nationalist convictions, while also distancing itself from and ridiculing the party's enemies.

Jobbik's 2018 manifesto, 'Hungarian Heart, Common Sense, Clean Hands', retains largely the same programmatic elements, but it scales back on nationalist discourse. As in 2010 and 2014, positions pertaining to the environment can be found in several places. However, the separate section dealing with environmental protection specifically has been shortened and, strikingly, has been almost entirely stripped of references to 'harmony' and 'national thinking', while the rhetoric of exaggeration and indignation has also been cut back. What remain are technical proposals, a commitment to more effective control and harsher punishment, and a ubiquitous critical tone against the government:

> Domestic environmental protection is bleeding from numerous wounds: since 2010, the Orbán government completed what its left-wing predecessors started. The Environmental Protection Authority – and through it the entire supervisory system – has been fatally weakened. It is "thanks" to this that abandoned hazardous waste deposits are popping up across the country, along with chemical plants that continue to pollute the environment even as they are being liquidated. In the eyes of the cabinet, environmental protection is a necessary evil, which must be diminished as much as possible.
>
> (Jobbik 2018: 54)

Moreover, the section devoted to the 'eco-social national economy' is shorter and much less noticeable than before. Anti-globalist rhetoric is also scaled back: e.g. the text refers to the need to 'mitigate the adverse effects of global economy' rather than eradicating it. As nationalist elements have been relatively de-emphasized, the strong critical tone has gained in salience and it seems that the single most important use of the environmental agenda is to highlight the incompetence of Jobbik's political rivals.

That said, the party has not abandoned its nationalist ideology, which is still an integral part of Jobbik's environmentalism. Indicatively, in a press conference presenting the 2018 environmental program as part of Jobbik's new manifesto, the head of the party's environmental cabinet, an environmental engineer by training and the face of environmentalism at Jobbik, MP Lajos Kepli, said the following:

> (...) unlike the Fidesz government, for us, the protection of the environment is important; it is important what kind of environment we leave to our children, grandchildren, to the future generations of Hungarians, that the environmental state and living conditions in the Carpathian Basin continue to improve in the coming years.[2]

The text equates the environment with national territory, where past and future generations of Hungarians come together, ensuring the continuity of the 'nation'. Note the use of the term 'Carpathian Basin', invoked by the Hungarian right to signify the unity of all Hungarians in their 'true home': a home defined by geomorphological boundaries and therefore more authentic than the official borders recognized by international law; a natural area on the Earth's surface embracing a 'nation' that historical calamity has dispersed onto the territories of alien states.

Priorities

Priorities, more than positions, reveal the true weight of any political program. Here, Jobbik's platform diverges from mainstream environmentalists. This is especially striking in the field of energy: Unlike LMP and other left/liberal parties, Jobbik is in favour of renewable energy sources only as long as they do not increase overall costs for consumers (Trechsel et al. 2014).[3] In fact, Jobbik is strongly pro-coal and pro-nuclear. It has supported the government's plan to extend, in cooperation with Russia, Hungary's nuclear power plant located in Paks City (commonly referred to as 'Paks 2'). In its 2014 manifesto, it bashed 'deep-Green' overreactions against the deal, pledging to examine whether the alternative solutions proposed by 'extremist' groups really bring the promised benefits to consumers and taxpayers (Jobbik 2014: 53). In other instances, Lajos Kepli also rallied in favour of coal mines to save jobs and to ensure energy independence.[4] Similarly, while Jobbik wants to reduce traffic levels and improve air quality, it is unwilling to also increase taxes for such purposes (Trechsel et al. 2014). This is hardly surprising, since Jobbik claims to represent those 'left behind': the Hungarian family struggling to make ends meet, the coal miner who will be out of job, the small business owner who cannot afford increased transportation costs. But it also reflects an ideological stance: a strong economy is the requisite of self-sufficiency, which serves the ultimate national goal, freedom. Environmental protection is subordinate to this ideal.

Jobbik's priorities, however, converge with LMP's in a key policy field: the sale of Hungarian land to foreigners. Both parties oppose this (Trechsel et al. 2014), arguing that land is a special asset and that local communities should be able to own the sources of their livelihood. However, while LMP is against the sale of land to foreign investors as members of the extremely wealthy (LMP 2018: 35), Jobbik objects to the fact of foreign ownership itself (Jobbik 2018: 41). One can hardly imagine a more transparently nationalist position than to assert that the nation's soil should be held under domestic control.

Jobbik has also gone beyond 'cheap talk' by taking action, i.e. involving itself in environmental activism through moves like founding its own environmental NGO, Green Answer (Zöld Válasz). Established in 2011, Green Answer scaled up its activities from 2017 onwards. Further, on its website and social media, Jobbik reports some fieldwork, like handing out flowers to locals and inspecting sewage treatment facilities. While Jobbik has not consistently linked environmentalism to migration (as, e.g. the British National Party has done; see Forchtner and Kølvraa 2015: 211), one noteworthy video documents the fusion of the two issues.[5] In it, Lajos Kepli, and other Jobbik politicians and volunteers are collecting 'the incredible amount of waste' that 'migrants' and NGOs allegedly left behind as they 'illegally' passed into and marched through the country. This is a rare instance where environmentalism directly links to Jobbik's staunch opposition to immigration, and where the latter is depicted as a disaster for both nature and nation.

Jobbik's environmental agenda in the Hungarian information environment

We have seen that Jobbik's environmental program is comprehensive. But how does it appear in the broader media environment? As previously mentioned, environmentalism is not an issue Jobbik is typically associated with – the scholarly literature barely ever mentions it. In fact, especially prior to 2010, Jobbik reached the mainstream news only in a very narrow range of topics: nationalism, anti-corruption, anti-communism and national security (Gessler and Kyriazi forthcoming). Though it has been increasingly able to present a fuller agenda, nonetheless, environmentalism continues to be marginal in the party's discourse. For this reason, I opted for a purposeful sampling of Jobbik's environmental communication, focusing on its patterns and configurations, rather than on its salience and visibility in the Hungarian information environment. Specifically, I analyzed audio-visual media data available on Jobbik's website as well as Jobbik's Facebook page dedicated to environmental issues. I present the results of these analyses below.

Audiovisual medially produced texts

For this inquiry I used 78 recordings uploaded to Jobbik's official website: 2 videos that can be found on the page dedicated to the party's environmental program[6] and 73 videos and 3 radio interviews on Lajos Kepli's Jobbik page.[7] I have not included parliamentary speeches also uploaded on these pages given that my aim was to examine environmental communication that reaches a broad audience. I have also disregarded material in which Lajos Kepli discusses issues other than the environment. The 78 videos have a total duration of approximately 685 minutes and were produced between January 2010 and March 2018 (the cut-off point of data collection for this paper). They are neatly cut to the relevant parts of TV programs or press conferences. Thirty-two are televised interviews or debates, 3 are radio interviews, 17 are reportages, 22 are press conferences and the remaining 4 are conversations or meetings recorded by the party. Most videos have been uploaded directly to Jobbik's website or YouTube. There are also numerous appearances at N1TV, an internet channel with close ties to Jobbik and other right-leaning networks, but few in mainstream channels.

The protagonist of most of these videos is Lajos Kepli, appearing either alone or with other Jobbik politicians and, more rarely, with representatives of other parties. It is unclear whether the videos posted on the Jobbik website are a full catalogue of all the media appearances related to environmentalism or only a selection (though they appear to be the former). Despite the problem of representativeness, this material constitutes a rich data source regarding Jobbik's environmental communication.

In analyzing these texts, I sought to reveal how the average viewer would interpret these segments: the global impression they convey on both the verbal and the visual level (Figueroa 2008), and the topics they discuss. I allowed for themes – i.e. recurring units of meaning (Guest et al. 2012) – to emerge freely from the material in an iterative process of data collection, categorization and analysis. This was fairly straightforward since most videos have a clear main topic (communicated in the description of the video) and discursive signposts that indicate when a new topic is introduced.

Predictably, the topics that Jobbik is able to communicate to wider audiences are much more limited than the detailed agenda outlined in its manifestos. Almost 90% of the themes discussed can be assigned to one of the five major substantive categories (presented in Figure 11.1). The remaining is about miscellaneous issues, such as the party's general environmental program, environmental education, etc. This suggests that not only is Jobbik's environmentalism not particularly visible in the broader information environment, but also, even when environmentalism gains some visibility, the variety of topics discussed remains small. Even so, put in a time perspective, there has actually been an expansion of both the thematic range and frequency of Jobbik's environment-related media appearances, especially from 2015 onwards.[8]

Until then Jobbik's environment-related appearances revolved around one major topic: the Ajka alumina sludge spill, commonly referred to as the 'Red Mud Disaster'. This industrial accident happened in October 2010 at an alumina plant in Ajka (Western Hungary), when the dam of a caustic waste reservoir collapsed. As a result, toxic waste flooded nearby localities, killing ten people, injuring 150 and causing severe material and environmental damage. The director of the plant and 14 other employees were arrested over charges of negligence,

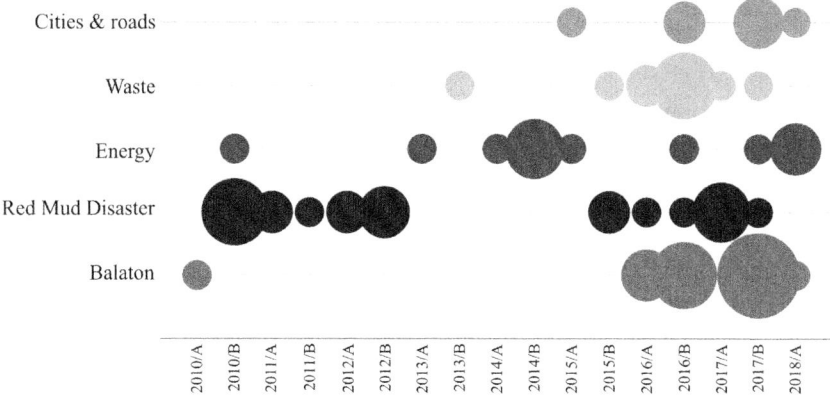

FIGURE 11.1 Major themes in Kepli Lajos' media appearances 2010–2018 (six-monthly). (Source: Author's elaboration.)

violations of waste management and environmental pollution, but were acquitted in January 2016 (which explains the reappearance of this theme around that time in Figure 11.1). The case is still under investigation and no convictions have been made as yet. Following the accident, a parliamentary committee was set up to investigate the omissions contributing to the spill, headed by Lajos Kepli. He linked the accident to legal and administrative defects, suspicious privatization contracts that disregarded environmental stipulations, and the impunity of big business. Early on he even floated a conspiracy theory that the dam was purposefully blown up.[9]

The videos also discuss some of the most obvious environmental problems and dangers: waste collection and management in the wake of the creation of a new waste collection agency (criticized for being inefficient) and energy, in particular the Paks 2 deal. This offers Jobbik a platform to advocate for Hungary's energy independence while also condemning the government's dealings with Russia, described as suspicious. There is also some preoccupation with the built environment, primarily road traffic and cities, and another, rather esoteric issue: Jobbik's pledge to eliminate daylight saving time. But the issue that has drawn as much attention as the Red Mud Disaster is Hungary's largest lake, the Balaton. Kepli, who is originally from the Balaton-region, spends a lot of airtime talking about the lake and its surroundings, mainly condemning the sale of waterfront parcels and summer campsites to businessmen reportedly tied to the Prime Minister.[10]

A constant element in this material is the unrelenting criticism of the administration, 'economic interest groups close to the government' and/or 'political circles'. An oft object of condemnation is Lőrinc Mészáros, a wealthy entrepreneur, Viktor Orbán's childhood friend and former mayor of the Prime Minister's hometown, who in Jobbik's discourse has come to epitomize the corrupt 'establishment'. The texts often refer to large sums of money, public procurement contracts and EU structural funds, conveying an impression of continual exploitation and misappropriation. Even when the accusations are not explicit, suggestive stylistic elements and rhetorical figures insinuate wrongdoing: corruption, embezzlement, shady business deals. These include the use of rhetorical questions, ellipses or irony ('How surprising!', 'quite amazing'), exaggeration ('incredible tempo') and ghastly metaphors ('wrangling over scraps'). Another characteristic of these texts is provocation, achieved through the use of a highly emotive vocabulary ('outrageous', 'shame'), puns and neologisms ('Orbán nostra', a word play with the Prime Minister's name and the Sicilian criminal syndicate 'Cosa nostra') and the use of slang ('buddies').

The demeanour of Kepli and the characters in the videos are, nonetheless, restrained and pedantically polite. Kepli is usually well dressed and formal, presenting himself as a serious, well-informed, articulate person. He often cites statistics and scientific studies and uses expert terminology ('ecological footprint', 'normal background radiation'). Whether Kepli's assertions truly reflect the mainstream scientific consensus is another question. This seems dubious at times, e.g. when he argues that the extension of Hungary's nuclear power plant would carry fewer environmental hazards than building solar energy infrastructure as LMP advocates.[11] Moreover, while Jobbik, unlike other (populist) far right actors, is not a climate-change denier (Lockwood 2018), nonetheless, this issue is conspicuously absent from the party's environmental communication, perhaps due to its low political salience in Hungary (McCright et al. 2015).

Social media

Jobbik is very active on social media (Bartlett et al. 2012). The party's official page has about half a million followers (Fidesz's official page has half as many). In the summer of 2017, 'Jobbik Environmental Protection' (*Jobbik Környezetvédelem*) was launched as a Facebook 'community' dedicated to the protection of the nature and the environment. This section analyses the content of 207 posts that have been uploaded since the page was launched on 31 July 2017 until the 2018 elections (the last post considered is from the 5th of April). Activity on the page has been relatively low with approximately 2000 total page likes. It is difficult to judge whether this is because of the marginality of environmentalism in the party's agenda or because the page is still new. In any case, the post/day rate shows a growing trend (Figure 11.2). Most entries are reposts from other internet sources: Facebook and online news outlets, with general and environmentalist profiles and belonging to the mainstream (e.g. *Magyar Nemzet*, HVG) as well as the fringe (*Alfahír*). Numerous posts also come from the large environmentalist portal, greenfo.hu, and Green Answer, Jobbik's environmental NGO.

Overall the page exhibits a congenial style. Its profile picture is a heart-shaped wordcloud combining red, green and white (the Hungarian national tricolour) with the words that stand out being 'Jobbik', 'environmental protection', 'energy',

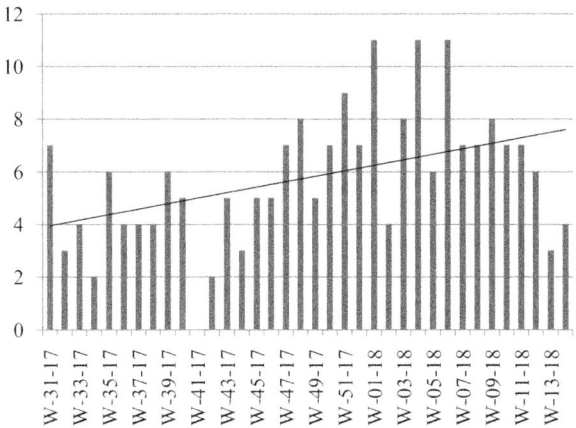

FIGURE 11.2 'Jobbik Environmental Protection' posts per week (31 July 2017–5 April 2018). (Source: Author's elaboration.)

'motherland', and 'green'. Its cover photo until recently depicted a pensive Lajos Kepli looking straight into the camera along with a quote attributed to Ghandi: 'We must be the change we wish to see in the world'. After the April elections, the cover photo was changed to a painting of the parliament building and the Hungarian flag, giving the page a more nationalist flare. This was motivated by the heightened political mobilization around the election and is somewhat incongruent with the style of the page, which generally avoids ultranationalist language and topics. It does, however, have a critical edge: 40% of all posts contain explicit or implicit criticism, targeting mostly the government and authorities (30%), but also corporations, plants and businesses as well as other states in the region. In addition, ten posts have only one dominant topic: the criticism of government and the authorities.

Like with the audiovisually produced texts discussed in the previous section, I performed thematic analysis to map out and evaluate the range of topics present on the Facebook page. Figure 11.3, listing the most frequent topics on Jobbik Environmental Protection (about 70% of all posts), indicates that most issues overlap: energy, the Balaton and the Red Mud Disaster are also often discussed on Facebook. However, there are notable differences, with the Facebook page dedicating much more space to pollution, waste, social questions (consumer protection, the abolition of daylight saving time, etc.) and crime (illegal dumping and construction). Beyond these major themes, a wide variety of issues are also present, from river ecosystems to climate and potable water. The page tries to engage followers in more light-hearted ways, too, e.g. by bringing attention to witty inventions, cute animals and beautiful landscapes.

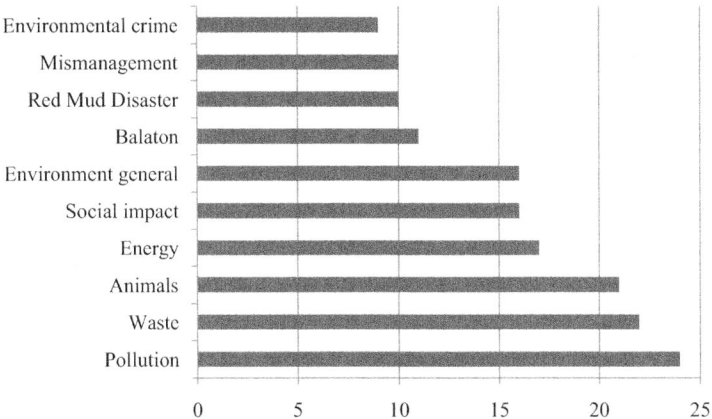

FIGURE 11.3 Major themes in 'Jobbik Environmental Protection' Facebook page. (Source: Author's elaboration.)

That said, the posts that prompt the largest response have explicit populist elements to them. By far the most popular post on the website (548 likes and reactions, 203 shares and 90 comments at the time of writing) is a call for action to protect the Balaton:

> Instead of making Balaton, Central Europe's largest lake and our national treasure, accessible and giving it back to the people, they [the Government] are selling it out day by day, inch by inch, to the modern landlords and straw men. Don't let the Balaton fall prey to robbers – to become the garden pond of Mészáros. Don't let greedy interests ruin this unique environmental area. Lake Balaton belongs to everyone!
>
> *(18 January 2018)*

The text is accompanied by a graphic of Lőrinc Mészáros sitting at a table preparing to slice up and eat Lake Balaton served up on his plate; sharp light comes down on him from above, and the text reads in all capital letters: 'Let's protect the Balaton!'

This post epitomizes how Jobbik fuses national populism and environmentalism in its communication. Lake Balaton serves as an identity marker, an idealized landscape that is entwined with the character of the nation (see: Forchtner and Kolvraa 2015). The vocabulary of populism is used to describe the people's 'enemies', few yet powerful: 'robbers', 'landlords' and 'straw men'. Note how the text plays with different scales: Balaton, a majestic natural treasure, 'Central Europe's *largest* lake' (my emphasis), is reduced to an insignificant 'pond' when it ceases to be a communal possession and becomes private property. This process is described as a 'sell-out' and 'robbery' not only literally – i.e. acquiring land around the Balaton with shady contracts to make the rich even richer – but also figuratively: that which belongs to everyone cannot be legitimately bought and sold on the market (Weiner 1992). Lőrinc Mészáros symbolizes politically well-connected greed that threatens to devour all that is invaluable: nature, the environment and the nation itself.

Conclusion: Jobbik, a party like all others?

In conclusion, Jobbik's environmental program is extensive and multifaceted, combining principled-ideological and instrumental-strategic components. Nationalist ideology and environmentalism intersect in numerous ways, so much so that the Hungarian far right has claimed to be the only 'true' environmentalists. Indeed, the protection of nature and the environment as the physical manifestation of the 'homeland' has been a part of Jobbik's program from the start, addressing a wide array of issues, from the sale of land to foreigners, to renewable energy and endangered species. Yet, when it comes to priorities, the autarchy and economic preponderance of the nation tends to come before the protection of the environment. Moreover, little of Jobbik's environmentalism is visible in the public sphere beyond a narrow set of issues. Environmental concerns are typically linked to the issues Jobbik already 'owns': the fight against corruption and mismanagement, anti-globalism and anti-capitalism, a strict law and order agenda.

Further, while earlier electoral manifestos saturated environmentalism with dense symbolic meaning that mirrored the party's ultranationalist ideology, this is much less the case today. As Jobbik's ability to present a fuller agenda has increased, exaggerated expressions of nationalist ideology have been somewhat de-emphasized. Stripped from its references to organic nationalism and cosmic harmony, Jobbik's environmental communication is currently almost entirely geared towards anti-establishment critique. Beyond this strategic issue linkage, environmentalism also aids Jobbik in conveying a sense of competence and professionalism and to go beyond identity-politics, which likely contributes to the party's recent efforts to present a more moderate image.

Notes

1. All translations from Hungarian are my own. All electronic sources were last consulted on 8 August 2018.
2. www.jobbik.hu/videoink/jobbik-bemutatja-kornyezetvedelmi-es-energetikai-programjat
3. Coded recording of the political position of the parties.
4. www.jobbik.hu/videoink/miert-jobb-nemet-szen-magyarnal-harc-banyakert
5. www.jobbik.hu/videoink/onkentes-munkat-vegzett-jobbik-asotthalmon
6. www.jobbik.hu/programunk/kornyezetvedelem
7. www.jobbik.hu/kepviseloink/kepli-lajos
8. Note that the overall media attention devoted to environmental issues has also grown slightly in the past few years (WWF 2016).

9 'The kolontár Committee did not learn much', *Index*, 4 October 2011. Online at: https://index.hu/belfold/2011/10/04/vorosiszap_vizsgalobizottsag/
10 See, e.g. 'The circle of Mészáros and Tiborcz take the best plots of Lake Balaton', *Index*, 3 August 2017. Online at: https://index.hu/gazdasag/2017/08/03/meszaros_tiborcz_balaton_kemping/

References

Bartlett, Jamie et al. (2012). *Populism in Europe: Hungary*. London: Demos. www.demos.co.uk/project/1219/ [8 August 2018].
Berg, Marni (2000). Red and Green: Twenty Years of Environmental Activism in Hungary, *Problems of Post-Communism*, 47(2): 46–56.
Bíró-Nagy, András, Boros, Tamás and Vasali, Zoltán (2013): More radical than the radicals: The Jobbik Party in international comparison. In: Ralf Melzer and Sebastian Serafin (Eds). *Right-Wing Extremism in Europe*. Berlin: Friedrich-Ebert-Stiftung, pp. 229–253.
Botcheva, Liliana (1996): Focus and effectiveness of environmental activism in Eastern Europe: A comparative study of environmental movements in Bulgaria, Hungary, Slovakia, and Romania, *Journal of Environment & Development*, 5(3): 292–308.
Carmin, JoAnn and Fagan, Adam (2010). Environmental mobilisation and organisations in post-socialist Europe and the former Soviet Union, *Environmental Politics*, 19(5): 689–707.
Feischmidt, Margit and Pulay, Gergő (2017): 'Rocking the nation': The popular culture of neo-nationalism, *Nations and Nationalism*, 23(2): 309–326.
Figueroa, Silvana (2008): The Grounded Theory and the Analysis of Audio-Visual Texts, *International Journal of Social Research Methodology*, 11(1): 1–12.
Forchtner, Bernhard (2019): Far-right articulations of the natural environment: An introduction. In: Bernhard Forchtner (Ed). *The Far Right and the Environment: Politics, Discourse and Communication*. London: Routledge.
Forchtner, Bernhard and Kølvraa, Christoffer (2015): The nature of nationalism: Populist radical right parties on countryside and climate, *Nature and Culture*, 10(2): 199–224.
Forchtner, Bernhard, Kroneder, Andreas and Wetzel, David (2018): Being skeptical? Exploring far-right climate-change communication in Germany, *Environmental Communication*, 12(5): 589–604.
Gessler, Theresa and Kyriazi, Anna (2019): "Hungary – a Hungarian crisis or just a crisis in Hungary?" In: *European Party Politics in Times of Crisis*, edited by Swen Hutter and Hanspeter Kriesi. Cambridge: Cambridge University Press, pp. 167–188.
Griffin, Rogers (2000): Interregnum or endgame? The radical right in the 'post-fascist' era, *Journal of Political Ideologies*, 5(2): 163–178.
Guest, Greg, MacQueen, Kathleen and Namey, Emily (2012). *Applied Thematic Analysis*. Los Angeles: Sage.
Inglehart, Ronald et al. (Eds) (2014). *World Values Survey: Wave 5:2005–2009*. Madrid: JD Systems Institute.
Jobbik (2010): Radikális változás. A Jobbik országgyűlési választási programja a nemzeti önrendelkezésértés a társadalmi igazságosságért. (Radical change. Jobbik's parliamentary election program for national self-determination and for social justice). www.jobbik.hu/sites/default/files/jobbik-program2010gy.pdf.
Jobbik (2014): Kimondjuk. Megoldjuk. A Jobbik országgyűlési választási programja a nemzet felemelkedéséért (We say it. We solve it. The electoral program of Jobbik). www.jobbik.hu/sites/default/files/cikkcsatolmany/kimondjukmegoldjuk2014_netre.pdf.

Jobbik (2018): Magyar szívvel, józan ésszel, tiszta kézzel. A Jobbik 2018-as választási programja (Hungarian heart, common sense, clean hands. Jobbik's 2018 Electoral Program). www.jobbik.hu/magyar-szivvel-jozan-esszel-tiszta-kezzel.

Karácsony, Gergely and Róna, Dániel (2011): The Secret of Jobbik. Reasons Behind the Rise of the Hungarian Radical Right, *Journal of East European and Asian Studies*, 2(1): 61–92.

Kovarek, Dániel and Farkas, Attila (2017): Analysing Jobbik's move towards moderation in light of the candidates fielded in single-member districts, *Politikatudományi Szemle*, XXVI(1): 31–54.

Kyriazi, Anna (2016): Ultranationalist discourses of exclusion: A comparison between the Hungarian Jobbik and the Greek Golden Dawn, *Journal of Ethnic and Migration Studies*, 42(15): 2519–2538.

Lehet Más a Politika (2018). Politics can be different. Electoral program. www.lmp2018.org/valasztasi-programunk [8 August 2018].

Lockwood, Matthew (2018). Right-wing populism and the climate change agenda: Exploring the linkages, *Environmental Politics*, 27(4): 712–732.

McCright, Aaron, Dunlap, Riley and Marquart-Pyatt, Sandra (2015): Political ideology and views about climate change in the European Union, *Environmental Politics*, 25(2): 338–358.

Molnár, Virág (2016): Civil society, radicalism and the rediscovery of mythic nationalism, *Nations and Nationalism*, 22(1): 165–185.

Mudde, Cas (2014): The far right and the European elections, *Current History*, 113(761): 98–103.

Mudde, Cas (2016). *On Extremism and Democracy in Europe*. Abingdon; New York: Routledge.

Olsen, Jonathan (1999). *Nature and Nationalism: Right-Wing Ecology and the Politics of Identity in Contemporary Germany*. New York: St. Martin's.

Tóth, András and Grajczjár, István (2015): The rise of the radical right in Hungary. In: Péter Krasztev and Jon van Til (Eds). *The Hungarian Patient: Social Opposition to an Illiberal Democracy*. Budapest; New York: CEU Press, pp. 231–261.

Trechsel, Alexander, Garzia, Diego and De Sio, Lorenzo (2014). euandi (Expert Interviews). GESIS Data Archive, Cologne. ZA5917 Data file Version 1.0.0, doi:10.4232/1.12138.

Varga, Mihai (2014): Hungary's "anti-capitalist" far-right: Jobbik and the Hungarian Guard, *Nationalities Papers*, 42(5): 791–807.

Vona, Gábor (2009): Part 35: The eco-social national economy, *Szebb Jövő*. https://szebbjovo.hu/35-resz-oekoszocialis-nemzetgazdasag/ [8 August 2018].

Weiner, Anette (1992). *Inalienable Possessions: The Paradox of Keeping-While-Giving*. Berkeley: University of California Press.

Wodak, Ruth and Meyer, Michael (Eds) (2009). *Methods of Critical Discourse Analysis*. 2nd ed. Los Angeles: Sage.

World Wide Fund for Nature Hungary (WWF) (2016). Who is interested in the protection of nature and the environment? http://wwf.hu/media/file/1478074142_WWF_infografika_-01-01.jpg [8 August 2018].

12

IS BROWN THE NEW GREEN? THE ENVIRONMENTAL DISCOURSE OF THE CZECH FAR RIGHT

Zbyněk Tarant

Introduction

Czech national identity is deeply connected with romanticized visions of the Bohemian and Moravian countryside. The 'natural beauty' is a frequently glorified theme of the national legends. In their most popular version, collected by Alois Jirásek, the mythical Forefather Czech describes the land through reference to the Bible, as a 'Promised land, rich in honey' (Jirásek 1936: 11). The beauty of the landscape was praised in the works of nineteenth century romanticists and Czech national revivalists, such as Božena Němcová and Karel Hynek Mácha. Bedřich Smetana's famous symphonic poem, 'My Country' (*Má vlast*), contains musical depictions of the landscape (Moldau) and the legends associated with it. The very national anthem, which comes from the same period, speaks about a paradise on Earth: 'Water roars across the meadows / Pinewoods rustle among crags / The garden is glorious with spring blossom / Paradise on earth it is to see / And this is that beautiful land, The Czech land, my home ...'. One particular physical testimony of this close bond between landscape and national consciousness exists even today. It is the dense, sophisticated system of hiking trails, the development of which is claimed by both Czechs and Germans, as well as the entire hiking culture, whose roots were established in 1889 (Czech Tourist Club), making it possible to 'conquer the land by foot'.[1] A brief look at the horizon in most places of the country reveals that the landscape is bursting with observation towers that, once again, serve the purpose of enabling one to view the Motherland and hold its beauty in the palm of one's own hand.

While the Czech far-right is well aware of the aesthetic role of nature in the national consciousness, this chapter argues that its contemporary environmental policies have been mostly developed against the background of real, serious environmental issues. Having been suppressed by the Communist regime for more than four decades after the World War Two, the Czech far-right had to rebuild its entire leadership structure and political agenda, using either the inspiration of historical 1930s Czech fascism (Kelly 1999; Mareš 2011) or foreign influences, such as the White-Power Skinheads. The contemporary Czech far-right scene is extremely divided, with about two dozen various active groups, parties, movements, networks and youth organizations, ranging from populist right-wing organizations through ultra-conservative Christian movements, nativist and Czech chauvinist movements, up until the neo-Nazis and Fascist groups (Bastl et al. 2011).

The following overview will be focused on the more significant post-1989 movements that have either entered Parliament or have played an important role in the development of the domestic far-right discourse in general. Specifically, the sample includes the only two far-right populist parties that have entered the Czech Parliament so far: the 1990s Republican Party (*Sdružení pro republiku – Republikánská strana Československa*, SPR-RSČ) and the contemporary Freedom and Direct Democracy (*Svoboda a přímá demokracie*, SPD). Any overview of the Czech far-right would not be complete without reference to the Workers' Party (*Dělnická strana*, DS) – a party with neo-Nazi links that has managed to utilize social conflict with the Romany minority in North Bohemian towns to acquire several seats in the local municipalities. The Czech far-right often operates in a semi-legal or semi-official fashion and its definition cannot be limited to political parties and movements only. To a great extent, it is also a social movement in the sociological sense, as well as a subculture in the anthropological sense. This is, why this chapter also needs to reflect, at least briefly, on the skinhead subculture of the 1990s, as well as the movement of Autonomous Nationalism of the early 2000s, which was instrumental in the transformation of the entire visual style of the Czech far-right at the time and had an influence on its 'eco-activism' as well.

This chapter is based on a content analysis of journals and websites published by the selected far-right movements in Czechoslovakia/the Czech Republic from 1990 until 2017. Some of the source material, especially the more recent, is available online; however, the older publications are accessible only in the newspaper sections of the Czech libraries (*Republika*, *Týdeník Politika*) or personal archives (skinhead zines). In order to cover a representative sample and evaluate the importance of environmental topics in the far-right discourse in general, an analysis of entire websites or all issues of a journal was performed where possible. The source material included 200 digitized issues of the ultra-conservative *Týdeník Politika*, about 60 issues of the *samizdat* skinhead zines, 550 issues of *Republika* (Republican Party's official bulletin), entire website of the Autonomous Nationalists (defunct in 2010, offline mirror was used), 56 issues of the Workers' Party bulletins (*Workers' Journal* and *Workers' Youth*) and entire offline mirror of the Freedom and Direct Democracy website. The type of analysis accorded with the nature of the source material – hand-coding in the case of physical copies or automated keyword searches in the case of archived websites and issues scanned

Is brown the new green? The environmental discourse of the Czech far right 203

by using optical character recognition (OCR). Due to the diverse nature of the analysed material, descriptions of the units of analysis for each of the parties and movements is given in their relevant sub-sections.

The source material was coded for remarks on environmental issues, such as ecology, natural diversity, environment, pollution, recycling, waste management, global warming, ozone layer etc. The aim was to find out how the most influential far-right movements and political parties respond to these issues, or how they incorporate these topics into their own discourse. It will be argued that outside the points in which the environmentalist speech directly merges with the far-right ideology, the Czech far-right also discussed genuine 'mainstream' policies of environmental protection, even though it never managed to complete such a comprehensive policy and put it into practice.

The eco-pioneers of the Czech far-right

It is not certain who was the first to introduce environmentalist themes into the discourse of the reborn far right in post-Communist Czechoslovakia. There are, for example, only half a dozen random, isolated remarks in the almost one hundred issues of the influential anti-Communist, later ultra-conservative and anti-Semitic, *Politics Weekly* (*Týdeník Politika*), published between 1991 and 1992. Such random remarks included a translation of the political program of the French National Front (*Durand 1992*) or a remark on 'ecological fanatics' in the context of the Roman Club conspiracy theory (*Čistota rasy jako... 1992*). The same can be said about the skinhead 'zines'[2] of that time, although research is hindered by their *samizdat* nature and the lack of a comprehensive central archive. These home-made bulletins were copied and distributed anonymously at the white-power concerts and most of them were extremely short-lived, usually disappearing after the publication of only one or two issues. However, a sampling of about 60 various Czech skinhead 'zines' from the 1990s to the mid-2000s shows that there were only random, insignificant remarks regarding 'environmentalism' or 'ecology'. Most of them focused primarily on white-power music, antisemitism, the ideology of National Socialism, the history of Nazi Germany, Nordic mythology, anti-Communism and anti-Capitalism. A somewhat different approach can be observed in the magazines that have a more neo-Pagan focus. These often contained descriptions of romanticized nature and its natural laws, with 'Judeo-Christian scum' being regarded as a chaotic and destructive force – a force of chaos that had killed the ancient gods, banned the old customs and cut down the thick forests (*V nitru matky přírody... 2005*). Such visions, which the contemporary Czech far-right often links to Maximiani Portaz (*Sávitrí Déví*), have continued to appear among the more intellectual and spiritually oriented neo-Nazi supporters until the present day, such as in the case of neo-Nazi blogs *Deliandiver* or *Sarmatia*.

One of the pioneers of far-right environmentalism might well have been the civic association *Patriot Front* (*Vlastenecká Fronta*, VF), established in 1993.

Although it was staffed by the same White-Power Skinheads, it defined itself in terms of a political movement rather than a subculture and its program already contained a section on 'ecology' at the time of its establishment (Lylová 2000: 422–439). In an expanded iteration of the program from 1999, it admitted that the Czech post-Communist legislation on nature protection was actually well-written, even though it was not properly enforced. The ecological program of the movement contained concrete recommendations, such as a continuation of the recultivation program, improvements in the housing stock on city peripheries, a ban on animal testing, support for ecological education, as well as potential benefits for those energy producers who made use of more efficient or sustainable technologies (*'Akční program Vlastenecké fronty'*).

Republican Party

Another far-right movement that manifested a certain degree of awareness about environmental issues in the 1990s was the right-wing populist Republican Party (*Sdružení pro republiku – Republikánská strana Československa*, SPR-RSČ), led by the former Communist censor, Miroslav Sládek. The party, which had neo-Nazi connections, was represented in Parliament between 1990 and 1998 and was the leading far-right force in the country during that time. A small section on 'Ecology' could already be found in its 1991 political program. The entire section consisted of a short and vague statement:

> 'Nature is a living organism, whose natural balance is influenced by all the actions of mankind. Legal, technical and economic measures must then point to the maximal protection of nature, because our responsibility is not only to ourselves and our living standards, but also to all future generations'
>
> *(Úkoly současné a perspektivní … 1991: 7)*

A later version of the party's program from March 1992 included more concrete measures, such as tax deductions for citizens who were forced to live in polluted environments or local referendums linked to municipal and regional environmental policies (*Teze volebního programu* 1992: 2). In other cases, 'ecology' was regarded as a pretext for introducing protectionist measures against foreign companies (*SPR-RSČ vyhlašuje…* 1992: 2). Several party members were either environmental experts or activists, who formed the Environmental Committee, responsible for its 'ecological program'. When discussing environmental issues, the Party used the correct terminology and valid definitions of the key terms. Its detailed 'ecological program', presented with slogans such as: 'SPR-RSČ for sustainable living' (Hrůzová 1998: 4), included practical measures regarding air, water and soil pollution in Czechia, again presented using correct terminology and awareness (*Zásady ekologické politiky* 1998: 5). On the other hand, environmental language was sometimes used as an ideological instrument, for example during the debate about joining NATO. In such occasions, isolated allusions to nativist environmentalism had appeared, for example in an article, according to which German NATO jets were 'destroying our forests' (*NATO ničí naše lesy* 1997: 3). The term 'nativist environmentalism' originally referred to an environmental policy of protection for local ecosystems from the threat of invasive or introduced species, but it also has its political and social dimension in far-right thought – the protection of both the natural and also the social environment from the foreign Other. As the case with 'NATO destroying our forests' shows, this rhetoric does not have to be limited to immigration, but it can also target international organizations.

Apart from the overviews of the party's political program, articles or references to environmental issues were very rare. Out of more than 550 issues of *Republika*, published between 1990 and 2004, there were only about a dozen isolated instances, usually connected to concrete contemporary issues such as air pollution. In these rare instances, the articles went into a certain depth and even included interviews with environmental experts (*Jak řešit zlepšení ovzduší...* 1991: 7). One such article, on air pollution, was further accompanied by quotations from Al Gore's *Earth in the Balance* (*Energie a poškozování ovzduší* 1997: 5). Diverse opinions were raised regarding nuclear energy. The party seemed sceptical with regard to the early 1990s Temelín nuclear plant project in South Bohemia, seeing it as a potential nuclear disaster (*Ještě k Temelínu* 1994: 7), yet it changed its opinion in 1998 when Tomáš Vandas, later known as a leader of the Workers' Party, called for Temelín's completion (Vandas 1998: 3).

The 'Golden era' of the Republican Party ended after the 1998 elections, with the party suffering a wipe-out, losing all of its parliamentary seats. The party's internal crisis was further deepened by suspicions that its leader, Miroslav Sládek, had embezzled the party's finances. Several of its organizations rebelled against the central leadership, forming the backdrop for new projects, such as the neo-Nazi Workers' Party, the small nationalist National Party, the small ultraconservative movement, Law and Justice (now National Democracy), and others.

The eco-activism of the Autonomous Nationalists

The Autonomous Nationalists (*Autonomní nacionalisté*) was a group of neo-Nazi activists, inspired by a similar network of Autonome Nationalisten (AN) in Germany (Kalibová 2011: 250–259). In a similar way to the German model, the group borrowed certain elements from their anti-Fascist opponents. This included developing a decentralized structure based on the creation of semi-autonomous cells, as well as adopting black-bloc clothing and street-art. These influences were further combined with elements borrowed from the Casapound Italia, together with squatting, a 'third way' neo-Fascist discourse (Ezra Pound, Julius Evola, Otto Strasser) and a focus on eco-activism. The advent of Autonomous Nationalism meant a significant innovation in the tactics and visual style of the Czech far-right, which had until then been represented mostly by the sub-culture of White Power Skinheads, including key personalities of the local subculture, such as Filip Vávra (for his memoirs from that era, see Vávra 2017). Autonomous Nationalism became the dominant stream of the Czech far-right between 2008 and 2010, replacing National Corporativism in its position as a unification platform for other far-right streams in the country.

The Autonomous Nationalists pushed the environmentalist agenda one step further by using eco-activism as a direct form of advertisement. It was not strong in terms of general numbers. Its central, now defunct website, Nacionaliste.com, contained about 510 articles in March 2010, about two dozen of which dealt with issues of ecology, environmental protection and animal rights.[3] However, the rhetoric, when manifested, was rooted much more deeply in the movement's ideological core. One of the ideological posters of the Autonomous Nationalists from March 2009 contained an image of a hooded AN activist hugging a sleeping baby fox. This was accompanied by the slogan: 'Join us in nature protection!'[4] The main website of the Autonomous Nationalists invited its readers to 'calculate their ecological footprint' (*Ekologická stopa – spočítejte si…* 2007). Local cells of Autonomous Nationalists, such as the one in Kutná Hora, bought dog food for animal shelters (*Akce solidarity Kutná Hora… 2011, Podpora útulku… 2010*) and built bird feeders in the forests. They also engaged in discussions about tree felling in public spaces (Noneman 2007).

It was also the Autonomous Nationalists who were to introduce the practice of 'eco-action' to Czechia, whereby a group of activists would document themselves cleaning garbage from a chosen area (e.g. *Letní eko-akce* 2008). About two dozen such 'eco-actions' took place between 2007 and 2010, with several more appearing under the banner of the Workers' Party or Revolt 114 (114 stands for the first and fourteenth letter of the alphabet, i.e. '[A]utonomous [N]nationalists'). A report on 'eco-action' in Brno, South Moravia, from April 2012 contained a call to readers to clean up the trash in their own environment: 'Bend your back and help nature! You are doing it for yourselves, for your friends, for nature and for the heritage of future generations' (*Eko Akce Brno* 2012).

Unlike the ecological activists, however, who primarily used the images of collected trash to warn against pollution and littering, the Autonomous Nationalists sometimes attached a different message. For example, in an 'eco-action' in Modřany, conducted under the banner of Revolta 114, they attached a message to the garbage bags stating, 'Freedom to political prisoners' (*Eko Akce Modřany* 2011). In the discourse of the Czech neo-Nazis, 'political prisoners' or 'POWs' are activists jailed or put on trial for criminal acts, related to their far-right activism, such as incitement to violence, genocide denial, assault or manslaughter. It was neither the first nor the last case in which environmentalist activism acted only as a form of advertisement, with photographs involving bags of trash serving only as the medium for communicating the key message.

In an opinion piece, 'Autonomous Nationalism and Nature Protection', written for the central website of the Autonomous Nationalists, Filip Vávra attempted to explain why environmentalism should not be considered as simply a 'left-wing topic'. Vávra's arguments mostly centred on the themes of anti-consumerism and the protection of the national heritage. This former key figure of the 1990s Nazi skinheads warned that global warming was a real problem, although he combined it with ozone depletion and air pollution, suggesting that these were the main culprits. He called for the use of public transportation (preferably the train over the bus), as well as vegetarianism (Vávra 2007). Interestingly enough, vegetarianism was discussed in another piece on the same website, but in this case was rejected as an option. This other voice from within the Autonomous Nationalist group also criticized the Anti-fascists for talking of vegetarianism and animal rights while wearing heavy leather boots (*Ne, děkuji... Jsem vegetarián* 2007). The discussion was not resolved and the issue remained open (see Forchtner and Tominc 2017).

The websites of the Autonomous Nationalists were abandoned shortly after the movement's split with the Workers' Party in March 2010 (*Oficiální distancování...* 2010). Supporters of Autonomous Nationalism in Czechia continued their activities under the banner Revolta 114 (*Revolta 114*). Judging from their websites, their Facebook activity and their street rallies, the final groups of Autonomous Nationalists in Czechia disappeared in 2014, with most of their members joining other far-right projects, such as Identity Generation.

The activism of the Autonomous Nationalists coincided with the establishment of intellectually oriented blogs, such as *Deliandiver* and *Sarmatia,* which focus almost entirely on the translation of ideological texts from foreign languages, including about two dozen texts on environmentalism by the American white nationalist Greg Johnson. The practice of 'eco-activism' introduced to Czechia by this movement was adopted by the neo-Nazi Workers' Party (see below) and later also by the small fascist party, National Democracy (formerly Law and Justice). In a similar way to the Autonomous Nationalists, the latter presented photographs of collected trash, above which it placed its key message: 'No to invaders! Adieu to EU tyranny. Goodbye to aggressive NATO'. (Mrázek 2017).

The Workers' Party

The Workers' Party (*Dělnická strana*, DS) was established in 2003 by former members of the Republican Party. It was banned for propagating Nazism in 2009 and re-established shortly thereafter under a slightly different name, the Workers' Party of Social Justice (*Dělnická strana sociální spravedlnosti*, DSSS). Tomáš Vandas has been the leader since its establishment. In its most successful period, the party was capable of attracting between 1% and 1.5% of the vote in elections. It was never to gain parliamentary seats, yet it managed to acquire several seats in local municipalities and regional councils, especially in North Bohemia (*Czech Statistical Office*). The discourse of the party was to change over time, shifting from a more socialist perspective in the beginning to the explicit referencing of German National Socialism, to a more contemporary right-wing populist approach (Tarant, 2009: 84–101; Mareš 2012). One can easily see this changing discourse by comparing any of the printed issues of the party's bulletin, *Workers' Journal* (*Dělnické listy*), from 2003, 2008 and 2017. The bulletin appears roughly as a quarterly publication, accompanied by about a dozen websites of various sizes,[5] and will be used here as a sample for exploration of the party's opinions on environmentalism.

Between 2003 and 2017, the Workers' Party published 49 issues of its printed bulletin.[6] An additional 7 issues of *Youth Voice* (*Hlas mládeže*), an irregularly published bulletin of the party's youth platform Workers' Youth (*Dělnická mládež*), were published during the same period.[7] Altogether, about 980 articles were published in the *Workers' Journal* from 2003 to 2017. Of these 980, about 20 articles, distributed at regular intervals throughout 12 of the 49 issues, contained keywords such as 'ecology' or 'environment'. Additionally, these keywords have been found in 2 out of the 7 issues of *Youth Voice*. When compared with the main focus of the party, which is mostly connected to identity politics, anti-Gypsyism and anti-immigration, environmental topics represent a rather marginal yet interesting issue, from which one can learn more about the internal diversity of far-right thought.

About half of the occurrences represent marginal remarks, shorter than one sentence. About one quarter of the remarks were of a pejorative nature, criticizing, for example, 'ecological fanatics' for hindering highway construction (Vandas 2004: 3), the Green Party's policies, referring to them as pseudo-ecological, or ecological grants, which they regarded as being a waste of money (Vandas 2008: 1). On the other hand, the party expressed its agreement with the state program for supporting the replacement of old boilers with more efficient ones (Komárek and Machová 2014: 3). The party called for the provision of support for the economies of local communities in order to limit the concentration of the workforce in large cities since they believed this created 'too much ecological load' (Müller 2013: 3). Regarding the issue of territorial coal-mining limit agreements, the breaking of which could lead to the forced evacuation of at least two towns in North Bohemia, the party called for a referendum in each of the respective towns (Komárek and Machová 2014: 3). The party was also extremely critical of the remarks made by the Minister of the Environment in 2010, where he spoke about 'the necessity to open up nature to the public', which the party saw as an attempt to open the doors for the exploitation of natural reserves by developers and lumberjacks (Vandas 2007: 1). Somewhat surprising are the plans to establish 'agricultural communes' in order to boost production (Komárek and Machová 2012: 3), which is a practice usually associated with socialism and communism (such as kibbutz, kolkhoz etc.).

The party has been reluctant to recognize the role of mankind in the global climate change phenomenon. In a reference to the climate change denial made by the Czech president, Václav Klaus, the vice-chairman of the party wrote: 'I am not an unconditional supporter of Václav Klaus' theses, nor am I supporter of Al Gore's. I think the truth is somewhere in the middle.' (Štěpánek 2010: 3).

In his op-ed for *Workers' Journal*, the leader of the party, Tomáš Vandas, while admitting that mankind had a negative impact on nature, which was responding in kind, labelled global projects to curb climate change as an attack on the freedom of trade and an expression of the views of 'fanatical eco-terrorists': 'Ecology is being used in a power struggle, to curb opinions that do not follow the mainstream view.' (Vandas 2007: 1). The party's vice-chairman, Jiří Štěpánek, wrote that he had 'no problem' with nuclear energy, seeing it as the only viable alternative to fossil fuels (Štěpánek 2010: 3). Let us remember that Štěpánek was a former member of the Republican party, which, as revealed above, originally opposed and later supported nuclear energy. Such a change of opinion might have been shaped by the international criticism raised against the Czech nuclear power plants (namely by the neighbouring Austria), which in effect became seen as a sort of 'cherished national treasure' that has to be defended against 'dictate' of foreign actors.

A slightly alternative voice appeared in an article entitled 'Avatar, we refuse to see'. Responding to the hype surrounding the 2009 James Cameron science-fiction movie, the article warned against the continuing decline of forested territories in the country and connected the issue to a more philosophical contemplation on consumerism and respect for nature.

> 'The term avatár comes from Hinduism. In Sanskrit, it means 'descent' or, metaphorically, 'incarnation'. To be more exact, the avatar of the tree in front of a house is not composed merely of wood, cellulose, lignin, organic and inorganic compounds. The tree has its own meaning and purpose, because it is there. The previous sentence might be problematic for the atheist mind of a materialist 'religion'. Because, if we accept the fact that the tree in front of the apartment bloc does not lack a soul, then it means that our fridge, full of apples, meat and milk, is something like a shrine of fallen gods. Or, perhaps, their gift to us.'
>
> *(Zbela 2010: 1–2)*

The practice of 'eco-action', started by the Autonomous Nationalists, was continued by the Workers' Party as well, especially by its youth organization. Headlines of such reports usually contain a call to follow the good example: 'Bend your back and help nature. Brno offers an example.' (Dudák 2012: 3). More recently, possibly in an attempt to imitate the example of the boy-scouts, the Workers' Youth set off for the Central Bohemian Highlands to admire the region that had inspired Czech romanticism, as well as to plant trees and remove trash (Aubrecht 2017).

The last issue of the *Workers' Journal* that was available at the moment of writing this text, contained the party's program for the November 2017 elections. Its last point is dedicated to environmental policies, which include support for rail transport, increased penalties for environmental pollution, increased protection for animal rights, support for alternative fuels, a ban on agricultural land sales to foreign corporations, as well state sanctioned restrictions on the exploitation of natural resources by foreign corporations (*Programové priority* 2017: 3).

Freedom and Direct Democracy

The leading far-right voice in the Czech Republic is currently the parliamentary party Freedom and Direct Democracy (*Svoboda a přímá demokracie*, SPD), led by Tomio Okamura (Bauerova 2017). The party can be described as being right-wing, populist and nativist, similar to Geert Wilders's Freedom Party in the Netherlands. The two key points on its program are anti-immigration and 'Swiss-style' direct democracy. In the November 2017 elections, the party received more than half a million votes, or 10.64% (Election Results 2017), the highest level of support for a far-right party during the entire post-Communist history of Czechia. It currently holds 22 of the 200 seats in the Czech Parliament. The party mostly focuses on Islam and immigration, together with some traditional populist themes such as anti-corruption.

Due to its rather monothematic nature, it does not seem to have any significant environmentalist agenda. Words like 'environment' or 'ecology' do not appear in its official program at all. The string 'natur*' did appear twice in the summaries of the party's program, but only in the context of the 'protection of natural resources from being exploited by foreign companies.' (Svoboda a přímá demokracie 'Program'... 2017). The term 'environment', in the Czech form *'životní prostředí'* ('living environment'), appears in the party's discourse in two major contexts. The first is related to its anti-corruption discourse: 'we do not want tunnelling [i.e. financial embezzlement] of education and environment'. The second is an actual remark on the environment that comes from the party's program overview for the November 2017 elections: '[We want] quality food and a healthy environment for all' (Okamura 2017). This was the only trace of environmentalism in the party's program and, quite surprisingly, this remark from the program summary did not make it to the official version of the program.[8]

The dominant image of an environmentalist on the party's website is an ecological activist, someone who blocks legitimate economic development. There are only random appearances of environmentalist issues, such as in the case of recycling, which the party appears to support. The vice-chairman of the party, Jaroslav Holík, is himself a recycling entrepreneur (*Předsednictvo hnutí* 2017). This seems to be the closest the party has come to having any sort of environmentalist agenda (Okamura 2016). It was also a safe thing for the populist party as the Czechs have already adopted recycling, with recycling rates reaching well above 60% (Waste management indicators 2017). Wider issues, such as animal rights, do briefly appear either in the case of halal slaughter (Okamura 2016) or in the case of dog-breeding facilities, where the party responded to some horrendous cases of maltreatment that were reported in the mainstream media (*Petice proti množírnám*).

Overall, this brief overview reveals that environmentalism is almost non-existent on the agenda of the Freedom and Direct Democracy party. Moreover, about one half of the several existing remarks on the issue do appear in a pejorative connotation. It can be argued that Freedom and Direct Democracy attempts to avoid this issue as its target audience consists of those who prefer to use the term 'eco-terrorism' when talking about environmental activism. Polls show that despite its right-wing populist image, they actually receive a significant portion of their votes from former Communist and Social Democratic voters (Škop 2017), many of whom represent the conservative electorate that has rejected the environmental agenda.

Conclusion

Any political movement has to make decisions on how much space, time and effort it wishes to dedicate to specific issues. In general, the Czech right-wing populist parties have dedicated little space to the issue of the environment and environmentalism does not seem to be at the core of their agenda, although there have been attempts to experiment with it or to respond to some particular issues. Election results show that the Czech far-right parties are strongest in regions that suffer from environmental issues, but these same communities are also economically dependent on mining and industry. The far right is thus forced to carefully balance its discourse by criticizing 'eco-terrorism' on the one hand, while demonstrating its concern for the daily issues experienced by the electorate, which also includes environmental issues. This necessity of maintaining a delicate balance between two seemingly incompatible priorities can be seen as one of the reasons behind some of the contradictory statements on particular issues, such as nuclear energy, global climate change and vegetarianism. Internal discussions about these issues are still continuing, thus giving a testimony of the evolving, dynamic face of the Czech far-right.

It can be argued that foreign influences, such as the Autonome Nationalisten or Casapound Italia in the mid-2000s, helped to transfer some of the new-right ideas from the West to the Czech scene, which included more elaborate far-right environmentalist concepts and strategies. These continued to exist alongside the older, more practical approach of the 1990s. Since then, the Czech movements and parties have kept experimenting with various concepts and communication strategies, but they have not settled with single one of them. There are some neo-Nazi blogs that translate Western far-right environmentalist texts, but these ideas are not being further reflected in the programs and policies of the far-right parties themselves. While the parties can sometimes surprise by their well-informed opinions and policy recommendations regarding local and regional ecological problems, they have no established environmental policy *per se*. Their programs are often limited to rather vague statements and some of their proclamations sound more like 'random shouts' rather than an expression of an organic, systematic ideology. Movements associated with neo-Nazi ideology, such as the Autonomous Nationalists, have manifested somewhat stronger interest in the spiritual dimension of nature. The Workers' Party, which has neo-Nazi links, does not dedicate significantly more space to environmental issues in its agenda, compared to the 1990s Republican Party or the late-2010s Freedom and Direct Democracy, yet it tends to enrich the 'practical' or 'materialist' approaches of the right-wing populists with the more spiritual ones to a somewhat higher degree. Cases of experimentation with the eco-activist communication strategies continue to appear, such as in the example 'eco-actions', where bags of collected garbage serve as a medium for communicating a completely unrelated political message, such as 'No to immigration', or 'Freedom to POWs'.

Some of the eco-experiments of the Czech far-right tended to head in the direction of 'nativist environmentalism', but in other cases, nativism seems to have directly conflicted with environmentalism. In fact, the Czech far-right discourse is constantly alternating back and forth on the spectrum between 'nativist environmentalism' and 'nativist anti-environmentalism'. The far-right can be expected to raise nativist objections in issues, such as immigration (the Czech one did so surprisingly rarely), but the same nativism, combined with conservatism and scepticism leads to denial of global environmental issues, even if they might pose direct threat to the national security. It is because solutions to these global issues would also require policies developed by some foreign 'Other' (e.g. the United Nations, the European Union, international ecological organizations) being imposed on the sovereign nation (see Forchtner and Kølvraa 2015). Where the Agenda 21 asks governments and individuals to 'think globally, act locally', the far-right prefers to 'think nationally, act locally', indicating that the environmental awareness of the Czech far-right seems to prevent them from seeing beyond national boundaries.

Notes

1 The connection between the hiking culture and nationalism has only recently received scholarly attention, thanks to the works of Shay Rabineau, who noted the significance of the hiking trail system, imported from Central Europe, for the Zionist hiking culture in Israel. See: Rabineau (2014).
2 'zin' or 'zine' was a 1980s and 1990s term used for samizdat magazines ("magazine' → 'zin').
3 The main website of the movement, Nacionaliste.com, is currently defunct, but available via internet archive: *Nacionaliste.com*. https://web.archive.org/web/20090430084320/; www.nacionaliste.com:80/ [4 April 2018].
4 Formerly available at: www.nacionaliste.com/storage/1238340075_sb_ekologie.jpg. Currently accessible at: *Autonomní nacionalisté Kutná Hora*. https://ankutnahora.wordpress.com/page/2/ [4 April 2018].
5 www.delnickastrana.cz, www.delnicka-strana.cz, www.dsss.cz, www.delnickelisty.cz, www.delnickamladez.cz etc. The archive of the oldest versions (2003–2008) can be visited via the Wayback machine: https://web.archive.org/web/20090301051139/; www.delnickastrana.cz/.

6 A partial archive, containing issues 26–49, is still available at: www.delnickelisty.cz/archiv-tisku [4 April 2018].
7 The archive of *Youth Voice* (*Hlas mládeže*) is available at: http://hlas.delnickamladez.cz/archiv-tistenych-vydani [4 April 2018].
8 Google search 'islám* site:www.spd.cz'. Analysis mine.

References

'Energie a poškozování ovzduší.' *Republika* 2/1997, 5.
Akce solidarity Kutná Hora – Podpora slabších rodin – psí útulek. (2011): Autonomní nacionalisté Kutná hora. https://ankutnahora.wordpress.com/2011/01/23/akce-solidarity-kutna-hora-podpora-slabsich-rodin-psi-utulek/ [4 April 2018].
Aubrecht, Martin (2017): Pochod Dělnické mládeže na vrch Radobýl s vysazováním dřevin. *Dělnická mládež*. www.delnickamladez.cz/pochod-delnicke-mladeze-na-vrch-radobyl-s__vysazovanim-drevin [4 April 2018].
Autonomní nacionalisté Kutná Hora. https://ankutnahora.wordpress.com/page/2/ [4 April 2018].
Bastl, Martin, Mareš, Miroslav, Smolík, Josef and Vejvodová, Petra (2011): *Krajní pravice a krajní levice v ČR*. Prague: Grada Publishing.
Bauerová, Laďka Mortkowitz (2017): How a Tokyo-born outsider became the face of Czech Nationalism. *Bloomberg*. www.bloomberg.com/businessweek [4 April 2018].
Čistota rasy jako základní podmínka (1992). *Týdeník Politika* (97/1992), p. 5.
Czech Tourist Club (KČT). www.kct.cz/cms/czech-tourist-klub-kct.
Dudák, Jiří (2012): Ohněte svůj hřbet a ulehčete přírodě. V Brně jdou příkladem. *Hlas mládeže*. 6/2012: 3.
Durand, Charles J. (1992) Co chce J. M. Le Pen – Program Národní fronty. *Týdeník Politika* (63/1992), p. 2–4.
Eko Akce Brno (2012): *Revolta.info*. https://revolta114.blogspot.cz/2012/04/ekoace-vol-6-brno.html [4 April 2018].
Eko Akce Modřany (2011): *Revolta.info*. https://revolta114.blogspot.cz/2011/03/eko-akce-modrany.html [4 April 2018].
Ekologická stopa – spočítejte si jak zatěžujete planetu. (2007): *Autonomní nacionalisté*. www.nacionaliste.com/view8f52.html?nazevclanku=ekologicka-stopa-%96-spocitejte-si-jak-zatezujete-planetu&cisloclanku=2007030004 [4 April 2018].
Election Results. (2017): *Czech Statistical Office*. https://volby.cz [4 April 2018].
Forchtner, Bernhard and Kølvraa, Christoffer (2015): The nature of nationalism: 'populist radical right parties' on countryside and climate, *Nature & Culture*, 10(2): 199–224.
Hrůzová, Vlasta (1998): SPR-RSČ za trvale udržitelné žití. *Republika* 36/1998: 4.
Jak řešit zlepšení ovzduší (1991): *Republika* 37/1991: 7.
Ještě k Temelínu (1994): *Republika* 22/1994: 7.
Jirásek, Alois (1936): *Staré pověsti české*. Prague: Státní nakladatelství 1936.
Kalibová, Klára (2011): Autonome Nationalisten in Tschechien. Autonome Nationalistem – Neonazismus in Bewegung. In: Jan Schendler and Alexander Häusler (Eds). *Autonome Nationalisten: Neonazismus in Bewegung*. Wiesbaden: Springer. pp. 250–259.
Kelly, David Donald (1999): The Czech fascist movement, 1922–1942. PhD disserstation, University of Nebraska, Lincoln. AAI9425288.
Komárek, Václav and Machová, Hana (2014): Hornictví, odbory a limity.' *Dělnické listy* 42/2014: 3.
Komárek. Václav and Machová, Iveta (2012): 'Myšlení lidí je stále stejné.' *Dělnické listy* 37/2012: 3.

Letní eko-akce (2008): *Nacionaliste.com*.www.nacionaliste.com/viewb6c5.html?nazevclanku= letni-ekologicka-akce&cisloclanku=2008070016 [4 May 2011].

Lylová, Hana (2000):Vlastenecká fronta – Studie o působení pravicově-extremistické organizace v českém prostředí, in: *Czech Journal of Political Science*, 4: 422–439.

Mareš, Miroslav (2011): Czech extreme right parties an unsuccessful story, in: *Communist and Post-Communist Studies*, 44: 283–298.

Mareš, Miroslav (2012): Right-wing extremism in the Czech Republic. *International Policy Analysis*. Fridrich Ebert Stiftung. http://library.fes.de/pdf-files/id-moe/09347.pdf [4 April 2018].

Mrázek, Jan (2017): Aktivisté ND Chomutov uklízeli přírodu ve svém okolí. *Národní Demokracie*.http://narodnidemokracie.cz/aktiviste-nd-chomutov-uklizeli-prirodu-ve-svem-okoli/ [4 April 2018].

Müller, Milan (2013): Chci radikální změnu.Volím DSSS! *Dělnické listy*, 41/2013: 3.

NATO ničí naše lesy (1997): *Republika* 25/1997: 3.

Ne, děkuji... Jsem vegetarián (2007): *Autonomní nacionalisté*. www.nacionaliste.com/view8818.html?nazevclanku=ne-dekuji-jsem-vegetarian&cisloclanku=2007090009 [4 May 2011].

Noneman, František (2007): K čemu stromy. *Nacionaliste.com*.www.nacionaliste.com/vieweb80.html?nazevclanku=k-cemu-stromy&cisloclanku=2008020001 [4 May 2011].

Oficiální distancování Autonomních nacionalistů od Dělnické strany sociální spravedlnosti (2010): *Nacionaliste.com*. htwww.nacionaliste.com/viewe1e5.html?nazevclanku=oficialni-distancovani-autonomnich-nacionalistu-od-delnicke-strany-socialni-spravedlnosti&cisloclanku=2010030004 [4 May 2011].

Okamura, Tomio (2016): Politické aktivity muslimů. *Svoboda a přímá demokracie*. www.spd.cz/novinky/tomio-okamura-politicke-aktivity-muslimu [4 April 2018].

Okamura, Tomio (2017): Volební program hnutí SPD 2017. *Svoboda a přímá demokracie*. www.spd.cz/novinky/tomio-okamura-volebni-program-hnuti-spd-2017?type=join&title=Přidejte+se+k+nám&sm=1&do=openModal [4 April 2018].

Podpora útulku a eko akce (2010): Autonomní nacionalisté Kutná hora. https://ankutnahora.wordpress.com/2010/06/14/podpora-utulku-a-eko-akce/ [4 April 2018].

Programové priority – Suverenita nebo zánik (2017): *Dělnické listy* 49/2017: 3.

Rabineau, Shay (2014): Marking and mapping the nation: Simun Shvilim and the creation of Israel's Hiking Trail Network. PhD dissertation: Brandeis University.

Rozhovor Roberta Starka s Gregem Johnsonem na téma ekofašismu (2015): *Deliandiver*. https://deliandiver.org/2015/10/rozhovor-roberta-starka-s-gregem-johnsonem-na-tema-ekofasismu.html [4 April 2018].

Sobota, Daniel (2009): Národ sobě? Zapomeňte.' *Dělnické listy* 30/2009: 3.

SPR-RSČ vyhlašuje 17 naléhavých opatření na záchranu životního prostředí v naší zemi (1992): *Republika* 9/1992: 2.

Svoboda a přímá demokracie. 'Program' (2017): www.spd.cz/program.

Škop, Michal (2017): Babiš a Okamura vysáli levici, od TOP 09 se přebíhalo k ODS, míní analytik. *iDnes.cz*. https://zpravy.idnes.cz/presuny-hlasu-volici-strany-volby-2017-ekologicka-inference-pac-/domaci.aspx?c=A171023_153934_domaci_ale [4 April 2018].

Štěpánek Jiří (2010): Příroda ve stínu peněz. *Dělnické listy* 29/2010, 3.

Tarant, Zbyněk (2009): Workers' Party – From Socialist Nationalism to National Socialism. In: Věra Tydlitátová and Alena Hanzová (Eds). *Anatomy of Hatred – Essays on Anti-Semitism*. Pilsen: Západočeská Univerzita v Plzni. pp. 84–101.

Teze volebního programu (1992): *Republika* 9/1992, 2.

Úkoly současné a perspektivní (Programový dokument) (1991): *Republika* 9-10/1991: 7
V nitru matky přírody (2005): *Heathen* 1/2005: 16.
Vandas, Tomáš (1998): Temelín se musí dostavět. *Republika* 17/1998: 3.
Vandas, Tomáš (2004): Vyvlastňování, nebo zlodějina. *Dělnické listy* 7/2004: 3.
Vandas, Tomáš (2007): Popírači. *Dělnické listy*, 14/2007: 1.
Vandas, Tomáš (2008): Nový styl. *Dělnické listy* 16/2008: 1.
Vávra, Filip. (2017): *Těžký boty to vyřešej hned! - Skinheads v Praze na konci 80. a začátku 90. let* Prague: Fiva.
Vávra, Filip. Autonomní nacionalismus a ochrana přírody (2007): *Autonomní nacionalisté.* www.nacionaliste.com/viewdeb2.html?nazevclanku=autonomni-nacionalismus-a-ochrana-prirody&cisloclanku=2007040009 [4 May 2011].
Waste management indicators (2017): *Eurostat.* http://ec.europa.eu/eurostat/statistics-explained/index.php/Waste_management_indicators [4 April 2018].
Zásady ekologické politiky (1998): *Republika* 18: 5.
Zbela, Martin (2010): Avatar, který nechceme vidět. *Dělnické listy*, 26/2010: 1–2.

13
BEYOND THE 'GERMAN FOREST'

Environmental communication by the far right in Germany[1]

Bernhard Forchtner and Özgür Özvatan

> Nature protection is homeland protection
> *National Democratic Party of Germany (NPD 2010: 15)*

> '"ENVIRONMENTAL PROTECTION IS NOT GREEN", but essential for us, our children and our country.'
> *Christoph Hofer (2007: 1), editor of the far-right quarterly Environment & Active*

Introduction

In the beginning, there was the forest – at least, this is one of the stereotypical associations readers might have when considering concerns over the environment by the far right in Germany. Indeed, the forest remains an 'ethno-scape' (Smith 2009: 50) where Germans and 'the land' are symbiotically interwoven in some far-right imaginaries; though even the German far right is neither solely nor primarily concerned with the 'German forest' anymore. Rather, their contributions to the discourse about the environment and the embeddedness of (wo)man in nature have long also addressed, for example, biodiversity, waste and pollution, climate change and genetically modified organisms (GMOs). In fact, the German case stands out when studying far-right environmental communication due to not only substantial amounts of primary material, but also because this material has been analysed comparatively extensively. In addition, the German case is telling as 'being green has emerged as a matter of national identity in Germany' (Uekötter 2014a: 2), due to the significance of the country's National Socialist past, and given that forerunners of today's far right have been involved in nature protection since the nineteenth century (e.g. Dominick 1992; Riordan 1997; Rollins 1997).

Against this background, this chapter provides both an overview of the historical trajectory of far-right environmental communication in Germany and a comparative empirical analysis of such communication today. More specifically, we ask: How does the German far-right communicate the environment, and in what way do differences and similarities exist between anti-liberal, (nominally) democratic (populist) radical-right actors and the anti-democratic extreme right? We thus cover a continuum, ranging from increasingly mainstream, anti-liberal actors to *organic* ethnonationalist, *völkisch* ones. As such, we do not provide an in-depth case study on one far-right actor or on one environmental issue, but offer a broad overview, covering a range of actors and issues in the German far right.

To illuminate diverse communication through which the far right has discursively constructed the natural environment, making it, at times, a significant part of its social imaginary and (idealised) subjectivity, we draw eclectically on the conceptual toolkit of the Discourse-Historical Approach (DHA, Reisigl and Wodak 2016) in Critical Discourse Studies (CDS). We systematically investigate material published between 2013 and 2016, which ranges from the radical-right weekly *Junge Freiheit* (*JF*, Young Freedom) to the extreme-right monthly *Zuerst!* (*Z!*, First!), the extreme-right ecological quarterly *Umwelt & Aktiv* (*U&A*, Environment & Active) and the monthly *Deutsche Stimme* (*DS*, German Voice) – the party newspaper of the extreme-right *National-Demokratische Partei Deutschlands* (NPD, National Democratic Party of Germany). In addition, we draw on manifestos by, first, the populist radical-right *Alternative für Deutschland* (AfD, Alternative for Germany – the latter has found a particularly welcoming reception in the *JF*) and, second, the extreme-right NPD which complements data from *DS*.

As such, we illustrate that an environmentalist or even ecological outlook is not non-political as the protection of the natural environment can be pursued with different arguments, some quite inclusive and universalist, others rather exclusive and particularistic.[2] In the words of Olsen (1999: 28): far-right ecology 'is a distinct right-wing form of modern environmentalism, one that is grounded in hostility to Enlightenment universalism characteristic of ultra-nationalism'. Although politics is usually a strategic endeavour, we emphasise that such stances are not always simply instrumental attempts to connect to a broader audience and increase appeal. Rather, we consider the natural environment to be also a window which offers another view of far-right ideology, another opportunity to understand such actors and their differences.

In the next section, we offer a historical overview of how the German far right has discursively constructed the environment since the nineteenth century. The 'Data and method' section then elaborates the data and method underlying our empirical analysis of contemporary articulations, which we subsequently present in 'Analysing the discourse about the environment in Germany'. Finally, the 'Conclusion' offers remarks and reflects on further steps.

Nature and nation: *Heimatschutz* in the past and the present

The root of far-right concerns over the environment is usually located in the Romantic response to the Enlightenment and rapid social change. It is here that the German forest emerged as a powerful symbol of Germandom. Indeed, the forest became a long-lasting symbol of the connection between nature and nation, such as when Ernst Moritz Arndt (1815: 401f) claimed that 'German men must nowhere lack trees', warning that a 'naked and forestless Germania will no longer be Germania'. Similarly, Wilhelm von Riehl (1854: 32) claimed that '[w]e must preserve the forest, not only to prevent our oven from getting cold during winter, but also so that the pulse of the life of the people [*Volkslebens*] keeps beating warm and cheerfully, so that Germany remains German'.

As the *völkisch* movement, with ideologues such as Paul Anton de Lagarde, gathered pace during the late nineteenth century, *Heimat* (homeland) and nature were key points of reference in their anti-urban, anti-industrial-capitalist and anti-Semitic agenda. Meanwhile, Ernst Haeckel founded the science of ecology, a science of all the relations of organisms to their environment. Importantly, Haeckel was not only a scientist, but also 'an ardent German nationalist' and founder of the Monist League, 'designed to promulgate his biological-ecological anthropology and scientific proposals for reordering society and politics' (Olsen 1999: 68). It is against this background that '*Umwelt*' (environment, literally 'surrounding-world') is at times replaced by the more ecologist '*Mitwelt*' (co-world, for example, Sojka 2014: 35 in U&A). Another source of far-right inspiration is Hermann Löns, a social Darwinist, anti-democrat, anti-urbanist and anti-materialist who, being a passionate hunter, claimed that the latter could reunite the degenerate with the *Volksgemeinschaft* (Franke 2017: 1074f). Löns furthermore argued in favour of creating nature reserves so as to preserve supposedly typical, untouched German landscapes. He famously promoted the foundation of Germany's first large nature reserve on Lüneburg Heath. His continuous relevance for the far right is visible in an article in U&A which claims that Löns was a holistic thinker who combined nature- and homeland-protection instead of working towards 'the dissolution of cultures in a coffee-brown mixture without identity, a collection of consuming creatures without tradition and without any spiritual substance' (Kast 2010: 37).

Already the late nineteenth century witnessed the rise of *Heimatschutz* (homeland protection), forcefully articulated by Ernst Rudorff (1994[1897]). With nature protection as a key part, it resulted in the foundation of the Association for Homeland Protection (*Bund Heimatschutz*) in 1904; and yet, *Heimatschutz* was regularly more concerned with aesthetic and cultural goals than with environmental issues. Such a focus on the aesthetic has been praised as providing opportunities to mobilise for conservationist goals (Rollins 1997). However, even if this would be correct, *Heimatschutz* was also about claims such as: '['the German homeland' provides] the roots of our ['German folkdom'] strength', which must be preserved 'unweakened and untainted' in the face of 'homeland-alien internationalism' (Rudorff 1994[1897]: 76f). While the movement was not exclusively *völkisch*, it was significant for the reproduction of *völkisch* ideas (Wolschke-Bulmahn 1996: 533; see also Dominick 1992: 86).

Many conservationists hoped that National Socialist rule would support their cause – and thus joined the Nazi ranks (e.g. ibid.: 113). Now, 'Volk, racism and conservation became integrally linked' (Brüggemeier et al. 2005: 8). One of the most famous supporters was Paul Schultze-Naumburg who had co-founded the *Bund Heimatschutz* and already joined the Nazi party in 1930. Other key figures included Walther Schoenichen, who was the *Reich's* chief conservationist until 1935 and had, during the 1920s, spoken of the purity of the German 'race' and the Jewish 'Other'; Alwin Seifert, who had warned of desertification, supported, like Rudolf Hess, biodynamic agriculture and played a key role in the planning of the *Autobahn*; and Hans Klose who worked for Hermann Göring and was responsible for the *Reichsnaturschutzgesetz* (RNG, Reich nature protection law, Uekötter 2014b) of 1935.[3] Indeed, the RNG and the Reich animal protection law of 1933 have become *lieux de mémoire* in the extreme right.

These laws and interests by some National Socialists, such as the Reich Peasant Leader and Minister of Agriculture (1933–1942) Richard Walther Darré's focus on, for example, 'blood and soil' and organic farming, have at times led to an emphasis on a 'green' agenda within the Nazi party (most famously by Bramwell 1985, see Gerhard 2005 for a critical discussion). And indeed, Pois (1986) rightly speaks of National Socialism as a religion of nature, worshipping 'the natural', nature's laws, and advocating an anti-transcendent, i.e. anti-Judeo-Christian, view of (wo)man as part of nature. However, while National Socialist concern for the natural environment should not be ignored, it should also not be exaggerated (Staudenmaier 2011: 126) as its deeds include land reclamation, dam-building and war mobilisation. While 'nature' was worshipped, nurturing the environment was ultimately contradicted as 'the Third Reich subordinated nature protection to economic development, war preparation, and racist expansion' (Lekan 2005: 94; see also Lekan 2004: 204–251; Blackbourn 2006: 266–280; Uekötter 2006: 30–43; Radkau 2008: 264f). Indeed, racist expansion meant that the 'ideal' German landscape constructed in 'the East' (*Generalplan Ost*) relied on genocidal efforts, as Slavs were viewed as being incapable of proper husbandry (Blackbourn 2006: 239ff; Wolschke Bulmahn 2005). A link between historical attempts to claim land and far-right activities today are so-called *völkisch* settlers who view themselves as a continuation of the *Artamanen*. The latter hoped to revive Germandom through farming during the 1920s and 1930s before being integrated into the Hitler Youth (Linse 1983; HBS 2012; oekom 2012).

After 1945, this past remained alive and influential (Franke and Pfenning 2014). The aforementioned Seifert co-prepared (and signed) a key document of German post-war nature protection, the *Grünen Charta von der Mainau* (Green Charter of the Mainau) in 1961 (Eissing 2014). Klose, responsible for the RNG (see above), ran the newly founded *Bundesanstalt für Naturschutz und Landschaftspflege* (Federal Institute for Nature Conservation and Landscape Management) and was followed, in 1954, by another former National Socialist, Gert Kragh. Only partly revised, the RNG was applied in many federal states until 1976. Furthermore, the *Weltbund zum Schutz des Lebens* (World Union for the Protection of Life), referred to by Radkau (2008: 270) as 'the germ cell of the movement against nuclear power', was founded in 1958 by the Austrian Günther Schwab, a former member of the Nazi party and the *Sturmabteilung*. The organisation's long-time chairman, Werner Haverbeck, was a National Socialist, too, who founded the far-right think tank *Collegium Humanum* in 1963 (banned in 2008) and co-signed the first version of the *Heidelberger Manifest* in 1981. The latter characterised the nation in terms of a system which had to be kept in balance, resulting in the apparent need to oppose immigration.

It was also during the 1970s and 1980s that modernisation attempts by the German *Neue Rechte* (New Right), aiming to establish a more intellectual far right not directly connected to National Socialism, drew on, for example, (a) scientists like Konrad Lorenz and Irenäus Eibel-Eibesfeldt who, amongst other things, opposed population growth and (b) linked national identity to the environment. Henning Eichberg, for example, took an ethnopluralist position, arguing that immigrants might lack an environmental consciousness because they were uprooted and could thus not have a substantive relation to the land (Olsen 1999: 99). This line of argumentation persists, as when Schretter (2016), in *U&A*, diagnoses a 'change in mentality' towards weakening nature awareness (*Naturbewußtsein*) in the younger generation characterised by increasing immigration.

Another actor remembered by the far right is Herbert Gruhl, initially a member of the centre-right *Christlich Demokratische Union Deutschlands* (CDU, Christian Democratic Union of Germany). Gruhl subsequently played a role in the formation of the Greens, before leaving the latter due to their move to the left, and co-founded the *Ökologisch-Demokratische Partei* (ÖDP, Ecological-Democratic Party) in 1981/82 (Geden 1996: 83–105). The ÖDP, too, experienced infighting over ultra-nationalistic tones which led to the foundation of the *Unabhängigen Ökologen Deutschlands* (Independent Ecologists of Germany) in 1991 by Gruhl and others. These others included Baldur Springmann, an early member of the ecological movement in Germany and well-known figure on the extreme right, and Wolfram Bednarski, former member of the CDU who, today, represents the AfD at the communal level and publishes in *JF* and *U&A* where he links animal and human migration (see below). It was also during the late 1980s and early 1990s that the party *Die Republikaner* (The Republicans) briefly established themselves by pursuing an anti-immigration agenda – without ever entering the German national parliament. Their (ephemeral) success stimulated studies discussing the party's framing of environmental protections as a patriotic duty and its use of ecology as a metaphor for the organic, national whole (Jahn and Wehling 1991: 73–90).

Since the mid-2000s, the extreme right has become more visible again in their engagement with the environment (HBS 2012; oekom 2012; Heinrich et al. 2015). The NPD, which was on the rise during the 2000s, made contributions, and so has the ecological quarterly *U&A*, founded in 2007. We will now turn to such contemporary environmental communication.

Data and method

Our corpus comprises articles published between 2013 and 2016 (providing us with a view of the current status of the discourse about the environment) in four publications plus programmatic writings of two far-right parties, the populist radical-right AfD and the extreme-right NPD, going beyond the 2013-16 period. Through this selection, we cover key sections of the far-right continuum, thus getting insights into a variety of far-right perspectives on the environment (these perspectives are not sharply separated as ideological and personal overlaps exist). The four publications we systematically analyse are the weekly newspaper *JF* founded in 1986, more precisely: its *Natur und Technik* (Nature and Technique) section. *JF* was long under observation by the North Rhine-Westphalia Office for the Protection of the Constitution for its 'extremist' views. The newspaper has become less radical, though *völkisch* nationalism is still present, and it will help to analyse the 'bourgeois' part of the far right in general and the AfD in particular. In fact, Gebhardt (2018: 115) calls the *JF* the party's 'unofficial party newspaper'. Concerning the environment, the AfD (2016: 84), established in 2013, acknowledges the 'responsibility' that 'we' have for an 'unspoilt and diverse environment' – a claim immediately followed by the assertion that the costs of 'nature conservation should not be to the detriment of mankind'.

To the right of the populist radical-right AfD, we consider the extreme-right NPD, its manifestos and the official, monthly party newspaper *DS*, which has been published since 1976. Like the NPD, the *DS* openly voices *völkisch* ideas and provides a platform for historical revisionism. The NPD was founded in 1964 to unify the splintered extreme right and while their party programme of 1973 did mention environmental protection, this happened through the prism of a healthy nation (NPD 1973). In contrast, their more radical youth organisation, *Junge Nationaldemokraten* (JN, Young National democrats), presented *12 Thesen zur Ökologie* (12 Theses on Ecology, JN 1976) which argued against 'quantitative growth' and for environmental protection trumping industrialisation. Subsequently, the NPD's environmental profile was strengthened: the party manifesto of 1985 (NPD 1985: 11f) speaks of 'materialist values' resulting in the 'destruction of nature and environment', 'alienation' and 'uprooting' of individuals. Instead, (wo)man has to be recognised as being part of nature leading to calls for, for example, the 'polluter pay principle'. The party has never entered the German national parliament and, following a relatively successful period during the 2000s in which the party openly welcomed neo-Nazis and entered the state parliaments of Saxony and Mecklenburg-Vorpommern, has, again, greatly diminished.

Closely connected to the NPD, both ideologically and personally (oekom 2012), is Germany's most prominent far-right ecological magazine, *U&A*. Published quarterly since 2007, it covers environmental issues such as nuclear energy, GMOs and biodiversity, the history of green thinking and its far-right origins, as well as areas such as gardening and (traditional) handicrafts.

Finally, we include the monthly extreme-right magazine *Z!*, since it can be viewed not only as a publication which attempts to speak to the entire far-right spectrum, but also (and so far unsuccessfully) attempts to address a broader audience. In late 2009, it replaced *Nation & Europa*, for decades one of the main extreme-right publications in Germany.

As mentioned earlier, we adopt a discourse-analytical approach inspired by the DHA in CDS (Reisigl and Wodak 2016). As such, we 'regard (a) macro-topic relatedness, (b) pluri-perspectivity and (c) argumentativity as constitutive elements of discourse' (ibid.: 27). We are therefore interested in contributions by different streams of the far right to the discourse about the environment and, consequently, focus on what, within the DHA, is referred to as topic analysis and analysis of argumentation. That is, we, first, consider the quantitative salience of topics within the discourse about the environment and code the evaluation of these topics (positive-neutral-negative, see Table 13.1). Coding was done by both authors separately and divergent cases were settled following a discussion. Second, we highlight some of the salient *topoi* (understood as conclusion rules which take an 'if-then' form; see Reisigl and Wodak 2016) so as to illustrate key structures underlying far-right environmental communication. That is, we carve out an inconclusive list of typical (often implicit) patterns of justification present in far-right environmental communication in Germany.

We collected all articles featuring environmental issues as 'macro structures' via which language users understand and summarise texts, for example through a headline and/or lead paragraph (van Dijk 1991: 131). In line with the principles of the DHA (Reisigl and Wodak 2016: 32), coding followed an abductive strategy: we drew on a set of macro-topics developed during one of the author's projects (see Endnote 1), thus approaching sources with a general understanding of what to look for. However, as we read the data, we inductively extended and renegotiated the list of topics. For instance, articles previously coded as 'Urban Sprawl', were subsumed under 'Forest', 'Environmental Protection' ('EP') and others. We then coded the stance taken by the authors of the respective articles vis-à-vis this topic, its evaluation (positive-neutral-negative), the results of which we turn to now.

Analysing the discourse about the environment in Germany

Looking at Table 13.1, it is apparent that *Z!* and *DS* constitute a small part of the corpus. Most of the 456 articles gathered are published in *JF* (n = 241), which is a weekly and thus more likely to feature a larger number of articles, and *U&A* (n = 172) which, as a publication explicitly concerned with ecology, was expected to provide many articles. Interestingly, environmental issues are evaluated rather similarly across the sources. While this similarity was not expected, the distribution of topics was neither. Yes, topics such as 'EP', which comprise a wide range of issues, are prone to be extensive, but this does not change the relatively low number of articles concerning issues prominently present in the wider public, such as climate change and GMOs. Turning to the data now in more detail, let us begin with two further surprises: 'Forest' and 'Overpopulation'.

TABLE 13.1 Overview of macro-topics in far-right environmental communication and their evaluation[4]

	Biodiversity	Material pollution	Environmental Protection	Meta	Energy-Conventional	Alternative	Transformation	Climate change	Genetically modified organisms	Forrest	Over-population	Total
JF	49	45	35	15	23	20	19	19	4	5	7	241
(+)-(0)-(-)	46-2-1	7-7-31	35-0-0	15-0-0	2-14-7	3-2-15	0-5-14	5-8-6	0-1-3	4-1-0	0-0-7	117-40-85
Z!	3	3	1	3	3	3	1	4	2	2	0	25
(+)-(0)-(-)	3-0-0	1-0-2	0-1-0	3-0-0	0-2-1	0-1-2	0-0-1	0-2-2	0-0-2	2-0-0	0-0-0	9-6-10
DS	1	0	3	2	2	1	5	1	3	0	0	18
(+)-(0)-(-)	0-0-1	0-0-0	2-1-0	2-0-0	1-0-1	0-0-1	0-1-4	0-0-1	0-0-3	0-0-0	0-0-0	5-2-11
U&A	56	25	33	15	5	3	2	1	16	12	4	172
(+)-(0)-(-)	55-1-0	0-1-24	33-0-0	15-0-0	0-1-4	1-0-2	0-0-2	1-0-0	0-0-16	12-0-0	0-0-4	117-3-52
Total	109	73	72	35	33	27	27	25	25	19	11	456
(+)-(0)-(-)	104-3-2	8-8-57	70-2-0	35-0-0	3-17-13	4-3-20	0-6-21	6-10-9	0-1-24	18-1-0	0-0-11	248-44-160

First, and considering Table 13.1, 'forest' is not a key site of environmental concern and (national) identity formation in our corpus, although it is very positively evaluated. As mentioned earlier, the forest is the site through which the link between the natural environment and German nationalism was forged during the Romantic period. Following the Nuremberg Laws, signs were erected, saying 'Jews are not welcome in our German forest' (Zechner 2009: 34). Imort (2005: 67f), while stating that, in purely ecological terms, National-Socialist forest laws where mostly beneficial, adds that the forest was represented 'as an analogy of the German Volk and state and as the cornerstone of National Socialist ideas of race, community, and eternity'. The historical salience of the forest has been forcefully formulated by Canetti who described it as a key crowd symbol:

> The crowd symbol of the Germans was the army. But the army was more than just the army; it was the marching forest. In no other modem country has the forest-feeling remained as alive as it has in Germany. The parallel rigidity of the upright trees and their density and number fill the heart of the German with a deep and mysterious delight.
>
> *(1981: 173)*

However, although the forest is still significant for and appreciated by the far right – *Z!* claims that even the green-alternative generation of the 1980s was 'German enough' to feel a special affinity for the forest (Warncke 2014: 67), and *U&A* has 'The Germans and their forest' on its cover (3/2014), discussing this intimate relationship – it is not an empirically dominant theme anymore.

Similarly, overpopulation, once a major topic – see Paul Ehrlich's *The Population Bomb* (1968), the 1972 report by the Club of Rome and, in Germany, Gruhl's *Ein Planet wird Geplündert* (1975, The plundering of a planet) – is less prominent now. It is, however, still and unanimously viewed as a problem and *U&A* has the topic on the cover of issue 2/2014. Its author, Heiko Urbanzyk, who is also a prolific contributor to *JF*, conducts a lengthy review of Alan Weisman's *Countdown* (2013), praising the author for his discussion of immigration in the context of overpopulation (the danger of underpopulation of the ethnic host) and an economy not based on growth. As such, the far right's (neo-Malthusian) stance on overpopulation is a first indicator of the presence of the topos of sustainability ('if we do not restrain ourselves, then our environment is doomed') in far-right contributions; in this case concerning the claim that more people imply a higher ecological foodprint.

One of the key issues of our times, climate change, which could replace overpopulation as a salient carrier of apocalyptic constructions (Jahn and Wehling 1991: 77), rarely discussed as the macro-topic (but present in other discussions, for example concerning energy sources). Though this topic is at times prominently displayed – *Z!* (2/2013) has climate change on the front page, asking 'World climate apocalypse. Has man influence over global warming?' – the far right 'scandalises' the topic in a rather different way. Instead of pointing to the danger of climate change, and in line with existing research on far-right climate-change communication in Germany and beyond (Gemenis et al. 2012; Forchtner and Kølvraa 2015; Forchtner et al. 2018; Lockwood 2018; Forchtner forthcoming), a significant extent of scepticism towards anthropogenic climate change is articulated. Indeed, even though the findings are fairly balanced, this is still comparatively sceptical when put in relation to the European mainstream.

Examples of scepticism towards anthropogenic causes include a report in *DS* (Richter 2016) on Udo Voigt's, previously party leader of the NPD and Member of the European Parliament until recently (2014–2019), activities in the European Parliament, saying that the climate has changed 'countless times' and will not be affected by human intervention. This claim is also present in *Z!* (Tiedemann et al. 2013) which employs a common array of ideas when speaking of 'climate craze', climate change as having 'dimensions of a religion, illustrated by the fanaticism of parts of its followers' and by depicting Al Gore as 'climate pope'.

Such arguments also characterise the less extreme, populist radical-right AfD, which is the only party voicing climate-change scepticism in the German national parliament. As such, the party's manifesto for the European election in 2014 (AfD 2014: 19) speaks of uncertainties concerning anthropogenic CO_2 emissions, while the party programme (AfD 2016: 78; see also AfD 2019: 79) claims that '[c]limate changes have occurred as long as the earth exists'; that '[c]arbon dioxide (CO^2), however, is not a harmful substance, but part and parcel of life' which would benefit the growth of plants; and that the claim of the Intergovernmental Panel on Climate Change that anthropogenic CO_2 emissions 'will result in catastrophic consequences for mankind (...) is based on computer models that, however, are not backed by quantitative data and measured observations'.[5]

Subtler are those articles which do not straightforwardly deny anthropogenic causes, and which we coded as 'neutral', but which feature a polemical tone and engage in process (concerning knowledge generation and decision-making processes) and response scepticism (concerning policy responses), instead of evidence scepticism (the denial of climate change happening at all, is anthropogenic, and/or might not be a bad thing) (see van Rensburg 2015 for more on this typology of scepticism). Indeed, 'neutral' articles oppose 'One-sided alarmism' (Keller 2016) and ridicule supposedly 'religious fervour' by those favouring greenhouse-gas neutrality (Glaser 2016).

Overall, there are a number of significant *topoi* present in climate-change-related contributions, including the *topos of (quasi-religious) irrationalism* ('if those arguing in favour of climate change are driven by (quasi-religious) irrationality, then their agenda should be rejected'), the *topos of scientific untrustworthiness* ('if established scientists/science working on climate change do(es) not follow proper scientific procedures, then their warnings concerning climate change cannot be trusted') and what might be called the *topos of 'we' first* ('if cutting back emissions causes too much harm to the national economy/the 'little guy', then we should not do it'). As such, all three *topoi* realise the thin-centred ideology of populism (Mudde 2007: 23) according to which 'the pure people' stand antagonistically to 'the corrupt elite', and that the general will of the people has to be the basis of politics. In this case, this leads to a rejection of allegedly (liberal) left elites and the irrationalism they spread to confuse 'common sense', an opposition to mainstream scientists which help to uphold the 'global order', and a portrayal of 'the little guy' as being punished by emission reduction targets (e.g. green taxes) beyond *our* 'reasonable' share.

Although 'Material Pollution' and 'EP' are extensively reported, we only touch upon them briefly due to their common-sense character, internal diversity and clear-cut evaluation. 'Material Pollution' deals with various sorts of waste, as in plastics in the oceans and pollution caused by cars, and is negatively evaluated. 'EP', which is positively evaluated, captures concrete environmental issues and the reaction to them which could not be allocated to more specific categories, for example an article on the dangers of extracting raw materials in the deep oceans. Concerning both topics, arguments regularly resemble those in mainstream publications, such as when Häusler (2014) reports approvingly of the International Coastal Clean-up Day in *Z!*, including shocking photos of plastic waste. Overlapping with the Czech case (see Tarant 2019 in this volume), the German extreme right has also engaged in hands-on eco-activism. The youth wing of the NPD, the JN (2008: 17, 19), collected waste and more recently, this has been displayed by the neo-Nazi party *Der III. Weg* (2017; The III. Way) in its criticism of non-environmentalist behaviour by littering comrades. Both topics draw on the *topos of sustainability*, a *topos* not only present in these areas, though it is clearly visible here and connected to *our* land. Moreover, it is not limited to the far right, but, in its liberal variant, has long played a key role. However, at least in some articulations, in particular by *U&A*, far-right criticism of unsustainable practices might be more fundamental than in liberal interpretations of sustainable development (see Hurd and Werther 2016: 164) due to the actors' organic ethnonationalism.

As visible in Table 13.1, the most salient topic is 'Biodiversity', a topic very positively evaluated. It includes biodiversity more generally, but also endangered and so-called 'invasive' species (on the role of this theme in mainstream and far-right texts, see Forchtner 2019a; in our corpus, about one-fifth of the articles on 'Biodiversity' predominantly address this subtopic). The literature has long pointed to analogies between animals and plants on the one hand, and the social world on the other in far-right communication, although the extent to which this is the case todays is surprising. Here, we cannot discuss why biodiversity is also defended staunchly by left-wing environmentalists; but for the far right, it has to do with the richness of the community's land, its unique character, and the pressure the latter experiences from materialist, so-called globalist forces and the desire to maximise consumption. JN (1999: 12) identified these as the 'main enemy', something also articulated by the NPD (2010: 35). Most clearly, this is visible in discussions of 'invasive' species. This facilitates thinking about the particularities of *our* ecosystem, of animals (including humans) and plants as part of an organic whole, particularly well and resembles what Dudek (1984: 98f) described as 'ecological racism' (the *Volk* as a biological system).

For example, Keller (2013) begins his article on a 'homogenised world' by pointing to a 2004 quote by Angela Merkel ('The multicultural society has grandiosely failed'), before warning against 'biological invasions' by animals and plants facilitated by globalisation ('the tremendous force of delocalisation and uprooting'). It is furthermore visible in an article concerned with the return of the wolf to Germany (see Kølvraa 2019 on cultural perceptions of 'the wolf' in this volume), in which Wolfram Bednarski (2014: 33), who has worked with Gruhl, acts as deputy chairman of the Herbert-Gruhl-Society e.V. and is a local politician for the AfD, states that 'Germany has an immigration problem – the wolf, however, does not belong to it' (In contrast to Bednarski, the AfD has voiced scepticism towards the return of 'the wolf' to Germany, see Höhne 2019.). Although not necessarily as outspoken as these two, many relevant articles view human communities, that is, national communities, as being part of nature, part of unique ecosystems. Therefore, and in line with ethnopluralism, to protect the particularity of *our* environment is also to protect *our* culture.

As such, ecosystems have to remain pure. Following Douglas' (2002: 2) work, opposition to pollution requires an 'effort to organise', to put matter in the right place (ibid.: 140), that is, its 'natural' place. While this *topos of naturalness* ('if nature organised X in a certain way, then we should not interfere') is certainly not limited to this topic, it is clearly visible in such contributions. According to the far right, the consumerist, globalist and materialist *zeitgeist* is responsible for matter being put.

A further topic which similarly draws on the *topos of naturalness* is 'GMOs'. The latter is very negatively evaluated across the investigated publications and has played a visible role in far-right communication. For example, the NPD group in the legislative assembly in the federal state of Mecklenburg-Vorpommern (NPD 2007) raised the issue on the front page of its first bulletin ('GMO muck [*Genfraß*] – no thanks!'). This is not only a regional position, it is also reflected in the party's national programme which calls for a ban on genetically modified food and seeds (NPD 2010: 35f). Opposition to GMOs in agriculture is connected to biodiversity and the survival of 'indigenous cultivated plants and their seeds' (ibid.), and is equally visible in *JF* (Urbanzyk 2014). The latter has criticised new dimensions in genetic manipulation, 'synthetic biology', as nature would soon be produced artificially in the lab without any need for actual nature. It is in this context, that imagining catastrophe is visible. For example, *U&A* (2015) frames the introduction of biotechnology in the Ukrainian agricultural sector as the next 'ecological disaster after the catastrophe of Chernobyl'. In contrast, the AfD (2016: 86) speaks of the need for more research in the area which might result in approving genetically modified products, while calling for clearly defined rules, for the preservation of seed diversity and against dependencies on 'multi-national mega-corporations'.

Turning to energy, we divided the material into three areas to increase transparency, although they will usually intersect: articles on conventional energy sources (e.g. fracking and nuclear power), alternative energy sources (e.g. sun and wind) and German energy transformation.

Concerning, first, conventional energy, the investigated actors are not wedded to them. Arguments in favour of conventional energy sources point to, for example, their importance for keeping Germany running, thus drawing on the *topos of autarky* ('if a particular energy sources brings the nation closer to autarky, its use must be encouraged'). At times, this is combined with the construction of the elite prohibiting discussion, as in *JF* (Manns 2013): 'pleading for (secure) nuclear power plants is not covered by freedom of opinion anymore. Who stands up for that, apparently commits a crime.' Yet, criticism dominates, pointing to, for example, environmental risks (e.g. Fukushima) and dependency on foreign powers as well as multinational corporations. This mirrors positions taken by, the British National Party and the Danish People's Party (Forchtner and Kølvraa 2015).

Thus, these actors are not necessarily advocating the status quo, but alternative energy sources too are evaluated rather negatively. At times, they are welcomed for environmental reasons and the promise of autarky (see also Forchtner et al. 2018: 597f); here too, the *topos of autarky* is present – a subtype of a broader *topos of sovereignty* ('if X strengthens/weakens the nation's sovereignty, then it has to be welcomed/rejected'). The NPD (2010: 37f) is in favour of 'domestic energy sources and renewable energy', though adding that a 'strategic energy alliance with Russia should be sought'. Still, building coal power plants is not called for and the country's post-nuclear energy policy requires Germany to remain at the forefront of developing renewable energy (ibid.). In contrast, the AfD (2019) argues in favor of coal and nuclear energy, and remains sceptical towards towards alternative energy sources. At the same time, opposition to alternative energy sources is regularly justified via environmental concerns, mainly in relation to the harm caused by wind turbines to bird populations (AfD 2016: 79; Rademacher 2015). Criticism also concerns aesthetic reasons (see Forchtner and Kolvraa 2015), a point presented in a now infamous speech by the best-known face of the *völkisch* wing of the AfD, Björn Höcke, in 2017. Speaking of 'a memorial of shame in the heart of their [Germans'] capital' and demanding a '180-degree turn in the politics of memory', Höcke (2017) also lamented that '[o]ur once beautiful homeland is increasingly misshapen due to ugly structures, wind turbines and chaotic populating'. These (aesthetic) concerns are also present in one of the many contributions to *JF*, 'Ugly landscapes', from Volker Kempf (2016), chairman of the aforementioned Herbert-Gruhl-Society e.V. and local AfD politician. The supposed threat from wind turbines to the nation's cultural landscape, and thus its communal identity, is also strongly present in the AfD's party programme (2016: 79, 85). This *topos of national beauty* ('if X threatens the beauty of the land, then X has to be prohibited') has a clear historical trajectory, as aesthetics were already significant in the *Heimatschutz* movement.

Conflicts over energy sources are interwoven with the negatively evaluated German energy transformation. While the overall goal is not necessarily rejected, the project is viewed as endangering the country's strong economy and/or individual consumers ('the little guy'), as prices are supposedly bound to rise. This harks back to the *topoi of autarky/sovereignty* and the *topos of 'we' first*, and is also connected to arguments about the alleged threat from communism/socialism (*topos of planned economy*; 'if economic freedom is restricted, then unfreedom will follow'). Keller (2015), in *JF*, claims that big investors use the project to speculate on German soil, including the monoculturalisation of German land (maize), making it unaffordable for individuals and small local businesses. The AfD (2016: 78) demands a secure, affordable and 'environmentally compatible' energy supply and takes a stance against patronisation, for example concerning the demand for insulation and increased energy efficiency. Such populist opposition is in line with the findings of Engels et al. (2013), who found a correlation between opposition to Germany's energy transformation and renewables and climate-change scepticism.

We close with a look at what is the most abstract theme, 'Meta', which covers principles of far-right (wo)man-nature relations, key figures in the history of far-right environmental thought, and the alleged aberration of left-wing Greens. Instead of being concerned with particular issues of the day, articles in this category deal with, for example, the reconstruction of how love for one's homeland and national landscape is a motivating force for environmental protection (Erdmann 2016). Consequently, it is claimed big supranational organisations are doomed to fail in this regard (ibid.: 32) and the article closes by citing Rudorff approvingly:

> The world is not only becoming uglier, more artificial, more American with each day, but with our pushing and hunting after illusions of alleged happiness we, at the same time, constantly undermine on and on and on the soil which carries us.
>
> *(ibid.: 33)*

Not only is it in passages like this that the German far right's anti-Americanism, to the extent that the United States represents liberal modernity, comes to the fore, but such references furthermore connect present-day far-right thinking to this tradition's forefathers. Stories of 'great men' prevail in criticism of the Greens and the party's transformation from an initially 'conservative' or 'patriotic' force into a left-wing one. In particular, Gruhl is remembered as 'one of the first true "Greens"' (Rose and Schimmer 2013) and as a 'conservative bedrock' (*Zuerst!* 2013: 27). Contemporary Greens, that is, left-wing environmentalists, are supposedly unable to act truly in favour of the environment: *they* are alienated from the (home-)land and thus are not able to truly care for it, *they* are not able to connect the protection of homeland, environment, culture and identity (*topos of alienation*; 'if one is alienated from the land, then one cannot truly care for it'). From here, the link to themes, for example, opposition to the ideology of growth and overpopulation (and immigration), is easily drawn as nature, in such a world view, is not multicultural, but diverse. Each living being has its place in an ecosystem, they are not mixed, as in the stereotypical melting pot (the alleged goal of 'globalists' and 'one-world apologists'), but live in separation, based on the 'right to difference' (de Benoist and Champetier 1999), so as to protect the homeland. Along these lines, and in addition to *topoi* supporting populism, *topoi* which reveal (ethno)nationalist ideology are, for example, the *topos of national beauty, of naturalness, of sovereignty* and, at the collective level, the *topos of 'we' first*.

Conclusion

The case of far-right environmental communication in Germany has significant potential to contribute to the study of far-right environmental communication more generally. Not only happen such articulations in a wider discursive context which acknowledges the importance of environmental protection, but they are characterised by relevant historical and contemporary far-right concerns. The distribution of topics and their evaluation (Table 13.1) illustrate the significance of issues *directly* connecting land and people. In contrast to climate change, an inherently transnational, global and seemingly abstract topic, imagining the protection of the environment via thematising living organisms populating *our* land, discussions of concrete environmental threats and instances of material pollution is to imagine protection of the environment in terms of protection of particular, demarcated spaces, one's own landscape and culture. It is only those who are rooted in *our* land who are able to truly appreciate *Heimat*, and who will subsequently care for it. In consequence, those 'unrooted' and alienated, immigrants and the left for example, lack a holistic understanding of (wo)man and environment and are not able to sense its 'true' needs. Such a position is not universalist, but particularistic; it imagines one's own land as a site of difference, as a site through which difference is reproduced. It is the 'globalist' structure of today's world which threatens this land, this difference; and though this does not necessarily imply a rejection of international cooperation, solutions are thus anti-global, pro local/national, protecting the people's ecosystem against rationalist and materialist globalist forces.

This argument helps to understand the stance taken towards the thesis of anthropogenic climate change, a topic characterised by a low number of articles and mixed evaluation. It is here, and this sentiment is also visible in energy-related discussions, that different ideological pillars seem to collide, pitching care for the natural environment against, for example, populist sentiments vis-à-vis a liberal, globalist regime and its representatives, as well as alleged restrictions on the national economy.

Indeed, across the topics, a number of *topoi* which reproduce populist and (ethno)nationalist ideology are present – the *topos of (quasi-religious) irrationalism, of scientific untrustworthiness,* of *'we' first,* of *autarky,* of *sovereignty,* of *sustainability,* of *naturalness,* of *national beauty,* of *a planned economy* and of *alienation* – a list future research should complement. Particularly the *topoi* of *alienation, naturalness* (these two are closely linked), *national beauty* and *sustainability* justify the far-right's holistic worldview of the nation as a sacred organism which must be preserved. *Topoi* are present across the various topics and interact with and strengthen each other. In fact, some of them go clearly beyond the discourse about the environment and characterise far-right politics more generally (as in the case of the *topos of naturalness* and gender). Furthermore, some *topoi* unfold especially powerfully in the context of specifics topics, such as in the case of biodiversity and, again, the *topos of naturalness*. Here, the *topos* facilitates certainty through purity, purity which 'is the enemy of change, of ambiguity and compromise' (Douglas 2002: 200). The demand to keep *us* free of pollution is thus absolute. Moreover, the conflict between the stances we have mentioned is, at the level of argumentation, a conflict between *topoi*; either because a *topos* can justify both environmental protection and its opposite (e.g. the *topos of autarky*) or because some arguments are commonly directed against environmental policies (e.g., the *topos of 'we' first*).

Both ends of the far-right continuum, from the radical right to the extreme right (which tends to be more *organic* ethnonationalist), regularly claim 'Nature/environmental protection is homeland protection' (see the opening quote by the NPD and, for example, Friedhoff [2019], a member of the AfD, during a speech on the protection of biodiversity). Indeed, the investigated actors often draw upon similar arguments – though the extreme-right, organic-ethnonationalist *U&A* is most clearly driven by an 'ecological imaginary' (Forchtner 2019a), while the case of the radical-right AfD is less comprehensive and reminds one of what Kølvraa (2019 in this volume) calls an 'environmental imaginary'. Yet, regarding climate change, things are more complex: the AfD is clearly climate-change sceptic, the JF does, at times, accept anthropogenic climate change, while the more extreme *DS* und *Z!* are, again, outright climate-change sceptic. Beyond the case of climate change, similarities across the continuum might (partly) be due to personal overlaps, visible in that one of the most prolific authors in *JF*, Urbanzyk, is also a contributor to *U&A*. Besides further investigation into the messages conveyed through far-right environmental communication, it is these 'nodes' in the far-right network, that is, actors who operate across the far-right continuum, which keep alive and (re)tell stories about our relation to the natural environment. They too are thus useful foci to better understand the meanings and internal dynamics driving far-right contributions to the wider discourse about the environment.

While this analysis focuses on the period between 2013 and 2016, current debates, for example during the European Parliamentary Elections in 2019, confirm our findings. For instance, the AfD has kept pushing climate-change scepticism. Yet, how climate-change communication by the far right in Germany will develop, following the recent rise of the Greens and the centrality of climate change in public debate, remains open. In fact, calls to accept the existence of anthropogenic climate change were raised shortly after the election by the AfD's youth organization in Berlin and by individuals belonging to the far right on social media. While these calls have so far been rejected, this illustrates that meanings can change.

Notes

1 This work was supported by FP7 People: Marie-Curie Actions [grant number 327595]. All quotes from German sources have been translated into English by the authors.
2 As mentioned in the Introduction to this volume (Forchtner 2019b), we use 'environmental' for pragmatic reasons, thus not connoting only superficial concern for the environment. However, at times we will use 'ecological' to stress a particularly holistic perspective.
3 On Seifert's relation to the cultivation of herbs in the concentration camp Dachau, see Kopke (2012).
4 We exclude animal welfare to do with pets and ritual slaughter for pragmatic reasons; though articles dealing with animals are part of our corpus, for example concerning 'Biodiversity' and in discussions of bird populations and wind turbines.
5 For another view on the AfD, see Schaller and Carius (2019: 84).

References

AfD (2014): *Mut zu Deutschland. Für ein Europa der Vielfalt. Programm der Alternative für Deutschland (AfD) für die Wahl zum Europäischen Parlament am 25. Mai 2014.* Erfurt: AfD.
AfD (2016): *Manifesto for Germany. The Political Programme of the Alternative for Germany.* Berlin: AfD.
AfD (2019): *Europawahlprogramme. Programme der Alternative für Deutschland für die Wahl zum 9. Europäischen Parliament.* Berlin.
Arndt, Ernst Moritz (1815): Ein Wort über die Pflegung und Erhaltung der Forsten und der Bauern im Sinne einer höheren, d.h. menschlichen Gesetzgebung, in: *Der Wächter, 2. Band.* Köln: Heinrich Rommerskirchen. pp. 346–408.
Bednarski, Wolfram (2014): Rückkehr des Wolfes – ein Geschenk und eine zweite Chance, *Umwelt & Aktiv,* 4, 32–33.
Blackbourn, David (2006): *The Conquest of Nature: Water, Landscape and the Making of Modern Germany.* London: Vintage.
Bramwell, Anna (1985): *Blood and Soil: Walther Darré and Hitler's Green Party.* Bourne End: Kensal House.
Brüggemeier, Franz-Josef, Cioc, Mark and Zeller, Thomas (2005): Introduction. In: Franz-Josef Brüggemeier, Mark Cioc and Thomas Zeller (Eds). *How Green Were the Nazis? Nature, Environment, and Nation in the Third Reich.* Athens: Ohio University Press. pp. 1–17.
Canetti, Elias (1981): *Crowds and Power.* New York: Continuum.
de Benoist, Alain and Champetier, Charles (1999): The French New Right in the year 2000. http://home.alphalink.com.au/~radnat/debenoist/alain9.html [25 June 2017].
Der III, Weg (2017): Diskussionsbeitrag: Die Szene ist unser Unglück. https://der-dritte-weg.info/2017/07/19/diskussionsbeitrag-die-szene-ist-unser-unglueck/ [29 July 2018].
Dominick, Raymond H. (1992): *The Environmental Movement in Germany. Prophets and Pioneers, 1871–1971.* Bloomington: Indiana University Press.
Douglas, Mary (2002): *Purity and Danger.* London: Routledge.
Dudek, Peter (1984): Konservatismus, Rechtsextremismus und die <Philosophie der GRÜNEN>. In: Thomas Kluge (Ed). *Grüne Politik. Der Stand einer Auseinandersetzung.* Frankfurt/Main: Fischer. 90–108.

Eissing, Hildegard (2014): Wer verfasste die „Grüne Charta von Mainau"?, *Naturschutz und Landschaftsplanung*, 46(8): 247–252.
Engels, Anita, Hüther, Otto, Schäfer, Mike and Held, Hermann (2013): Public climate-change skepticism, energy preferences and political participation, *Global Environmental Change*, 23: 1018–1027.
Erdmann, Bastienne (2016): Wer die Heimat liebt, schützt sie. Ernst Rudorff und die Ursprünge des Umweltschutzes aus moderner Perspektive, *Umwelt & Aktiv*, 2: 31–33.
Forchtner, Bernhard (2019a): Nation, nature, purity: Extreme-right biodiversity in Germany, *Cultural Imaginaries of the Extreme Right. Special Issue of Patterns of Prejudice*, 53(3): 285-301.
Forchtner, Bernhard (2019b): Far-right articulations of the natural environment: An introduction. In: Bernhard Forchtner (Ed). *The Far Right and the Environment: Politics, Discourse and Communication*. London: Routledge..
Forchtner, Bernhard and Kølvraa, Christoffer (2015): The nature of nationalism: populist radical right parties on countryside and climate, *Nature & Culture*, 10(2): 199–224.
Forchtner, Bernhard, Kroneder, Andreas and Wetzel, David (2018): Being skeptical? Exploring far-right climate-change communication in Germany, *Environmental Communication*, 12(5): 589–604.
Franke, Nils M. (2017): Naturschutz als völkische Aufgabe. In: Michael Fahlbusch, Ingo Haar and Alexander Pinwinkler (Eds). *Handbuch der völkischen Wissenschaften. Akteure, Netzwerke, Forschungsprogramme. Teilband 2*. Oldenbourg: De Gruyter. pp. 1073–1079.
Franke, Nils M. and Pfenning, Uwe (Eds) (2014): *Kontinuität im Naturschutz*. Baden-Baden: Nomos.
Friedhoff, Dietmar (2019): Umweltschutz ist Heimatschutz!. www.afdbundestag.de/ umweltschutz-ist-heimatschutz-dietmar-friedhoff-afd-fraktion-im-bundestag/ [5. July 2019].
Gebhardt, Richard (2018): >Mut zur Wahrheit<? *Compact, Sezession* und *Junge Freiheit* – das publizistische Netzwerk der AfD. In: Alexander Häusler (Ed). *Völkisch-autoritärer Populismus. Der Rechtsruck in Deutschland und die AfD*. Hamburg: VSA. pp. 109–116.
Geden, Oliver (1996): *Rechte Ökologie. Umweltschutz zwischen Emanzipation und Faschismus*. Berlin: Elefanten.
Gemenis, Kostas, Katsanidou, Alexia and Vasilopoulou, Sofia (2012): The politics of anti-environmentalism: positional issue framing by the European radical right. *Paper prepared for the MPSA Annual Conference*, 12–15 April 2012, Chicago.
Gerhard, Gesine (2005): Breeding pigs and people for the Third Reich. Richard Walther Darré's Agrarian ideology. In: Franz-Josef Brüggemeier, Mark Cioc and Thomas Zeller (Eds). *How Green Were the Nazis? Nature, Environment, and Nation in the Third Reich*. Athens: Ohio University Press. pp. 129–146.
Glaser, Dirk (2016): Unser Gral leuchtet treibhausgasneutral, *Junge Freiheit*, 40: 16.
Häusler, Sven (2014): Inferno aus Plastikmüll, *Zuerst!* 12: 59–60.
HBS (Ed) (2012): *Braune Ökologen. Hintergründe und Strukturen am Beispiel Mecklenburg-Vorpommerns*. Berlin: NBS.
Heinrich, Gudrun, Kaiser, Klaus-Dieter and Wiersbinski, Norbert (Eds) (2015): *Naturschutz und Rechtsradikalismus. Gegenwärtige Entwicklungen, Probleme Abgrenzungen und Steuerungsmöglichkeiten*. Bonn: BfN.
Höcke, Björn (2017): *Rede von Björn Höcke live bei der Jungen Alternative AfD - Dresdner Gespräche*. www.youtube.com/watch?v=WWwy4cYRFls [20 July 2018].
Hofer, Christoph (2007): Vorwort: Der Grundgedanke dieses Magazins, *Umwelt & Aktiv*, 1: 1.

Hurd, Madeleine and Werther, Steffen (2016): The militant media of neo-Nazi environmentalism. In: Heike Graf (Ed). *The Environment in the Age of the Internet. Activists, Communication and the Digital Landscape.* Cambridge: Open Book Publishers. pp. 137–170.

Imort, Michael (2005): 'Ethernal forest – eternal *Volk*'. The rhetoric and reality of National Socialist forest policy. In: Franz-Josef Brüggemeier, Mark Cioc and Thomas Zeller (Eds). *How Green Were the Nazis? Nature, Environment, and Nation in the Third Reich.* Athens: Ohio University Press. pp. 43–72.

Jahn, Thomas and Wehling, Peter (1991): *Ökologie von rechts: Nationalismus und Umweltschutz bei der Neuen Rechten und den "Republikanern".* Frankfurt/Main: Campus.

JN (1976): *12 Thesen zur Ökologie.*

JN (1999): Nationalismus ist Naturschutz, *Der Aktivist*, 8: 11–19.

JN (2008): Aktionsberichteaus den Stützpunkten und Landesverbänden, *Der Aktivist*, 17(3): 13–19.

Kast, D. (2010): Hermann Löns – Heidedichter, Naturschützer und ungehörter Warner, *Umwelt & Aktiv*, 2: 34–37.

Keller, Christoph (2013): Vorboten der homogenisierten Welt, *Junge Freiheit*, 19: 13.

Keller, Christoph (2015): Hasen verschwinden, Störche wandern ab, *Junge Freiheit*, 16: 15.

Keller, Christoph (2016): Einseitiger Alarmismus, *Junge Freiheit*, 44: 16.

Kempf, Volker (2016): Häßliche Landschaft, *Junge Freiheit*, 36: 16.

Kølvraa, Christoffer (2019): Wolves in sheep's clothing? The Danish far right and 'wild nature'. In: Bernhard Forchtner (Ed). *The Far Right and the Environment: Politics, Discourse and Communication.* London: Routledge.

Kopke, Christoph (2012): Kompost und Konzentrationslager. Alwing Seifert und die „Plantage"im KZ Dachau. In: Annett Schulze and Thorsten Schäfer (Eds). *Zur Re-Biologisierung der Gesellschaft. Menschenfeindliche Konstruktionen im Ökologischen und im Sozialen.* Aschaffenburg: Alibri. pp. 185–207.

Lekan, Thomas (2004): *The Nation in Nature. Landscape Preservation and German Identity 1885–1945.* Cambridge: Harvard University Press.

Lekan, Thomas (2005): "It shall be the whole landscape!" The Reich Nature Protection Law and regional planning in the Thirds Reich. In: Franz-Josef Brüggemeier, Mark Cioc and Thomas Zeller (Eds). *How Green Were the Nazis? Nature, Environment, and Nation in the Third Reich.* Athens: Ohio University Press. pp. 73–100.

Linse, Ulrich (1983): *Zurück o Mensch, zur Mutter Erde. Landkommunen in Deutschland 1890–1933.* München: dtv.

Lockwood, Matthew (2018): Right-wing populism and the climate change agenda: exploring the linkages, *Environmental Politics*, 27(4): 712–732.

Manns, Michael (2013): Atomare Preisfrage, *Junge Freiheit*, 26: 13.

Mudde, Cas (2007): *Populist Radical Right Parties in Europe.* Cambridge: Cambridge University Press.

NPD (1973): Programm der Nationaldemokratischen Partei Deutschlands. In: Siegfried Hergt (Ed) (1977): *Parteiprogramme.* Leverkusen: Heggen. pp. 406–439.

NPD (1985): *NPD-Parteiprogramm. Nationaldemokratische Gedanken für eine lebenswerte Zukunft.*

NPD (2007): Genfrass – Nein Danke! *Der Ordnungsruf*, 1: 1.

NPD (2010): *ARBEIT. FAMILIE. VATERLAND. Das Parteiprogramm der Nationaldemokratischen Partei Deutschlands (NPD).* Berlin: NPD.

oekom (Ed) (2012): *Ökologie von rechts. Braune Umweltschützer auf Stimmenfang.* München: oekom.

Olsen, Jonathan (1999): *Nature and Nationalism: Right-wing Ecology and the Politics of Identity in Contemporary Germany.* Basingstoke: Palgrave.

Pois, Robert (1986): *National Socialism and the Religion of Nature.* London: Croom Helm.

Rademacher, Bernd (2015): Der Milan: rot von Blut, *Junge Freiheit*, 36: 15.
Radkau, Joachim (2008): *Nature and Power. A Global History of the Environment*. Cambridge: Cambridge University Press.
Reisigl, Martin and Wodak, Ruth (2016): The discourse-historical approach (DHA). In: Ruth Wodak and Michael Meyer (Eds). *Methods of Critical Discourse Studies*. London: Sage. 23–61.
Richter, Karl (2016): NEIN ZU KLIMAWAHN UND GENMAIS! *Deutsche Stimme*, 11: 10.
Riordan, Colin (Ed) (1997): *Green Thought in German Culture. Historical and Contemporary Perspectives*. Cardiff: University of Wales Press.
Rollins, William H. (1997): *A Greener Vision of Home: Cultural Politics and Environmental Reform in the German Heimatschutz Movement, 1904–1918*. Michigan: University of Michigan Press.
Rose, Olaf and Schimmer, Arne (2013): Geistiger Vater und Vordenker der Umweltbewegung, *Deutsche Stimme*, 28: 24–25.
Rudorff, Ernst (1994[1897]): *Heimatschutz*. St. Goar: Reichl.
Schaller, Stella and Carius, Alexander (2019): *Convenient Truths: Mapping Climate Agendas of Right-wing Populist Parties in Europe*. Berlin: adelphi.
Schretter, Helmut (2016): Naturbewusstseinsstudie im Lichte des Bevölkerungsaustauschs, *Umwelt & Aktiv*, 3: 41.
Smith, Anthony D. (2009): *Ethno-Symbolism and Nationalism*. Oxon: Routledge.
Sojka, Klaus (2014): Kirche und Tierschutz. Theologen und Tierbehandlungen, *Umwelt & Aktiv*, 4: 34–35.
Staudenmaier, Peter (2011): Right-wing ecology in Germany: Assessing the historical legacy. In: Janet Biehl and Peter Staudenmaier (Eds). *Ecofascism Revisited: Lessons from the German Experience*. Porsgrunn: New Compass Press. pp. 89–132.
Tarant, Zbyněk (2019): Is brown the new green? The environmental discourse of the Czech far right. In: Bernhard Forchtner (Ed). *The Far Right and the Environment: Politics, Discourse and Communication*. London: Routledge.
Tiedemann, Falk, Rehwaldt, Dorian and Diehl, Robert (2013): Weltklima-Apokalypse, *Zuerst!* 2: 8–15.
U&A (2015): Kornkammer für Gen-Getreide, *Umwelt & Aktiv*, 1: 26.
Uekötter, Frank (2006): *The Green and the Brown: A History of Conservation in Nazi Germany*. Cambridge: Cambridge University Press.
Uekötter, Frank (2014a): *The Greenest Nation? A New History of German Environmentalism*. Cambridge: MIT Press.
Uekötter, Frank (2014b): Die Autoritäre Versuchung: Das Reichsnaturschutzgesetz. In: Frank Uekötter (Ed). *Ökologische Erinnerungsorte*. Göttingen: Vandenhoeck & Ruprecht. pp. 86–100.
Urbanzyk, Heiko (2014): Saatgutmultis lassen nicht locker, *Junge Freiheit*, 9: 14.
van Dijk, Teun (1991): The interdisciplinary study of news as discourse. In: Klaus Bruhn-Jensen and Nick Jankowksi (Eds). *Handbook of Qualitative Methods in Mass Communication Research*. Abingdon: Routledge. pp. 108–120.
van Rensburg, Willem (2015): Climate change scepticism. *SAGE Open*, 5(2): 1–13.
von Riehl, Wilhelm (1854): *Land und Leute*. Stuttgart und Tuebingen: Cotta'scher Verlag.
Warncke, Xaver (2014): Die Deutschen und ihr Wald, *Zuerst!* 10: 66–67.
Wolschke-Bulmahn (2005): Violence and the basis of National Socialist landscape planning in the "Annexed Eastern Areas". In: Franz-Josef Brüggemeier, Mark Cioc and Thomas Zeller (Eds). *How Green Were the Nazis? Nature, Environment, and Nation in the Third Reich*. Athens: Ohio University Press. pp. 243–256.

Wolschke-Bulmahn, Joachim (1996): Heimatschutz. In: Uwe Puschner, Walter Schmitz and Justus Ulbricht (Eds). *Handbuch zur 'Völkischen Bewegung' 1871–1918*. DeGruyter: München. pp. 533–545.

Zechner, Johannes (2009): Vom Naturideal zur Weltanschaung: Die Politisierung und Ideologisierung des deutschen Waldes zwischen Romantik und Nationalsozialismus. In: Landschaftsverband Westfalen-Lippe (Ed). *Mythos Wald*. Münster. pp. 34–41.

Zuerst! (2013): "Grüner Selbstbetrug", *Zuerst!* 6: 26–27.

14
THE ENVIRONMENT AS AN EMERGING DISCOURSE IN POLISH FAR-RIGHT POLITICS

Samuel Bennett and Cezary Kwiatkowski

Introduction

In Poland only a very few electorally significant political parties approach environmental issues directly or propose any form of consistent green politics and 'climate change is still largely unimportant unless...it appears in the discussion as a threat to the Polish economy' (Jankowska 2016: 146). Instead, both discourse and social action on the environment is largely the preserve of non-political bodies, including far-right groups.[1]

The environment, then, is a relatively new site of political and discursive contestation in Poland and is not a salient electoral issue in and of itself; awareness and concern for the environment is only recently on the rise in Poland. According to a study conducted by ARC Rynek i Opinia, in 2008 only 15% of Poles were concerned with environmental issues (Sadura and Kwiatkowska 2008: 13). In a different study from the same year, 48% of respondents agreed 'the intensive development of Poland's economy justifie[d] the increase of Poland's CO_2 emission limits' (Jankowska 2011: 163–164). By 2014, however, there had been a noticeable change in the attitudes, with 86% of respondents acknowledging the seriousness of climate change (TNS Polska 2014: 59).

Situating this within the Central and Eastern Europe (CEE) geographical context, we can see that Poland's case is possibly an outlier. Environmental concerns and even 'eco-nationalist' politics (though not of a far-right nature) have been found in the Baltic republics – including Latvia (Galbreath and Auers 2009) and Estonia and Lithuania (Malloy 2009) – as well as Slovakia (Podoba 1998) during their transition from Soviet rule. Environmental concerns also emerged in nationalist discourses both in the region (after Chernobyl) (Turnock 2004), and further afield in Europe (cf. Plaid Cymru's green nationalism in Wales, Hamilton 2002).

Given its, by and large, peripheral salience in Poland, media coverage and political (re)action to the environment are largely issue-based. That there is little coherent elite-level discourse on, for example, climate change (other than public relations surrounding United Nations Climate Change summits), means that environmental issues can be, and have been exploited by far-right actors to promote their ideologies to a wider audience. Individual environmental 'flashpoints', like other issues in Polish politics – such as immigration (Krzyżanowski 2018) – have been politicised and have emerged rapidly as highly mediatised spaces of far-right discourse production, which has been, at times supported and propagated by individual members of the governing Law and Justice party (*Prawo i Sprawiedliwość*, PiS).

Whilst others in this volume work from theoretical perspectives of, for example, populism and Barthes' myth as master frames, we take a discourse-theoretical position. Following Reisigl and Wodak (2001: 36), we understand discourse as 'a complex bundle of simultaneous and sequential interrelated linguistic acts, which manifest themselves with and across the social fields of action as thematically interrelated semiotic, oral or written tokens, very often as texts that belong to specific semiotic types, i.e. genres'. These discourses are 'socially constituted and constitutive' which are 'situated within specific fields of social action' (Reisigl and Wodak 2009: 89). That is, discourses are shaped by the rules, constraints, and contexts of multiple social fields and, conversely social action is at least in part enacted through discourse.

In this paper, we argue that there has been a diffusion of far-right discursive frames and strategies into mainstream Polish political discourse, when reacting to environmental issues. Following Kallis (2013: 221) there has been a 'mainstreaming' of far-right discourse 'that involves previously taboo ideas, frames and practices, becoming the new "common sense" for growing sections of European politics and societies.'

In the following sections we give a brief historical overview of modern Polish politics since 1989 and how far-right groups have emerged within this context. We then move on to an explanation of our data and methodology, followed by a comprehensive analysis of the data. Lastly, in the conclusion we discuss our key findings.

Historical context
The evolution of Polish politics after 1989

The end of the communist era in Poland brought about the pluralisation of the political scene, which allowed newly surfacing (and resurfacing) far-right groups to enter the fray. In the years following, successive governments represented various ideological orientations, but the far right did not constitute a major part of mainstream politics, except for the League of Polish Families (*Liga Polskich Rodzin*, LPR), a nationalist coalition partner in the right-wing government of

2005–2007. The coalition soon dissolved due to personal conflicts and this was followed by the victory of the centre-right Civic Platform (*Platforma Obywatelska*, PO), which managed to build a positive self-image of a moderate, pro-European party. The positive sheen did not last forever, though, and PO's eight-year-long rule led to a widespread feeling of disappointment with the neoliberal status quo, experienced most drastically by those who had not gained economically in the post-EU accession years (especially inhabitants of impoverished villages and small towns, the so-called 'Poland B') and who seemed to have been more eager to vote for candidates who they saw as 'anti-establishment', including Law and Justice (CBOS 2015).

The socio-political climate of post-communist Poland can be best described as socially conservative and wary of the left, with left-wing political actors usually being associated with the previous Communist regime (Kurek-Bąk 2013: 64). The devotion to conservative social values has been further reinforced by the influence of the Catholic Church. Far from diminishing in more recent years, the church's social and political influence and opinion-setting role has in fact increased since 2015.

The post-2015 shift to the right

The 2015 elections brought about a massive rightward shift in the political scene, with PO's dedication to maintaining the status quo were challenged by the triumphant Law and Justice and smaller far-right groups. As Abbass et al. (2011: 2–3) write, 'right-wing extremism in Poland is based on small quasi-political organisations', including the National Radical Camp (*Obóz Narodowo-Radykalny*, ONR), a descendant of a pre-World War Two illegal fascist party of the same name; the All-Polish Youth (*Młodzież Wszechpolska*, MW), formerly associated with LPR; and many other, smaller movements. Such groups have been successful in engaging people in pro-nationalist activism; one of the manifestations of their success has been the Independence March, an annual far-right event taking place in Warsaw on Polish Independence Day (11 November). In 2017, the event was attended by approximately 60,000 demonstrators and was full of far-right slogans, including: 'Pure Poland, white Poland' and 'Refugees get out!' (Taylor 2017). The demonstration was praised by the Interior Minister, Mariusz Błaszczak (ibid.), which seems to indicate that the far-right discourse has, in some ways, become normalised or maybe even incorporated into mainstream political discourse.

There are also parties which are more concerned with economic matters, such as New Right (*Nowa Prawica*) and Liberty (*Wolność*, formerly known as KORWiN, and led by the 'enfant terrible' of the European Parliament, Janusz Korwin-Mikke). Although KORWiN did not manage to gain any seats in Parliament, it did manage to establish a strong community of supporters through active social media use and the voicing of radical opinions (e.g. that the EU is a 'communist' organisation). These parties differ from far-right organisations, but they exhibit certain similarities when it comes to Euroscepticism and xenophobia, though this could be seen as of a right-wing populist bent.

However the ideological change can also be observed in mainstream politics. Firstly, fascist politicians associated with the National Movement (*Ruch Narodowy*, a party connected with the MW) have gained seats in Parliament. Secondly, the ruling party has drastically changed its rhetoric and their initial moderate discourse prior to the 2015 elections has been gradually replaced with more aggressive, Eurosceptic and anti-immigrant language.[2] Additionally, 'individual PiS MPs support right-wing extremist initiatives' (Abbass et al. 2011: 3). For example, Stanisław Pięta, an MP from Bielsko-Biała, co-organised a march commemorating the 'cursed soldiers' (i.e. soldiers of the post-war anti-communist resistance movements) together with local far-right organisations (Pięta 2017). To some extent, then, far-right groups, including fascist ones, do not promote ideologies that differ greatly from those of mainstream parties and they are, arguably, just more radical espousals of them. Indeed, rather than being dismissive or adversarial to far-right groups, mainstream right-wing political actors can be said to take an 'accommodative' approach to them (Mudde 2010).

Methodology and data

Although, overall, little work has been done on the analysis of far-right discourses of the environment, a number of studies should be highlighted in order to situate the Polish case in its wider geographical and political context. Barcena et al. (1997) track the evolution of the relationship between ecological and national discourse. In the United Kingdom, the importance of 'the land' to British far-right groups has been found by Richardson (2017). Elsewhere, German far-right discourses on the environment have been studied by Forchtner (2019) regarding biodiversity in the extreme-right, and Forchtner et al. (2018), who looked at the climate-change communication of far-right actors, which complements pan-European studies on far-right actors and climate scepticism by, for example Lockwood (2018) and Gemenis et al. (2012).

The far-right landscape in Poland is a mixture of traditional political actors (established political parties and their members), grassroots social movements (fascist organisations) and para-political entities which possess a combination of characteristics of both of the former. For example, on the one hand Ruch Narodowy produces party manifestoes and has professional management bodies, but at the same time, its membership and support is drawn largely of members of far-right and fascist organisations such as Mlodzież Wszechpolska and ONR (with which it signed a shared-ideology agreement), and it is also a co-organiser of the annual Independence March. Given such complexity in the fields of social and political action, it is therefore unsurprising that the sites of these actors' discursive action are equally varied. Thus, we include material from a wide range of genres.

Our entry points into gathering the data were official Facebook (FB) pages for the main right-wing, far-right, and fascist groups in Poland. We analysed

47 posts which referred to environment-related issues. The exception to this is ONR, which is banned from Facebook, and so its official webpage was used instead, along with MW's, which only has regional chapter pages on FB.

When analysing social media data, it is also important to analyse hyperlinked texts because FB posts frequently included links to other official party material (webpages, manifestos). However, there is a second, and more vital reason for doing so: the need to account for different levels of hyper-intertextuality, through which 'texts can be transformed and given multiple and alternative interpretations' (Barros 2014: 1223). Elsewhere, Bennett (2018b) has termed this 'hyper-recontextualisation', which can give hints to underlying ideologies; the linked-to texts act as both 'proxies' for the actors' claims via the text's content and legitimations (authorisation) of/for these claims via the text's genre and authorship (see Kaiser 2019 in this volume who analyses networks of climate sceptics). As Barros (2014: 1225) argues, 'the battle for legitimation goes beyond the text' and in social media 'arguments can be supported or legitimated by links to texts and text producers that are deemed authoritative' (Bennett 2018b) to the intended recipient of the post (see below for methodology and categories of analysis). In addition to FB posts and official webpages, we also take into account various traditional print media outlets that are supportive of, or even are unofficial mouthpieces of the ruling PiS government. We do this to highlight the wider transference of far-right discursive strategies into popular media and show how mainstream political actors have internalised, or at the very least, have not disowned the use of far-right language.

Using a combination of the literature review above and the large body of critically discursive work on the far right (cf. Wodak 2015; Richardson 2017), we hypothesise that the types of discursive and rhetorical strategies used by far-right actors in Poland will be similar across different topics. Methodologically, we orient ourselves to a selection of analytical tools offered by critical discourse scholars who have analysed exclusionary or far-right language. Following Reisigl and Wodak (2001) and van Leeuwen (1996) we identify and analyse how social actors (individuals, groups, and institutions) are constructed in talk and text and to accomplish this, we focus on three categories of analysis. The first two are nomination and referential strategies, in which we look at how actors' agency, role, actions and qualities are indicated through different grammatical and lexical instantiations, including pronouns, modality, verbs and adjectives (ibid.). Referential strategies are also predicative strategies; as Reisigl and Wodak (2001: 45) point out, they frequently involve 'a denotatively as well as connotatively more or less deprecatory or appreciative labelling.'

Our third category of analysis is argumentation as a form of classical rhetorical persuasion which gives a text its cohesion and function. This can be done in two ways, either via employing convincing arguments or by 'influencing somebody suggestively and manipulatively by fallacies (Reisigl 2014: 70). Specifically, we are interested in the strategic use of topoi and pragmatic fallacies. In line with other critical discourse researchers (Krzyżanowski 2010; Bennett 2018a) we follow van

Eemeren and Grootendorst's (1992: 96) understanding of topoi as: 'a more or less conventionalised way of representing the relation between what is stated in the argument and what is stated in the standpoint'. In contrast, pragmatic fallacies violate generally accepted rules of argumentation. Whilst some argumentation schemes are context and content dependent and can often become formalised and typical of certain social practices, others are universal (Reisigl 2014). It can thus be presumed that argumentation schemes identified in previous analyses of far-right and racist discourse will also be evident in discourses on the environment, especially given that the actor profile is similar.

Analysis

As mentioned in the introduction, the environment is not a stable, salient topic in the Polish political sphere and so a lot of the discourse production was of a reactive nature. That said, we noticed two trends. Firstly, that far-right organisations embedded the environment in their fundamental ideological claims and secondly, they reacted to separate issues with highly symbolic actions that were both discursive and non-discursive.

1 As well as the spiritual aspects, we emphasise the great importance of national heritage, understood as a multigenerational material and linguistic achievement, strictly connected with devotion to the homeland, especially the natural environment (…) Maintaining this natural balance and mutual complementation is what we consider vital and, through its scope, relevant to the other spheres of life of the Polish nation. (ONR 2017)
2 Earth Day is celebrated annually on 22 April in order to promote pro-environmental behaviours. For us – as nationalists – this aim is even more important. As the generation which currently inhabits the Polish land, we are neither the first, nor the last, of its inhabitants. The land is the greatest wealth we, Poles, have. Therefore, let's take care of it so that our children receive our homeland in a good state and order. We also steer clear of pseudo-environmental theories concerning, among other things, global warming. Environmentalism is much more grassroots and depends on us (Młodzież Wszechpolska 2015)

In excerpt 1, from ONR's ideological declaration, we see the traditional language of *'Blut und Boden'*, which idealizes the natural environment and privileges it over urban living. This echoes other works on historical Nazism in the United Kingdom (Richardson 2017) and Germany (Cosgrove 2004). Nature is discursively connected to national identity through the use of *'ziemia ojczysta'* (homeland), but is closer in this context to the German *'Heimat'*, which implies a love of a typically rural, idyllic homeland, as denoted here by the strong emotional noun 'devotion'. The thread linking the two excerpts is the concept of inter-generational

The environment as an emerging discourse in Polish far-right politics 243

stability – 'multigenerational', 'we are neither the first, nor the last, of its inhabitants' – of the nation and its citizens extending temporally backwards into the past and forwards into the future. This serves to reify and concretise the nation and also to legitimise certain actions, be it carrying on a tradition and/or being custodians of something that will be passed down to future generations. The plural pronoun 'our' also moves the claim from an individual responsibility to a collective one.

In excerpt 2, the (symbolic) action is cleaning forests, which is a practice of far-right organisations familiar from other contexts (see chapters by Tarant 2019, as well as Forchtner and Özvatan 2019 in this volume). Interestingly, analysis of other posts on the MW website finds that regional chapters engage in the cleaning of other symbolic spaces, for example Polish war graves. Thus, clearing rubbish is less of an environmental issue and more one concerning heritage and national collective action. This belief might be pointed to in the final two sentences of excerpt 2. In it, a comparison is made between two types of environmentalism. In the first sentence the prefix 'pseudo' is attached to 'environmentalism' as a way to delegitimise claims of global warming, which is juxtaposed to 'our' environmentalism; one that happens at the micro-level. This comparison marks out ideological differences between 'us' on the right concerned with the national and 'them' on the left, who are worried about wider, internationalist, trends. Lockwood (2018) has also found that right-wing populists often couch their rejection of climate change within claims to its universal and cosmopolitan nature, whilst Barcena et al. (1997) note that radical Basque nationalists also use local ecological frames in their campaigns for independence.

There appears to be a right way and wrong way of caring about the environment, which leads us to our second tranche of examples that focus on the delegitimisation of alternative approaches to the environment. We understand it as a *conscious discursive strategy that intertextually refers to and builds on a complex network of other ideologically grounded ad hominem attacks* against 'environmentalists'. These excerpts all come from what could be termed non-far-right actors, or at least people who are not members of fascist, far-right organisations. Yet, they use strong language and far right discursive frames that are also present in mainstream political and social actors.

3 We only want to heal our country from certain diseases. The previous government implemented a leftist ideology: as if the world had to move in only one direction towards a mixture of cultures and races, a world of cyclists and vegetarians who only focused on renewable energy sources and which combats all forms of religion, it has nothing to do with traditional Polish values (Waszczykowski, *Bild* 2016)

4 The ideology of green, atheistic, materialistic and nihilistic neo-communism aims to negate God's role as the creator of Poland, Poles, animals and plants. The neo-Marxist strand goes further, from total opposition towards God to partial animalisation of a human being, in order to eventually lead to total

annihilation. This ideology – and I know what I am saying ... is a variety of green Nazism. The only difference between it and the Third Reich is that now no human race has a right to exist. (Guz, *Gazeta Wyborcza* 2017)

Excerpt 3 is from the former Minister for Foreign Affairs Witold Waszczykowski in an interview with the German newspaper *Bild*. Typical far-right frames are set up via the metaphor of 'diseases' (i.e. Jews as rats). According to Musolff (2012: 303) the metaphor implies that, such as with diseases that attack the human body, 'any socio-political elements of the nation that threaten its existence have to be eliminated.' These 'diseases' (here: abstract, yet damaging concepts, ideologies) include another common far-right trope, the mixing of races, implying a monolithic, Polish national identity. In subsequent parts, the speaker lists other 'diseases' that run counter to 'Polish values', including renewable energy, cycling and vegetarianism.

Excerpt 4 comes from Tadeusz Guz, a priest and philosopher who was speaking at a conference organised by the then Environment Minister (Jan Szyszko) and Tadeusz Rydzyk, the owner of Radio Maryja and TWRM (Catholic-nationalist media outlets). It continues, albeit not cohesively, the previous line of argument and creates a binary us/them ideological division that links left-wing ideologies with environmentalism. The overall argument is that care for the environment is a threat to the natural order of things (a 'Polish' world created by God) and, worse, 'total annihilation'. There is a later shift where Guz reframes the subject not as Communists, but as Nazis, comparing greens to the Third Reich. Such reference to totalitarianism is a form of *reductio ad absurdum* known as *reductio ad Hitlerum* (Strauss 1953), which is a way of attaching guilt by association and automatically delegitimising a claim, actor or action. There are three additional functions to the speaker's strategic reference to Communism and Nazism. Firstly, it references two historical totalitarian enemies of Poland in the twentieth century, which is an example of topos of history, i.e. because history has shown that a certain action has a certain consequence, a given, comparable action now should or should not be performed (Reisigl and Wodak 2001: 80). Secondly, it is also a topos of fear, i.e. its totalitarian nature threatens 'our' existence and so action should be taken. Thirdly, it is a legitimising *ad hominem* attack on opponents; rather than rationally addressing environmentalist's claims, the speaker resorts to using hyperbolic referential terms.

In the above excerpts disparate opposition groups are 'lumped together' or collectivised (van Leeuwen 1996) and a single (fictional) opponent is thus discursively constructed that can best be described as a folk devil representing a moral panic that needs addressing. On a societal level this construction is intended to play well with core PiS voters and traditional Polish churchgoers, and forms part of a larger, ideologically-led political discourse (Krzyżanowski 2018) that has support in the wider media.

Figure 14.1 is from the self-styled 'liberal-conservative' magazine *Najwyższy Czas!* (High Time!), originally established by Korwin-Mikke. The front cover

The environment as an emerging discourse in Polish far-right politics **245**

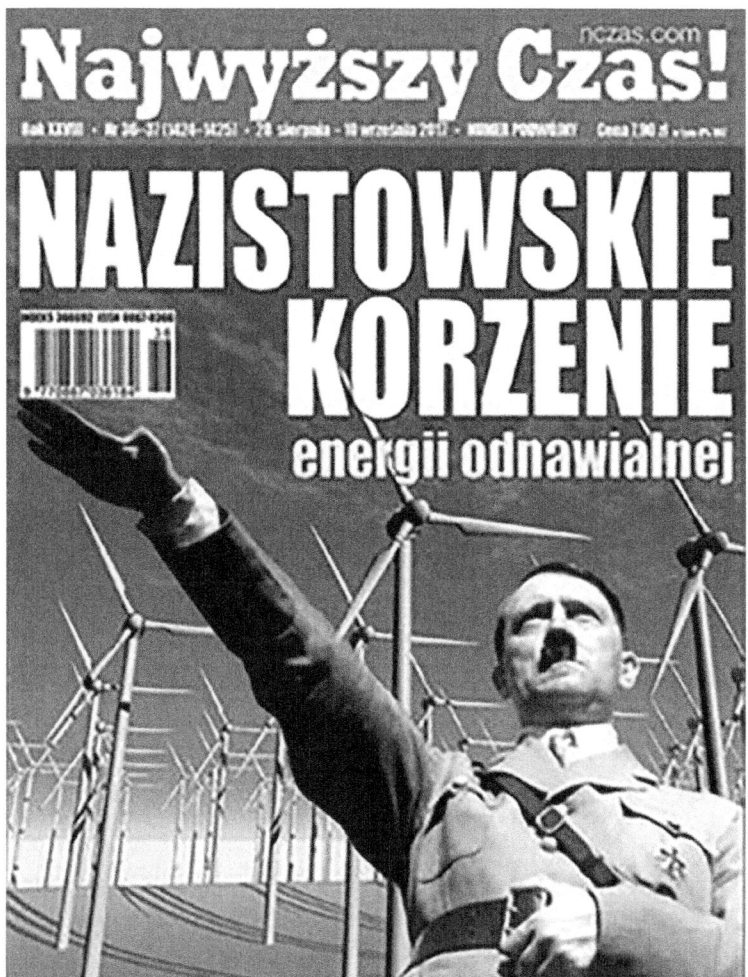

FIGURE 14.1 'Nazi Roots'.³

translates as: 'THE NAZI ROOTS of renewable energy' which tallies with excerpt 4. By contrast, a controversial headline in *Do Rzeczy* (a magazine editorially close to PiS) provides an example of how Hitler is discursively constructed of the left and thus separated from the ideological right. It reads: 'HITLER WAS A LEFTY: Contrary to Communist propaganda the Nazis had nothing in common with the Right'. In these front covers the negative connotations are transferred onto 'leftist' ideology to, first, delegitimise renewables (visually represented through windmills) as an acceptable form of energy and to, second, 'cleanse' (or maybe 're-Baptise'?) the right of its negative connotations.

We now move to analysing the discourse surrounding two specific environmental issues that have become publically salient recently: Smog and logging. The discursive strategies identified in general ideologies of far-right actors are also present in issue-specific discourse. Smog is a perennial problem in Poland, especially during winter months due to people heating their homes with coal, wood, newspaper, rubbish. In 2017, there was heightened media coverage of the issue, especially in the Małopolska (Lesser Poland) Voivodeship, where levels of dangerous particles reached seven times the healthy limit.

5. No complex analysis that would enable an effective battle against this phenomenon [smog] has been conducted yet. The decisions of local authorities are taken on the basis of speculations. Scientists have already warned that a ban on the use of coal for heating in Krakow will produce negligible results and it will be an additional burden for citizens. (Ruch Narodowy 2016)
6. Instead of speaking about Poland exceeding a billion złotys in debt, about struggling with smog in the cities and not providing money for counteracting it, we keep discussing the relationship between Marshal Kuchciński and MP Szczerba! I'm participating in a circus show. (Kukiz'15 2017)
7. Activists from the Kraków chapter of Młodzież Wszechpolska came to the rescue of acclaimed Polish figures (Młodzież Wszechpolska 2017)

In these excerpts we see three very different approaches to the issue of smog. The first (excerpt 5) is an example of a questioning of science. Previous research is delegitimised via the nominative 'speculations'. However, in the following sentence scientists are used to legitimise the negative impacts of regulating coal usage in Kraków and there is an additional instance of *argumentum ad populum* through reference to the potential consequences of the ban for citizens. By contrast, in excerpt 6, Paweł Kukiz politicises the issue of smog and uses it to make a populist, anti-establishment point. He employs an *ad hominem* attack – 'circus show' – to portray the Sejm (Parliament) as not focussing on the correct topics. 'Smog' here represents a serious problem for citizens that requires action rather than words. Lastly, in response to the smog in Krakow, the local MW chapter carried out a campaign where they 'rescued' (excerpt 7) important figures from Polish history by adorning them with anti-smog masks. In this activity we see echoes of the cleaning of forests and graves from earlier, i.e. grassroots activity to protect sites important for their worldview and which place the actors as custodians of Poland's heritage, albeit a heritage that they themselves define.

Finally, the issue of logging was also a highly mediatised issue in Poland and in EU member states. The controversy was two-fold. The first was that logging would begin in Bialowieża national park – a UNESCO heritage site and one of the last primeval forests in Europe – ostensibly to prevent the spread of disease in the trees, but Szyszko and his colleagues stood to benefit from sales of timber. The second element was the removal of restrictions on logging on private land.

8 Interestingly, those in favour of abortion shout 'my body, my concern' when fighting for the 'logging' of the unborn (…), but risk being 'cut' for the trees … (Partia Wolność 2017).
9 We must accept two assumptions (…) First, that it is man that is the subject of sustainable development, and so man has not only the right, but the duty to use natural resources. Second, that human development is not detrimental to the environment (Dybalski 2017).
10 The key fuel for the Polish electrical energy industry is, and will be, coal. Poland possesses the biggest resources in Europe of this black gold, which can help Poland ensure almost 200 years of sovereignty and energy security. (Ruch Narodowy 2016)

Responses from the far right were limited, but more mainstream right-wing actors did politicise the issue. Excerpt 8 comes from a Facebook post on Wolność's page. Similarly, to excerpts 3 and 4, the cohesion of the claim relies on a collectivisation of different actors, in this case women's rights protestors and environmental protestors in Białowieża trying to prevent logging. The claim tries to sarcastically imply the inconsistencies between support for the two causes. Following Musolff (2017) sarcasm '[flouts] the maxim of quality and thus [generates] conversational implicatures.' Reference to abortion is a strategic move to delegitimise the environmental protesters in the eyes of socially conservative supporters of Wolność. The accompanying image depicts a tree being cut down with a chainsaw along with the phrase 'my tree, my business'. This is a decontextualisation and recontextualisation of a common slogan seen during protests against anti-abortion laws in Poland occurring at around the same time. Thus, 'my body, my business', becomes 'my tree, my business', which supports Wolność's libertarian ideology. Furthermore, appropriation of the same rheme (whilst replacing the theme) also somehow challenges 'ownership' of it and thus, in a sense, disempowers, or tarnishes, the rhetorical impact of the original slogan.

Lastly, excerpt 9 is from Minister Szyszko in a comment about the logging law. The overall goal of the passage is to rationalise, and consequently legitimise logging through reference to the great chain of being (e.g. God→Man→Animals→Plants). There is a rhetorical call to collective action, which manifests in the strong pronoun + modal verb construction 'we must', the noun 'duty', and use of present tense, which reifies the claim and establishes it as stable, foundational 'truth' about the world. The same structure is also present in the final sentence where it is used to deny human impact on the environment. The environment is placed 'below' humanity and is constructed as a resource that should be exploited. This chimes with Ruch Narodowy's perspective on coal (excerpt 10), in which natural resources are discursively constructed as vital to the continuing existence of the Polish nation, despite elsewhere in their manifesto accepting there is a problem with smog (see excerpt 5). Similar discourses about the importance of natural resources to nationalist projects have been found in Denmark and the United Kingdom (Forchtner and Kølvraa 2015).

Discussion and conclusion

In bringing this chapter to a close, we want to now draw together the analysis and place it within the current socio-political context in Poland. Regarding the actual form and content of the discourse, four findings stand out. Firstly, far-right actors, especially MW, often took 'symbolic' (non-discursive) social action and the accompanying texts on websites or social media are recontextualisations of these social actions. Such actions are ideologically coherent with fascist doctrine, 'which conceive[s] politics as pure political activism, and as direct, immediate, decisive action' (Froio and Castelli Gattinara 2016: 1044). In this sense they are also visual performances, or 'ritual affirmations' (Schedler 2014: 241) of the group's collective identity, key values and, 'emotional orientations' (Fahlenbrach 2008: 99). This can be seen in some way as an example of Wodak's (2015: 20) idea of a 'right-wing populist *perpetuum mobile*' in which 'provocative utterances' are made and are reacted to by the media. However, as hybrid entities that straddle the divide between political parties and social movements, these actions are more than just performances, but populist moves which seek to highlight the supposed inefficiency of governments (at any level) in working in the best interests of 'the people'.

Secondly, mainstream actors, or least those linked to mainstream parties constructed a 'folk devil' as a *pars pro toto* for anyone opposing their ideology. This is part of a wider, and ever-increasing, us/them division in Polish political discourse, engendered from the top by Jarosław Kaczyński and mediated via state controlled television, religious media outlets such as Radio Maryja, and closely connected periodicals. There are 'good Poles' and 'bad Poles', or in Kaczyński's words 'the worst sort' who are 'guilty of national treason' (Czornak 2015). By being linked to totalitarianism, environmentalists are constructed as an internal enemy that challenges Polish values and endangers the nation, and so needs to be eradicated.

Thirdly, there was little ideological cohesion in far-right discourses of the far right in Poland. This can be explained by the disparate nature of the right that includes fascist groups, libertarians and right-wing populists. Environment issues are not a natural comfort zone for such ideologies and so discourse production on the environment largely occurs in reaction to specific issues (smog, logging, etc.). Political antagonism is mapped onto a non-party political issue, instrumental in making political points – i.e. attacking the EU for being overbearing, attacking the government for doing nothing, or attacking opponents for going against 'Polish values'.

The data, taken from different political and quasi-political organisations, indicates the wide spectrum of environmental positions. Other than PiS, these differences are, we argue, largely ideologically grounded rather than predicated on the potential for electoral success. In this sense at least, the environment is a stable, albeit background issue. Radical and extreme right groups – MW and ONR – have similar discourses that are close to 'traditional' fascist frames of interpretation. The noticeable difference between them is their approach to protecting the environment as MW is much more social active: their discourse is aligned with their praxis, whereas ONR's ventures into environmentalism remain at the ideological and declaratory stage. By contrast, Ruch Narodowy can be placed somewhere between these organisations and more the mainstream right-wing in Polish politics. Rather than this difference being visible at the level of ideology (their rejection of climate change along cosmopolitan/trans-national grounds echoes that of MW), it is present in the genre and register of their discourse, i.e. more formalised documents such as manifestos. Thus, the medium is an important part of the message. Climate change denial is also present in libertarian political discourses (Wolność and Nowa Prawica, as well as publications closely tied to them), but there is, unsurprisingly more emphasis on the individual freedom to use natural resources and any attempt to deny this is seen as 'meddling' from above, be it at a national or EU level.

This leaves the governing Law and Justice party, which remained largely absent in our data. The only relevant Facebook posts found concerned the signing of the Paris Climate Agreement, stating that Poland's interests should be protected by any agreement. In this, we see the balancing act that the government has at times tried to maintain with regard to its global reputation. Indeed, much of their domestic-facing words and deeds appear to be aimed at firming up support from the right, evidence of which can be seen in their European election victory in May 2019 when a confederation of far-right parties, led by Korwin-Mikke lost all of its four seats. At the same time though, their outward-facing rhetoric is essentially an attempt at "brand management" on the global stage. The mixed messages from PiS on the environment continue: at the Global Climate Action panel in late 2018 the Prime Minister, Mateusz Morawiecki claimed that Poland was a leader in the fight against global warming (Rzeczpospolita 2018). Yet, in June 2019, Poland (along with Czechia, Hungry and Estonia) vetoed the European agreement on carbon neutrality by 2050 and in doing so, Morawiecki claimed, "we have secured the interests of Polish citizens (Morawiecki: Polska liderem walki z globalnym ociepleniem 2019). However, in its dealings with the European Commission over logging in Białowieża, as well as its recalcitrance on other matters such as judicial reform and immigration, we see an ever-greater rejection of outside interference and a positioning of the environment as a national issue. Moreover, it is an issue that is secondary to economic development and energy efficiency, which is similar, at a policy level, with Ruch Narodowy. Thus, whilst official PiS party communication channels remained largely silent, mainstream actors and those close to them integrate, replicate, and leave unchallenged far-right language in their discursive action on the environment. Far from erecting a *cordon sanitaire* (Littler and Feldman 2017) around far-right actors, whereby they are 'either formally or informally deprived of a media platform' (ibid.: 512), mainstream right-wing actors have accommodated, adopted, and adapted far-right language.

With nationalism perceived as positive (CBOS 2016: 17) and far-right discourse being spread via colonised public media, the rightward shift in Poland is not surprising, and environmental discourse is another example of the diffusion of far-right frames into mainstream politics (Abbass et al. 2011; Krzyżanowski 2018) and public life in Poland.

Notes

1 Pro-environmental postulates were present during the Round Table Talks in 1989, but they did not leave any significant mark on the political life of the newly-neoliberal Poland, in which the effectiveness of the economy became a top priority (Charkiewicz 2008: 42).
2 Krzyżanowski (2018) notices this 'discursive shift' in immigration and provides examples of the far-right language strategically enacted by PiS.
3 http://nczas.com/2017/08/27/najwyzszy-czas-numer-podwojny-36-372017-odponiedzialku-w-kioskach-oraz-online/ [5 June 2018].

References

Abbass, Merin, Tvrdá, Kateřina, Walach, Václav, Rydliński, Bartosz and Nociar, Tomáš (2011): *Right-wing Extremism in Central Europe: An Overview*. Berlin: Friedrich-Ebert-Stiftung.
Absurdalne wystąpienie na konferencji Szyszki i o. Rydzyka. 'Ekologia odmianą zielonego nazizmu' (2017): *Gazeta Wyborcza*, 4 June 2017. http://wiadomosci.gazeta.pl/wiadomosci/7,114884,21910528,absurdalne-wystapienie-na-konferencji-szyszki-i-o-rydzyka.html [24 April 2018].
Barcena, Iñaki, Ibarra, Pedro and Zubiaga, Mario (1997): The evolution of the relationship between ecologism and nationalism. In: Michael Redclift and Graham Woodgate (Eds). *The International Handbook of Environmental Sociology*. Cheltenham: Edward Elgar. pp. 300–315.
Barros, Marcos (2014): Tools of legitimacy: The case of the Petrobras corporation blog, *Organizational Studies*, 35(8): 1211–1230.
Bennett, Sam (2018a): *Constructions of Migrant Integration in British Public Discourse: Becoming British*. London: Bloomsbury.
Bennett, Samuel (2018b): New 'crises,' old habits: Online interdiscursivity and intertextuality in UK migration policy discourses, *Journal of Immigrant and Refugee Studies*, 16(1–2): 140–160.
Centrum Badania Opinii Społecznej – CBOS (2015): *Kim są wyborcy, czyli społecznodemograficzne portrety największych potencjalnych elektoratów* [Who the voters are, i.e. the socio-demographic portraits of the largest potential electorates]. Warszawa: CBOS.
CBOS (2016): *Między patriotyzmem a nacjonalizmem* [Between patriotism and nationalism]. Warszawa: CBOS.
Charkiewicz, Ewa (2008): Zielony finał PRL-u [A green finale of the Polish People's Republic]. In: Przemysław Sadura (Ed). *Polski odcień zieleni: Zielone idee i siły polityczne w Polsce* [A Polish shade of green: Green ideas and political forces in Poland]. Warszawa: Heinrich-Böll-Stiftung. pp. 35–48.
Cosgrove, Denis (2004): Landscape and Landschaft. Lecture delivered at the 'Spatial Turn in History' Symposium, German Historical Institute, 19 February 2004, *GHI Bulletin*, 35, 57–71.
Czornak, Michał (2015): 'Najgorszy sort Polaków'? Co tak naprawdę powiedział Jarosław Kaczyński? [The 'worst sort of Poles'? What did Jarosław Kaczyński really say?], *wMeritum*, 14 December 2015. http://wmeritum.pl/najgorszy-sort-polakow-co-tak-naprawdepowiedzial-jaroslaw-kaczynski/130074 [23 April 2018].
Dybalski, Jakub (2017): Jan Szyszko o zrównoważowanym rozwoju: Polska to unikat [Jan Szyszko on sustainable development: Poland is unique]. *Rynek Infrastruktury*, 19 February 2017. www.rynekinfrastruktury.pl/wiadomosci/drogi/jan-szyszko-o-zrownowazonymrozwoju-polska-to-unikat-57394.html [28 April 2018].

Fahlenbrach, Kathrin (2008): Protest-Räume – Medien-Räume. Zur rituellen Topologie der Straße als Protest-Raum. In: Sandra M. Geschke (Ed). *Straße als kultureller Aktionsraum. Interdisziplinäre Betrachtungen des Straßenraums an der Schnittstelle zwischen Theorie und Praxis.* Wiesbaden: Verlag für Sozialwissenschaft. pp. 98–110.

Forchtner, Bernhard (2019): Nation, Nature, Purity: Extreme-right Biodiversity in Germany, *Patterns of Prejudice. Special Issue Cultural Imaginaries of the Extreme Right*, 53(2).

Forchtner, Bernhard and Kølvraa, Christoffer (2015): The nature of nationalism: Populist radical right parties on countryside and climate, *Nature and Culture*, 10(2): 199–224.

Forchtner, Bernhard and Özvatan, Özgür (2019): Beyond the 'German forest': Environmental communication by the far right in Germany. In: Bernhard Forchtner (Ed). *The Far Right and the Environment: Politics, Discourse and Communication.* London: Routledge.

Forchtner, Bernhard, Kroneder, Andreas and Wetzel, David (2018): Being skeptical? Exploring far-right climate-change communication in Germany, *Environmental Communication*, 12(5): 589–604.

Froio, Caterina and Castelli Gattinara, Pietro (2016): Direct social actions in extreme right mobilisations: Ideological, strategic and organisational incentives in the Italian neo-fascist right, *PArtecipazione e COnflitto*, 9(3): 1040–1066.

Galbreath, David J. and Auers, Daunis (2009): Green, black and brown: Uncovering Latvia's environmental politics, *Journal of Baltic Studies*, 40(3): 333–348.

Gemenis, Kostas, Katsanidou, Alexia and Vasilopoulou, Sofia (2012): The politics of anti-environmentalism: Positional issue framing by the European radical right. *Paper prepared for the MPSA Annual Conference*, 12–15 April 2012, Chicago, IL. http://dl.dropbox.com/u/4736878/MPSA.pdf [30 June 2018].

Hamilton, Paul (2002): The greening of nationalism: Nationalising nature in Europe, *Environmental Politics*, 11(2): 27–48.

Jankowska, Karolina (2011): Poland's climate change policy struggle: Greening the East? In: Rüdiger K.W. Wurzel and James Connelly (Eds). *The European Union as a Leader in International Climate Change Politics.* London: Routledge. pp. 163–178.

Jankowska, Karolina (2016): Poland's clash over energy and climate policy: Green economy or grey status quo? In: Rüdiger K.W. Wurzel, James Connelly and Duncan Liefferink (Eds). *The European Union in International Climate Change Politics: Still Taking a Lead?* London: Routledge. pp. 145–158.

Kaiser, Jonas (2019): In the heartland of climate scepticism: A hyperlink network analysis of German climate sceptics and the US right-wing. In: Bernhard Forchtner (Ed). *The Far Right and the Environment: Politics, Discourse and Communication.* London. Routledge.

Kallis, Aristotle (2013). Far-right 'contagion' or a failing 'mainstream'?: How dangerous ideas cross borders and blur boundaries, *Democracy and Security*, 9(3): 221–246.

Krzyżanowski, Michał (2010): *The Discursive Construction of European Identities: A Multi-level Approach to Discourse and Identity in the Transforming European Union.* Frankfurt am Main: Peter Lang.

Krzyżanowski, Michał (2018): Discursive shifts in ethno-nationalist politics: On politicization and mediatisation of the 'refugee crisis' in Poland, *Journal of Immigrant and Refugee Studies*, 16(1–2): 76–96.

Kukiz'15 (2017): Blog post, 10 January 2017. www.facebook.com/KlubPoselskiKukiz15/videos/572296589639263/ [24 April 2018].

Kurek-Bąk, Joanna (2013): Transformacja polskiego systemu partyjnego po 1989 roku [The transformation of the Polish party system after 1989], *Studia Politicae Universitatis Silesiensis*, 10: 53–82.

Littler, Mark and Feldman, Matthew (2017): Social media and the *cordon sanitaire*: Populist politics, the online space, and a relationship that just isn't there, *Journal of Language and Politics*, 16(4): 510–522.

Lockwood, Matthew (2018): Right wing populism and climate change: Exploring the linkages, *Environmental Politics*, 27(4): 712–732.

Malloy, Tove H. (2009): Minority environmentalism and eco-nationalism in the Baltics: Green citizenship in the making? *Journal of Baltic Studies*, 40(3): 375–395.

Młodzież Wszechpolska (2015): Kołobrzeg: Dbamy o środowisko [Kołobrzeg: We are taking care of the environment]. Blog post, 27 April 2015. http://mw.org.pl/2015/04/kolobrzeg-dbamy-o-srodowisko/ [24 April 2018].

Młodzież Wszechpolska (2017): Kraków: Akcja antysmogowa. Wszecpolacy chronią bohaterów przed smogiem [Kraków: Anti-smog action. The All-Polish Youth protect heroes from smog]. Blog post, 12 January 2017. http://mw.org.pl/2017/01/krakow-akcja-antysmogowa-wszechpolacy-chronia-bohaterow-smogiem/ [24 April 2018].

Morawiecki: Polska liderem walki z globalnym ociepleniem (2018). *Rzeczpospolita*, 4 December 2018. www.rp.pl/Rzad-PiS/181209813-Morawiecki-Polska-liderem-walki-z-globalnym-ociepleniem.html [24 June 2019].

Mudde, Cas (2010): The populist radical right: A pathological normalcy, *West European Politics*, 33(6): 1167–1186.

Musolff, Andreas (2012): The study of metaphor as part of critical discourse analysis, *Critical Discourse Studies*, 9(3): 301–310.

Musolff, Andreas (2017): Metaphor, irony and sarcasm in public discourse, *Journal of Pragmatics*, 109: 95–104.

Obóz Narodowo-Radykalny – ONR (2017): *Deklaracja ideowa* [Ideological declaration]. www.onr.com.pl/deklaracja-ideowa/ [24 April 2018].

Partia Wolność (2017): Blog post, 22 February 2017. www.facebook.com/jkm.wolnosc/photos/a.760499060693456.1073741828.757876837622345/1252725248137499/?type=3&theater [24 April 2018].

Pięta, Stanisław (2017): Marsz Żołnierzy Wyklętych w Bielsku-Białej [The march of the Cursed Soldiers in Bielsko-Biała]. Blog post. http://spieta.com.pl/marsz-zolnierzy-wykletych-w-bielsku-bialej/ [2 January 2018].

Podoba, Juraj (1998): Rejecting green velvet: Transition, environment and nationalism in Slovakia, *Environmental Politics*, 7(1): 129–144.

Polen-Merkel am EU-Pranger: Außenminister Waszczykowski verteidigt Vorgehen gegen Medien und Justiz (2016). *Bild*, 3 January 2016. www.bild.de/politik/ausland/polen/eu-kommissar-will-polen-unter-aufsicht-stellen-43997696.bild.html#fromWall [13 April 2018].

Reisigl, Martin (2014): Argumentation analysis and DHA: A methodological framework, in: Chris Hart and Piotr Cap (Eds). *Contemporary Critical Discourse Studies*. London: Bloomsbury. pp. 67–96.

Reisigl, Martin and Wodak, Ruth (2001). *Discourse and Discrimination: Rhetorics of Racism*. London: Routledge.

Reisigl, Martin and Wodak, Ruth (2009): The discourse-historical approach. In: Ruth Wodak and Michael Meyer (Eds). *Methods of Critical Discourse Analysis*. London: Sage. pp. 97–121.

Richardson, John E. (2017). *British Fascism: A Discourse-historical Analysis*. Stuttgart: ibidem Press.

Ruch Narodowy (2016): Suwerenny naród w XXI wieku: Program Ruchu Narodowego [Sovereign nation in the 21st century: National Movement's programme]. http://ruchnarodowy.net/wp-content/uploads/Program-Ruchu-Narodowego.pdf [13 April 2018].

Sadura, Przemysław and Kwiatkowska, Agnieszka (2008): Zielona polityka w społeczeństwie postpolitycznym [Green politics in a post-political society]. In: Przemysław Sadura (Ed). *Polski odcień zieleni: Zielone idee i siły polityczne w Polsce* [A Polish shade of green: Green ideas and political forces in Poland]. Warszawa: Heinrich-Böll-Stiftung. pp. 9–20.

Schedler, Jan (2014): The devil in disguise: Action repertoire, visual performance and collective identity of the Autonomous Nationalists, *Nations and Nationalism*, 20(2): 239–258.

Sobczak, Krzysztof (2019): Polska poparła weto w sprawie neutralności klimatycznej w UE [Poland supports veto on climate neutrality in EU]. *Prawo.pl*, 21 June 2019. www.prawo.pl/biznes/neutralnosc-klimatyczna-polska-zablokowala-decyzje-szczytu-ue,434508.html [24 June 2019].

Strauss, Leo (1953): *Natural Right and History*. London: The University of Chicago Press.

Tarant, Zbyněk (2019): Is brown the new green? The environmental discourse of the Czech far-right. In: Bernhard Forchtner (Ed). *The Far Right and the Environment: Politics, Discourse and Communication*. London: Routledge.

Taylor, Matthew (2017): 'White Europe': 60,000 nationalists march on Poland's independence day. *The Guardian*, 12 November 2017. www.theguardian.com/world/2017/nov/12/white-europe-60000-nationalists-march-on-polands-independence-day [2 January 2018].

TNS Polska (2014): Badanie świadomości i zachowań ekologicznych mieszkańców Polski: Badanie trackingowe – pomiar: październik 2014. Raport TNS Polska dla Ministerstwa Środowiska [A study of the environmental awareness and behaviours of the inhabitants of Poland: A tracking study from October 2014. A TNS Poland report prepared for the Ministry of Environment]. www.mos.gov.pl/g2/big/2014_12/fe749deb7e1414bf1c4afbc6548300f9.pdf [2 January 2018].

Turnock, David (2004): *The East European Economy in Context: Communism and Transition*. Taylor & Francis Group.

van Eemeren, Frans H. and Grootendorst, Rob (1992). *Argumentation, Communication and Fallacies*. Hillsdale, NJ: Lawrence Erlbaum.

van Leeuwen, Theo (1996): The representation of social actors. In: Carmen Rosa Caldas-Coulthard and Malcolm Coulthard (Eds). *Texts and Practices: Readings in Critical Discourse Analysis*. London: Routledge. pp. 32–70.

Wodak, Ruth (2015). *The Politics of Fear: What Right-wing Populist Discourses Mean*. London: Sage.

PART V
Beyond Europe

15

IN THE HEARTLAND OF CLIMATE SCEPTICISM

A hyperlink network analysis of German climate sceptics and the US right wing[1]

Jonas Kaiser

Introduction

Climate scepticism is, by any means, a fringe position in Germany that was not represented in the public discourse for the most part (Engels et al. 2013; Kaiser and Rhomberg 2016). Only in recent years have climate sceptic positions been picked up politically by the far-right party *Alternative für Deutschland* (AfD; Alternative for Germany) which is, for example, questioning mankind's influence on climate change. Similarly, prominent climate sceptic bloggers have covered the AfD's platform closely and even wrote about their experiences on their blogs.[2,3]

This connection is interesting for two reasons: on the one hand, a survey in Germany was not able to find a connection between political ideology and climate scepticism (Engels et al. 2013). While this is different in other countries like the United States where a conservative ideology is connected to climate scepticism (Leiserowitz et al. 2013; McCright et al. 2016), it signals that in the German case climate scepticism cannot be explained with phenomena like partisan ideology sufficiently. On the other hand, this might signal a change: a potential political appropriation of a minority belief. Indeed, in a study Kaiser and Puschmann (2017) have conducted, the authors were able to show a connection between German climate sceptics and the far-right (see also Forchtner and Kølvraa 2015; Lockwood 2018) and other niche beliefs that they called an alliance of 'antagonism,' that is, a community of societally excluded positions that band together based on their societal exclusion and/or overlapping views. And as the sceptics' attempts in reaching the mainstream predate the AfD's political success in recent years, German sceptics had to look to other places for support and inspiration – namely the United States where climate scepticism has been and still is more prominent politically and societally (Leiserowitz et al. 2013). There are, for example, numerous well-funded right-wing think tanks like the Heartland Institute, George C. Marshall Institute or the CATO Institute (Jacques et al. 2008; Elsasser and Dunlap 2013) as well as far-right conspiracy websites like Infowars.com that may offer resources, for example in the form of talking points, and support. This, then, shows that German sceptics are being pulled to the right in the climate sceptic discourse. The question is just how far right?

One important – if not the most important – tool for German sceptics to connect to fellow sceptics is the internet which not only allowed them to find resources but also to forge alliances (Schäfer 2012; Lörcher and Taddicken 2015; Kaiser 2017). As this book's focus is on the 'far-right', I will mostly focus on actors from the far-right (e.g. Forchtner et al. 2018) but will also highlight more prominent right-wing actors like the Heartland Institute.

In this study, I want to answer the following question: What role do international right-wing and far-right sites play for German climate sceptics? To answer this question, I will use an explorative approach: First, I first want to empirically understand which communities are relevant in the climate for German climate sceptics based on a hyperlink network analysis. Secondly, I will highlight how prevalent the connection between German climate sceptic websites and blogs and the international right-wing and far-right is – especially in the United States – to, finally, highlight the especially prominent far-right cases. Based on a hyperlink network analysis, which I conducted for this study (45,117 nodes/sites, 77,674 edges/sites) I will highlight the international actors climate sceptics refer to, and contextualize them in the German context. In doing so, this chapter will achieve two things: (1) I will shine a light on the role the international right-wing and far-right plays for German climate sceptics. (2) In profiling the prominent international right-wing and far-right sites, I will give context for the German climate sceptics and their connection to the right wing and far-right.

Theoretical background

The linking between websites and the ensuing formation of networks can be best understood against the theoretical background of the *networked public sphere* (Benkler 2006). The networked public sphere posits that the internet is able to facilitate and encourage the formation of online publics. For boyd (2008: 38) networked publics are 'constructed through networked technologies' and can be characterized by an 'imagined collective that emerges as a result of the intersection of people, technology and practice.' This highlights both infrastructural and perceived connections. Indeed, 'imagined collective' refer to Anderson's (2006[1983]) concept of 'imagined communities' that describes the feeling of belonging to a community based on an imagined shared identity. boyd argues that the feeling of an 'imagined collective' is similar to what users experience online and which contributes to the formation of a collective identity. Collective identity is, as Rauchfleisch and Kovic (2016) argue convincingly, one of the key functions of a public. However, the networked public sphere highlights not only the networked character of publics, that are all connected to each other on one level or the other (e.g. directly or via so-called weak links) and thus have the potential to influence the mainstream public sphere, but also allows marginalized and/or less-represented actors to coalesce and influence the agenda (Benkler et al. 2015; Kaiser 2017). In this sense, the internet allows them to, as Fraser (1990: 68) puts it, find 'spaces of withdrawal and regroupment [which] also function as bases and training grounds for agitational activities directed toward wider publics'.

As climate scepticism is rare among the German populace (Engels et al. 2013), the internet offers sceptics a way to find a shared space. Online, sceptics can find allies both domestically as well as internationally, strategize, and attempt to reach the broader public sphere. In an analysis of English-language sceptic blogs, Sharman (2014: 167) shows that some especially prominent blogs are not only acting as translators between scientific research and lay audiences, but, in their reinterpretation of existing climate science knowledge claims and critique of scientific institutions, are acting themselves as alternative public sites of expertise for a climate sceptical audience.

They thus function as alternative media for the climate sceptic public. In a similar study, Elgesem et al. (2015) first analysed the communities English-language climate blogs form with a hyperlink network analysis and, then, made use of topic modelling to identify the distinct topics that were present in these communities. Their main finding stresses a distinct polarization between sceptics and 'accepters' as most communities that they identified either consisted of sceptic or mainstream blogs. Those communities mostly dealt with political issues. In addition, they identified one community that focused on the topic of climate science and which functioned as a bridge between sceptics and mainstream. In an analysis of blog rolls and link lists, Kaiser (2017) mapped the German-language climate network and was able to identify three distinct communities: the administrative/mainstream, climate activist, and climate sceptic community. Based on a similar analysis, Kaiser and Puschmann (2017) show that the German-language climate network is highly polarized into climate sceptic sites which are mostly run by individuals on the one side and mainstream sites on the other which are mostly run by organizations like NGOs, universities or the government. In addition, we identify that the sceptic community also includes far-right sites. In a similar analysis, Häussler et al. (2017) look at the linking practices between German and US climate websites and conclude that 'hyperlink communication is responsive to the political context, and that countermovements, in particular, manage to reap the benefits from online communication mobilization efforts'. They, however, did not take a closer look at the role of the US right-wing for German climate sceptics.

Against the background of this book's topic and the theoretical assumptions and empirical findings laid out above, it is important to understand if a connection between the German climate sceptics and the US far-right exists and if so, what the connection looks like. As established in the introduction, the overarching research question is: What role do international right-wing and far-right sites play for German climate sceptics? To answer this question, I conducted a hyperlink network analysis on the basis of German climate sceptic sites. To tease out this connection, I will use an iterative explorative approach.

In a first step, I am interested in the network's general structure, that is, the communities that can be detected. Furthermore, I aim to identify the relevant international actors, and especially US sites, for German sceptics. I thus ask:

RQ1: What are the prominent communities in the climate network?

RQ2: What is the relationship between German and US sites and what are the prominent US sites within the German climate network?

But as studies like Elgesem, Steskal and Diakopoulos' (2015) have shown, we have to consider the topics that bring sites together, that is, if German sceptics link to sites that go beyond a validation of climate sceptic beliefs. In a next step, I am thus especially interested to find out:

RQ3: Are there websites from the US political (far-)right that that are relevant for German climate sceptics and if so which ones are these?

Finally, based on research by Kaiser and Puschmann (2017) and in line with this book's overarching theme, I am interested in the role far-right sites play for German climate sceptics.

RQ4: What role do far-right sites play for German climate sceptics (per indegree)?

To answer these questions, I will describe this study's method in the next section, to then present the results.

Methods

In line with research into the networked public sphere, I understand hyperlinks as indicators of association between websites and thus will analyse the communities that these linking practices form (for example Adamic and Glance 2005; Benkler et al. 2015; Kaiser et al. 2016; Kaiser 2017). To uncover the potential connections between German climate sceptics and the US right-wing, I relied on my prior research on the German-language climate hyperlink-network (Kaiser 2017). More specifically, I focused on the climate sceptic community that I had identified and selected every website within that community that had both 1 indegree and 1 outdegree, as these can be understood to be actively contributing to the formation of the community. This resulted into 130 unique websites, which I used as starting seeds for my web crawl. I then queried these websites in R and extracted every link in their html files. As some of the websites were unavailable or shutdown, I was able to capture links for 106 of those 130 websites. As I was not interested in scraping complete websites as this would have been both inconsiderate for the site owners and computationally intensive, I only selected websites that were linked to directly (that is, their domain; for example, *Spiegel*.de, the online presence of a main German-speaking news magazine) or on the first level of the website (e.g. Spiegel.de/*politik*). This resulted in 9,164 unique links. I then repeated the first step for the 9,164 links, which resulted in over 1.9 million links. These links included everything from the links I wanted to extract, to pdf files, individual YouTube videos or site analytics (that is, all links that are not particularly helpful in a network analysis). I further reduced the websites to their domain with the R package *urltools* and deleted duplicate edges (that is, duplicate links between, for example, *Spiegel*.de and *NYTimes*.com). Next, I deleted social media, search, ad, e-commerce, or analytic platforms (e.g. Facebook, Twitter, Google, etc.), as these tend to skew the network structure heavily.

The result was 45,117 pages (nodes) with 77,674 links (edges); the *full network* (see Figure 15.1). Next, I extracted the *core network* via a community detection algorithm (Blondel et al. 2008) and an indegree filter of 5. This way I could be sure that the sites within the network were linked to by different websites and could be considered relevant actors within the network (e.g. sites like the far-right Daily Stormer got filtered this way, as it got linked to only twice within the network). I also tested reducing the network with k-core decomposition (Alvarez-Hamelin et al. 2005), but as I was mostly interested in keeping the core network intact and removing the fringe communities (e.g. ad networks), I opted for a different approach. The core network can already be analysed visually as some more distinct communities like the climate one can be identified (Figure 15.1). Since I am especially interested in the potential relationship between the climate communities and the US right-wing, I removed some communities like the mass media one that drew too many inlinks and thus potentially skewed the community detection due to their central position within the network. This further reduction resulted in the *climate network*, which consists of 906 nodes. Of these nodes, 43 were part of the starting seed sites. The comparison with the starting seeds is fruitful for two reasons: First, it highlights the diversity of sites within the network, and second, it shows that although the network has been reduced quite heavily, the starting sites are still an integral part of it. To further understand the connection between German and US sites I queried the servers, that hosted the 906 websites with R to get their IPs, which then allowed me to get their location data. The visualizations of the networks were done in Gephi.

The crawl, IP gathering and analysis of the websites was conducted in April 2018. This is important, as hyperlinks are temporally dependent, that is the links are subject to change depending on the ongoing discourse (Ackland and Gibson 2004; Adamic and Glance 2005; Kaiser and Puschmann 2017). Although most results are consistent with prior research, it is important to keep the temporal aspect in mind.

FIGURE 15.1 Reduction of full network and comparison with seed sites (full network = 45,117 nodes, 77,674 edges, 106 seed sites; core network = 1,747 nodes, 13,155 edges, 61 seed sites; climate network = 906 nodes, 4906 edges, 43 seed sites).

Results

Communities

To make sense out of the climate network and to understand the role that right-wing sites play for German climate sceptics, I first identified the different communities using the Louvain algorithm (Blondel et al. 2008). Then, I looked at the most 'linked to' websites within the communities to get an understanding of the topics that are relevant for these communities (Table 15.1). I then labelled the communities based on these websites. For some communities, this

TABLE 15.1 Labelled communities per top five websites per indegree in each modularity class

Label	Modularity class	Website	Indegree
(Far-)right news & conspiracy	0	alles-schallundrauch.blogspot.com	31
		deutsch.rt.com	23
		epochtimes.de	21
		deutsche-wirtschafts-nachrichten.de	20
		volksbetrugpunktnet.wordpress.com	14
Far-right & conspiracy	1	kopp-verlag.de	40
		lupocattivoblog.com	27
		marialourdesblog.com	24
		derhonigmannsagt.wordpress.com	21
		hintergrund.de	21
Critical & left-wing blogs	2	heise.de	35
		nachdenkseiten.de	29
		de.sputniknews.com	26
		net-news-express.de	25
		globalresearch.ca	20
General conspiracy	3	wahrheiten.org	18
		wissensmanufaktur.net	17
		yoice.net	13
		zeitgeist-online.de	12
		infokrieg.tv	11
Economics, religion, & conspiracy	4	mmnews.de	21
		infokriegernews.de	13
		iknews.de	12
		conservo.wordpress.com	11
		michael-mannheimer.net	11
Anti-feminism	5	sciencefiles.org	18
		danisch.de	16
		gesetze-im-internet.de	11
		jungefreiheit.de	11
		de.wikimannia.org	10
Climate change	6	wattsupwiththat.com	29
		nature.com	19
		eike-klima-energie.eu	17
		realclimate.org	17
		ipcc.ch	16

was easier than for others. I will briefly describe the communities and give specific websites as examples, to make it more transparent how I picked the labels. Given that climate change scepticism is prevalent throughout the network, I will also highlight when one of these websites denies or questions man-made climate change.

The most distinct community in the network is the one I labelled '*climate change*'. It features both sceptic blogs like Wattsupwiththat (US) or EIKE (GER) but also mainstream sites like Nature, RealClimate or the Intergovernmental Panel on Climate Change (IPCC). Indeed, it would be wrong to label this community as sceptic, as it includes too many mainstream sites that deal with climate change and especially climate science. Other prominent sceptic actors are the right-wing Heartland Institute and the libertarian Cato Institute (Jacques et al. 2008; Elsasser and Dunlap 2013). As there are also news outlets in there, for example The Daily Caller or Slate but also the far-right site AmericanThinker. com, I picked the label climate change and not climate science. This is mostly in line with Kaiser and Puschmann's (2017) analysis of the German climate network, as well as with Elgesem et al.'s (2015) finding that climate science could be a bridge between mainstream and sceptics. Indeed, this highlights that sceptics are by no means part of an 'echo chamber' (Sunstein 2001), but rather are very aware of the mainstream's climate discourse, as their links to mainstream sources demonstrate. This does, of course, not imply that they are participating in the general climate 'discourse' but rather that they are aware of it; one can, for example, link to a climate journal to mock a new paper.

Another community that is relatively straightforward to identify is the '*anti-feminism*' community. Indeed, most sites in the community deal with what they call 'gender' question, that is, they criticize feminism, political correctness, and gender studies in particular. The most 'linked to' site within this community, Sciencefiles.org, is not only climate sceptic but also has, for example, a 'gender trash ranking' of German universities that have to be avoided, which is based on the number of professors in gender studies relative to professors in science theory. It is important to note, that the 'anti-feminism' is not apolitical but also far-right (one of the most linked to sites in that cluster is the far-right newspaper *Junge Freiheit*). Indeed, Ging (2017) highlights the misogyny in the so-called 'manosphere' and Kaiser and Rauchfleisch (2018) showcase their connection to the far-right in the United States.

The next community '*economics, religion, & conspiracy*' is both connected to the climate change community, the anti-feminism community, and the more densely connected political communities (Figure 15.2). It includes websites that focus on economic news, libertarian world views, conspiracy sites, as well as religious topics. Infokriegernews.de (literally translated Info Warrior News), for example, covers economic as well as global news, but also writes about the 'climate lie'. However, there are also blogs in the community like conservo.wordpress.com, which cover a wide variety of topics including religion (see also Ecklund et al. 2016 for a study on the connection between religion and attitudes towards climate change) and, as a result, the community also includes some Catholic blogs and Katholisch.de, the German Catholic church's official site.

The remaining four communities are somewhat harder to differentiate and, consequently, to label. Indeed, conspiracy theories about chemtrails, 9/11, or the New World Order are present to a varying degree in at least five communities

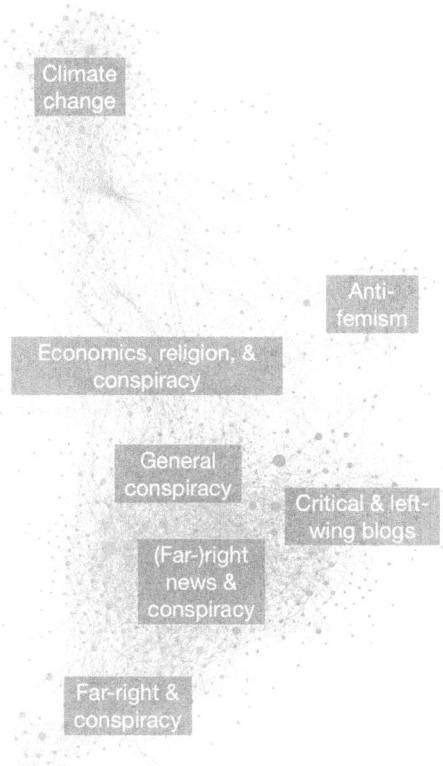

FIGURE 15.2 Labelled climate network by community.

within the network. Against this background, I labelled the '*general conspiracy*' community based on the lack of prominent media outlets and other topical foci in the most linked to sites. The site Wahrheiten.org, for example, lists nine 'lies', including the 'climate lie', 'vaccination lie', or 'BRD lie' (the conspiracy theory that Germany is still occupied and a company, not a country). Closely connected to 'general conspiracy' is the community '*(far-)right news & conspiracy*'. It features the German versions of Russia Today (RT) and Epoch Times; two prominent general news media outlets for the German far-right (Puschmann et al. 2016). In addition to these sites, however, there are also more extreme sites like Volksbetrugpunktnet.wordpress.com (literally translated Betrayal of the Volk dot net). The site, for example, features videos from the far-right Identitarian Movement or congratulates Victor Orbán for his actions against George Soros and the 'globalists'.[4] In addition, the author(s) also question(s) man-made climate change. Closely connected to this community is the '*far-right & conspiracy*' one. It features Kopp-verlag.de, a publishing company that publishes conspiracy books and which until 2017 also had a news section that was prominent in the German right-wing (Puschmann et al. 2016). It furthermore features blogs like the site Lupocattivoblog, which covers the 'climate terror' and questions that the Holocaust has ever happened.[5] The community also includes white nationalist sites from the United States like Theoccidentalobserver.com (motto: White Identity, Interests, and Culture) or counter-currents.com (see SPLCenter.org 2018).

Finally, the community that I labelled '*critical & left-wing blogs*' is the one where the label is, perhaps, the most inaccurate. The community includes, for example, a mainstream technology site with Heise.de, a critical blog by numerous authors with Nachdenkseiten.de, a Russian news site with Sputnikenews.com, but also 'the independent news agency' Net-news-express.de, where everyone can submit articles, the right-wing conspiracy sites Kenfm.de and Nuoviso.tv, but also the left-wing news sites Neues-Deutschland.de and Democracynow.com, and even the now shutdown left-wing extremist site Linksunten.indymedia.org (see also Figure 15.3). In sum, this community is highly diverse, both with regards to topics that are being discussed, as well as with the displayed political ideologies. Potential explanations for the diversity of this community are, for example, sites like Net-news-express.de, which link to a high number of different sites and thus pull them together, as well as popular sites like Heise.de, which are linked to from different sides.

To sum up, there are seven different communities in the climate network. Of these, one deals predominantly with the topic of climate change, while the others are more focused on politics and societal issues. It is noteworthy, that this initial analysis already demonstrates the breadth of different websites in the network, including (far-)right, (far-)left, and, indeed, (alleged) Russian misinformation sites (see also Gallacher et al. 2017; Figure 15.3). Especially the prominence of Russian sites that have been connected with misinformation campaigns are

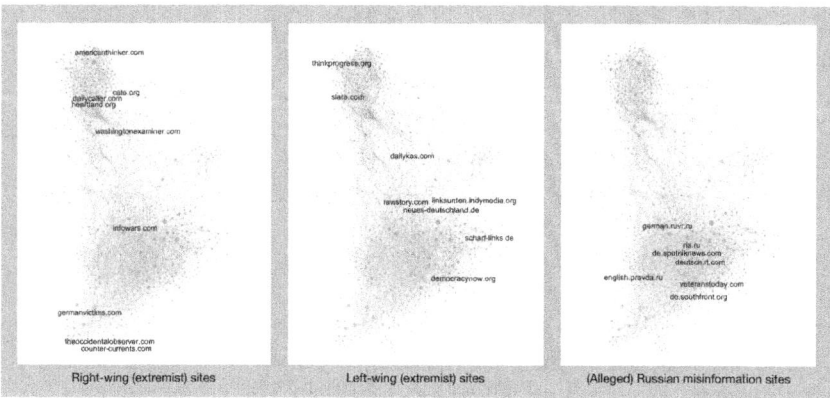

FIGURE 15.3 Selected sites for 'right-wing (extremist)', 'left-wing (extremist)', and '(alleged) Russian misinformation' sites.

noteworthy, as these offer a different take on the daily news than the mainstream media and are especially prominent on the fringe (Puschmann et al. 2016). With regard to climate scepticism, it is worth noting that sites like RT or Sputniknews are not climate sceptic per se. They both publish articles that promote the mainstream position,[6] but also feature sceptic talking points.[7]

Connection between German and US sites

In a next step, I aim to understand the relationship between German and US sites, to then identify particular prominent (far-)right US sites within the German climate network. As every method to identify a site's region has weaknesses, I visualized the climate network both based on their top-level domain (e.g. *.de* for Germany, *.com* for the United States) as well as their host locations (see Tables 15.2 & 15.3). Both tables show how intertwined German and US sites are in the climate network. Whereas there are 303 domains ending in .com and 258 ending in .de in the domain network, suggesting that there could be more US sites in the network than German ones, the host IP network shows the opposite.[8] There are 331 sites with their server in Germany and 268 sites in the United States. It has, however, to be noted that some German-language websites are actually hosted in other countries like the United States. Reasons for this can be innocent, pragmatic, or to avoid the German law. As this study deals with the far-right, it is worth noting that Holocaust denial, for example, is illegal in Germany. Take, for example, the far-right site Lupocattivoblog that I've mentioned above with regard to its Holocaust denial. It is hosted in the United States, its imprint refers to Uruguay,[9] but it is written in German. Although this serves as a cautionary tale, the host IPs can be used as a general heuristic approximation of a website's geographic location.

As Table 15.2 shows, over 50% of the websites with a server in the United States are part of the climate change community. This indicates that US sites are highly important within the climate change community, whereas the critical and left-wing blog communities, for example, mostly consist out of German sites.

However, when looking closer at the data, it is notable how prominent US sites are in the climate change community: ~55% of the sites are from the United States and only ~19% are from Germany (Table 15.3). In addition, the far-right and conspiracy community consists of ~31% from US sites and ~34% from German ones, while the (far-)right news community consists of ~25% from US sites and ~24% from German ones. The high number of 'other' regions, for example in the right-wing news community, can be explained with the numerous Wordpress- and Blogspot-based blogs. It is also worth noting that US sites are neither particularly prominent in the critical and left-wing community nor the anti-feminism one. This, then, might indicate that these communities deal with issues and topics for which the United States is less relevant.

TABLE 15.2 Share of websites in each modularity class with IPs in Germany or the United States (n = 906)

	Community (in %)							
Country code	(Far-)Right news & conspiracy	Far-right & conspiracy	Critical & left-wing blogs	General conspiracy	Economics, religion, & conspiracy	Anti-feminism	Climate change	Total (n)
DE	3.93	18.13	32.33	11.18	10.27	10.27	13.90	331
US	5.22	20.15	7.84	6.72	7.84	1.87	50.37	268

TABLE 15.3 Distribution of websites with IPs in Germany or the United States per modularity class (n = 906)

	Community (in %)						
Country code	(Far-)Right news & conspiracy	Far-right & conspiracy	Critical & left-wing blogs	General conspiracy	Economics, religion, & conspiracy	Anti-feminism	Climate change
DE	23.64	34.09	54.04	41.11	38.64	61.82	18.85
US	25.45	30.68	Oct-61	22.22	23.86	09-Sep	55.33
Other	50.82	35.23	35.35	36.67	37.5	29-Sep	25.82
Total	100	100	100	100	100	100	100
(n)	(n = 55)	(n = 176)	(n = 198)	(n = 90)	(n = 88)	(n = 55)	(n = 244)

German climate sceptics and the US (far-)right

In a next step I will focus on the connection between German climate sceptics and their connection to the US (far-)right (RQ 3 & 4, see p. 260). To do so, I will both draw from prominent (far-)right sites within the climate change community, as well as look at the links between the different communities (Table 15.4).

As highlighted above, some of the most prominent US right-wing sites in the climate change community are well-known: The Heartland Institute and, to some extent, the Cato Institute (see Figure 15.3). Several authors have looked at the publications of the two institutes, to understand how the 'merchants of doubt' (Oreskes and Conway 2010) operate (Jacques et al. 2008; McCright and Dunlap 2000; Elsasser and Dunlap 2013; Cann 2015). In their book, Oreskes and Conway (2010), show how Heartland has worked with the tobacco industry before and applies similar tactics to the climate change discourse. They write:

> The Heartland Institute is known among climate scientists for persistent questioning of climate science, for its promotion of 'experts' who have done little, if any, peer-reviewed climate research, and for its sponsorship

TABLE 15.4 Links between different modularity classes in percent

	Target						
Source	(Far-)right news & conspiracy	Far-right & conspiracy	Critical & left-wing blogs	General conspiracy	Economics, religion, & conspiracy	Anti-feminism	Climate change
(Far-)right news & conspiracy	45.82	20.07	13.38	8.36	7.69	1.34	3.34
Far-right & conspiracy	8.04	62.23	15.52	6.58	4.39	1.87	1.38
Critical & left-wing blogs	7.6	15.87	61.49	7.69	4.46	1.9	0.99
General conspiracy	1.58	14.78	17.94	54.88	2.64	0.53	7.65
Economics, religion, & conspiracy	3.96	10.06	13.11	1.52	60.67	3.96	6.71
Anti-feminism	3.45	1.15	8.43	1.92	3.45	76.25	5.36
Climate change	1	2.5	2.34	3.67	2.34	0.83	87.31

of a conference in New York City in 2008 alleging that the scientific community's work on global warming is a fake.

(Oreskes and Conway 2010: 439)

However, the Heartland Institute does not only focus on climate change, but also works on topics like immigration, criminal justice or health care and has received donations[10] from the Koch brothers; the Koch brothers are well-known for their funding of climate change denial (Oreskes and Conway 2010) but also of right-wing movements like the Tea Party (Parker 2018). Its role within the climate change community is thus not only that of a climate sceptic resource and ally, but also as a bridge to other topics.

Additional sites within the climate change community which might also function as a bridge, are the right-wing news sites The Daily Caller, Washington Examiner, and American Thinker. While Washington Examiner is a conservative daily, both The Daily Caller[11] and especially American Thinker[12] are further to the right and both feature conspiracy theories (Daily Caller to a lesser extent). And although this is already indicative of a connection between climate sceptics and the US right-wing, we have to keep in mind that left-wing sites like ThinkProgress.com or Slate.com are also part of the community (Figure 15.3).

I will thus focus now on the structural level, in order to understand the role the (far-)right plays for the climate change community (Table 15.4). When comparing the inlinks and outlinks between the communities, we can see that ~87% of links by sites within the climate change community stayed within the community. Against this background, sites within the climate change community only rarely link outside the community; only 1% of the links are directed at sites in the (far-)right news and conspiracy community, and 2.5% at the far-right and conspiracy community. This is, to some extent, reciprocal with ~3% outlinks from (far-)right news and conspiracy to climate change, ~1% from far-right and conspiracy. In that way, the (far-)right communities and the climate change one are among the furthest away within the network (see also Figure 15.2).

But these results have to be seen against the analysis above (see also Figure 15.1 for the comparison of new nodes and seed nodes). As I highlighted, numerous sites throughout the network talk about and deny climate change. Against this background, I suggest that there are two distinctly different climate sceptic communities in the network: one focusing on climate change and climate science and the other one dealing with right-wing and conspiracy sites, that discuss numerous different topics. Whereas the latter is deeply connected with sites from the right-wing and the far-right, the former is more mono-thematic, and albeit connected to right-wing sites, is not focused on politics per se. These differences are not surprising. As Forchtner et al. (2018) point out, the far-right is not a monolithic bloc but rather diverse in their positions, including climate scepticism (see also Kaiser and Rhomberg 2016; Kaiser 2017 regarding different forms of climate scepticism). And while the German climate scepticism discourse is mostly dominated by mono-thematic German and US sites, it also includes voices from the far-right.

Conclusion

In this analysis I investigated the connection between German climate sceptics and the US right-wing, especially the far-right. To do so, I conducted a hyperlink network analysis based on German climate sceptic sites. The main result is that, while there are undoubtedly connections between German climate sceptics and the US far-right, it has to be noted that US left-wing media outlets are also part of the network and, indeed, the climate change community. Looking at the results more granularly, there are three main takeaways:

1 The climate change community, although including political actors, is mostly focused on the topics of climate change and climate science. Particularly noteworthy is that there is no indication of an isolated sceptic community; a so-called 'echo chamber'. Indeed, climate sceptics seem to be very much aware of the mainstream discussions and, while most likely disagreeing, link to them nevertheless. This is in line with Elgesem et al.'s (2015) findings, that the topic of climate science might bridge between mainstream and sceptics.

2   German climate sceptics are very connected to other U.S. climate sites (especially sceptic ones). This suggests, that German climate sceptics are predominately interested in other climate (sceptic) sites. However, as I have shown, that there are also right-wing and far-right sites in the climate change community, like the Heartland Institute, Cato Institute, Daily Caller, or American Thinker. Yet left-wing sites like Slate or ThinkProgress are part of this community too. This also highlights that there is no potential 'political' echo chamber in the community. Indeed, while right-wing content is only one link away, so is left-wing content.
3   Climate change scepticism is very much connected with other conspiracy theories. Indeed, there are numerous German-language conspiracy theory sites in the network that, while not exclusively climate sceptic, are also promoting sceptic viewpoints. Conspiracy theories might, in this context, work as a bridge between climate change scepticism and the right wing. In addition, quite a few far-right sites (e.g. Wahrheiten.org, American Thinker, Counter Currents, etc.) I have mentioned above are also questioning climate change.

This leads me to the proposition that although sceptics are a minority in Germany, we have to address their differences and differentiate between mono-thematic sceptics that are mostly focused on climate change and climate science on the one hand and, on the other hand, political climate sceptics. And while the mostly mono-thematic actors might be reached with science communication, the political climate sceptics pose a different threat as they combine different conspiracy theories and often reject science and mainstream media sources. While this is no binary differentiation as the initial example of the climate sceptic blog Science-Skeptical.de clearly shows, it is worth noting that there *are* differences (see also the introduction to this volume by Forchtner 2019 and Lockwood 2018). In that sense it is a heuristic that is supposed to highlight the differences within the climate sceptic community and the climate sceptic discourse and emphasize that while some climate sceptic websites are clearly connected to the far-right, others are not. This has to be considered when planning communication strategies and outreach programs. Both the connection as well as the disconnect between the far-right and climate scepticism was at display during the AfD's European election campaign which was based around a climate sceptic message. And while the AfD gained 3.9% since 2014, their end result of 11% was nevertheless disappointing and highlights that climate scepticism might hurt the party's chances in reaching more voters.

Against this background it has to be noted that this analysis is based on a snapshot of weblinks. And while indicative of the networked public sphere, other factors like overarching topics, frames, shared imagery are lacking. Prior research thus should look into the connections between right-wing and climate scepticism based on these aspects. Additionally, this analysis lacks a metric of influence, that is, of the audience. It would be highly interesting to map this network against the overlap of site visitors. Finally, and as highlighted above, the community labels were manually chosen based on the most linked to websites. Future research could follow Elgesem et al.'s (2015) research and scrape the text of the websites and conduct a topic modelling analysis.

Notes

1 This work has been funded by the German Research Foundation (DFG; KA 4618/1-2).
2 www.science-skeptical.de/blog/kernpunkte-zur-klima-und-energiepolitik-aus-der-afd/0013952/ [30 April 2018].
3 www.eike-klima-energie.eu/2014/03/23/europawahlprogramm-der-afd-fordert-komplette-abschaffung-des-eeg-wissenschaft-vom-klimawandel-wird-als-sehr-unsicherheitsbehaftet-bezeichnet/ [30 April 2018].
4 https://volksbetrugpunktnet.wordpress.com/2018/04/16/soros-ungarn-victor-orban-macht-ernst-mit-seinem-kampf-gegen-die-globalisten/ [30 April 2018].
5 https://lupocattivoblog.com/2010/03/17/holocaustleugnung-und-die-abwesenheit-von-logik/ [30 April 2018].
6 For example, https://de.sputniknews.com/panorama/20170801316841497-klima-forscher-trockene-regionen-austrocknen-feuchte-feuchter-werden-klimawandel/ (Accessed 30 April 2018); https://deutsch.rt.com/newsticker/66972-weltbank-bericht-millionen-droht-umsiedeln-wegen-klimawandels/ [30 April 2018].
7 For example, https://de.sputniknews.com/wissen/20161107313252515-klimawandel-hauptschuldigen/; https://deutsch.rt.com/programme/der-fehlende-part/52127-michael-limburg-klimaschutz-ist-absurde-idee/ [30 April 2018].
8 For this analysis, I set the locations for the two prominent blogging platforms Wordpress and Blogspot as NA, as all sites would show up as U.S. sites. Given that there are 165 sites within the network which are hosted by either one of those platforms, this would have skewed the analysis heavily.
9 https://lupocattivoblog.com/impressum/ [30 April 2018].
10 http://conservativetransparency.org/recipient/heartland-institute/page/3/ [30 April 2018].
11 www.splcenter.org/hatewatch/2017/08/16/daily-caller-has-white-nationalist-problem [30 April 2018].
12 www.splcenter.org/hatewatch/2014/12/04/american-thinker-sinks-bottom-racist-barrel [30 April 2018].

References

Ackland, Robert, and Rachel Gibson (2004): Mapping Political Party Networks on the Www: Presented at *Australian Electronic Governance Conference*. University of Melbourne, 2004.
Adamic, Lada A., and Glance, Natalie (2005): The Political Blogosphere and the 2004 U.S. Election: Divided They Blog. In: *Proceedings of the 3rd International Workshop on Link Discovery*. Chicago, IL: ACM. 36–43.
Alvarez-Hamelin, J. Ignacio, Dall'Asta, Luca, Barrat, Alain, and Vespignani, Alessandro (2005): Large scale networks fingerprinting and visualization using the k-core decomposition. In: *NIPS'05 Proceedings of the 18th International Conference on Neural Information Processing Systems,* December 5–8, 2005, Vancouver, Canada. pp. 41–50. http://papers.nips.cc/paper/2789-large-scale-networks-fingerprinting-and-visualization-using-the-k-core-decomposition [July 30, 2018]
Anderson, Benedict R. O. (2006). *Imagined communities: Reflections on the origin and spread of nationalism* (Rev. and extended ed.). London, New York: Verso (Original work published 1983).

Benkler, Yochai (2006): *The Wealth of Networks – How Social Production Transforms Markets and Freedom*. New Haven, CT; London: Yale University Press.

Benkler, Yochai, Roberts, Hal, Faris, Rob, Solow-Niederman, Alicia, and Etling, Bruce (2015): Social Mobilization and the Networked Public Sphere: Mapping the SOPA-PIPA Debate, *Political Communication*, 32: 594–624.

Blondel, Vincent D., Guillaume, Jean-Loup, Lambiotte, Renaud, and Lefebvre, Etienne (2008): Fast unfolding of communities in large networks, *Journal of Statistical Mechanics: Theory and Experiment*, 2008: P10008.

boyd, danah (2008): *Taken Out of Context: American Teen Sociality in Networked Publics*. University of California-Berkeley.

Cann, Heather W. (2015): Climate change, still challenged: Conservative think tanks and skeptic frames. Presented at *Annual Meeting of the Western Political Science Association*, Las Vegas.

Ecklund, Elaine Howard, Scheitle, Christopher P., Peifer, Jared, and Bolger, Daniel (2016): Examining links between religion, evolution views, and climate change skepticism, *Environment and Behavior*, 49(9): 985–1006.

Elgesem, Dag, Steskal, Lubos, and Diakopoulos, Nicholas (2015): Structure and content of the discourse on climate change in the blogosphere: The big picture, *Environmental Communication*, 9(2), 169–188.

Elsasser, Shaun W., and Dunlap, Riley E. (2013): Leading voices in the Denier Choir: Conservative Columnists' dismissal of global warming and denigration of climate science, *American Behavioral Scientist*, 57(6): 754–776.

Engels, Anita, Hüther, Otto, Schäfer, Mike S., and Held, Hermann (2013): Public climate-change skepticism, energy preferences and political participation, *Global Environmental Change*, 23(5): 1018–1027.

Forchtner, Bernhard (2019): Far-right articulations of the natural environment: An introduction. In: Bernhard Forchtner (Ed). *The Far Right and the Environment: Politics, Discourse and Communication*. London: Routledge.

Forchtner, Bernhard, and Kølvraa, Christoffer (2015): The nature of nationalism: Populist radical right parties on countryside and climate, *Nature and Culture*, 10(2): 199–224.

Forchtner, Bernhard, Kroneder, Andreas, and Wetzel, David (2018): Being skeptical? Exploring far-right climate-change communication in Germany, *Environmental Communication*, 12(5): 589–604.

Fraser, Nancy (1990): Rethinking the public sphere: A contribution to the critique of actually existing democracy, *Social Text*, 25/26: 56–80.

Gallacher, Jihn D., Barash, Vlad, Howard, Phil, and Kelly, John (2017): *Junk News on Military Affairs and National Security: Social Media Disinformation Campaigns Against US Military Personnel and Veterans*. Data memo, Oxford. Accessed from https://comprop.oii.ox.ac.uk/politicalbots/wp-content/uploads/sites/89/2017/10/Junk-News-on-Military-Affairs-and-National-Security-1.pdf [30 July 2018].

Ging, Debbie (2017): Alphas, betas, and incels: Theorizing the masculinities of the manosphere, *Men and Masculinities*. doi:10.1177/1097184x17706401

Häussler, Thomas, Adam, Silke, Schmid-Petri, Hannah, and Reber, Ueli (2017): How political conflict shapes online spaces: A comparison of climate change hyperlink networks in the United States and Germany, *International Journal of Communication*, 11: 3096–3117.

Jacques, Peter J., Dunlap, Riley E., and Freeman, Mark (2008): The organisation of denial: Conservative think tanks and environmental scepticism, *Environmental Politics*, 17(3): 349–385.

Kaiser, Jonas (2017): Public spheres of skepticism: Climate skeptics' online comments in the German Networked Public Sphere, *International Journal of Communication*, 11: 1661–1682.

Kaiser, Jonas and Puschmann, Cornelius (2017): Alliance of antagonism: Counterpublics and polarization in online climate change communication, *Communication and the Public*, 2(4): 371–387.

Kaiser, Jonas and Rhomberg, Markus (2016): Questioning the doubt: Climate skepticism in German Newspaper Reporting on COP17, *Environmental Communication*, 10(5): 556–574.

Kaiser, Jonas, Rhomberg, Markus, Maireder, Axel and Schlögl, Stephan (2016): Energiewende's lone warriors. A hyperlink network analysis of the German Energiewende discourse, *Media and Communication*, 4(4): 18–29.

Kaiser, Jonas and Rauchfleisch, Adrian (2018): Unite the Right? How YouTube's Recommendation Algorithm Connects The U.S. Far-Right. *D&S Media Manipulation*. https://medium.com/@MediaManipulation/unite-the-right-how-youtubes-recommendation-algorithm-connects-the-u-s-far-right-9f1387ccfabd [30 July 2018].

Leiserowitz, Anthony, Maibach, Edward, Roser-Renouf, Connie, Feinberg, Geoff and Howe, Peter (2013): *Global Warming's Six Americas in September 2012*. New Haven, CT: Yale Project on Climate Change Communication.

Lockwood, Matthew (2018): Right-wing populism and the climate change agenda: Exploring the linkages. *Environmental Politics*, 27(4): 712–732.

Lörcher, Ines, and Taddicken, Monika (2015): »Let's talk about... co2-Fußabdruck oder Klimawissenschaft?« Themen und ihre Bewertungen in der Online-Kommunikation in verschiedenen Öffentlichkeitsarenen. In: M. Schäfer, S. Kristiansen, and H. Bonfadelli (Hrsg.), *Wissenschaftskommunikation im Wandel*. Köln: Herbert von Halem Verlag. 258–287.

McCright, Aaron M. and Dunlap, Riley E. (2000): Challenging global warming as a social problem: An analysis of the conservative movement's counter-claims, *Social Problems*, 47(4), 499–522.

McCright, Aaron M., Dunlap, Riley E., and Marquart-Pyatt, Sandra T. (2016): Political ideology and views about climate change in the European Union, *Environmental Politics*, 25(2), 338–358.

Oreskes, Naomi, and Conway, Erik M. (2010). *Merchants of Doubt: How a Handful of Scientists Obscured the Truth on Issues from Tobacco Smoke to Global Warming*. New York: Bloomsbury Press.

Parker, Christopher (2018): The Radical Right in the United States of America. In: Jens Rydgren (Ed.), *The Oxford Handbook of the Radical Right*. Oxford University Press.

Puschmann, Cornelius, Ausserhofer, Julian, Maan, Noura and Hametner, Markus (2016): Information laundering and counter-publics: The news sources of Islamophobic groups on Twitter. Presented at *International AAAI Conference on Web and Social Media*, Cologne, Germany, April 2016.

Rauchfleisch, Adrian, and Kovic, Marko (2016): The Internet and generalized functions of the public sphere: Transformative potentials from a comparative perspective, *Social Media + Society*, 2(2): 1–15.

Schäfer, Mike S. (2012): Online communication on climate change and climate politics: A literature review, *Wiley Interdisciplinary Reviews: Climate Change*, 3: 527–543.

Sharman, Amelia (2014): Mapping the climate sceptical blogosphere, *Global Environmental Change*, 26: 159–170.

SPLCenter.org. (2018, April 28): GREG JOHNSON. www.splcenter.org/fighting-hate/extremist-files/individual/greg-johnson [28 April 2018].

Sunstein, Cass (2001): *Republic.com*. Princeton, NJ: Princeton University Press.

16
ALT-RIGHT ECOLOGY

Ecofascism and far-right environmentalism in the United States

Blair Taylor

The rise of the so-called 'alt-right', or alternative right, has transformed the political landscape in the United States and challenged established political categories. For at least a generation, the political right has been understood primarily as a defence of the status quo: pro-capitalist, pro-state, pro-science and technology, and anti-environmental. By contrast, the alt-right draws its energy from a critique of the established order, liberal and conservative alike, not from its defence. Thus, it has resurrected older right-wing traditions of the antimodernist, revolutionary, and fascist right which have remained marginal in the North American conservative movement, often articulated in a white nationalist framework. While this milieu is highly rearticulated today and internally divided, much of it is animated by strains of reactionary thought which attack liberal democracy, the state, and the 'mongrelizing', amoral forces of global capitalism. It has adopted positions associated with the political left, for example against war and free trade and for (exclusionary) social protectionism. A variety of contemporary researchers of the far right have begun to examine these new political alignments and how they disrupt past understandings (Reid Ross 2017; Lyons 2018).

However, even these recent accounts often overlook the extent to which alt-right discourse draws from ecological discourse. Ecology is an increasingly important political vector for the rejection of traditional pro-business conservative positions by the constellation of esoteric, revolutionary, and traditionalist currents that comprise the alt-right. This chapter thus analyses how the alt-right deploys ecological discourse, rediscovering older Nazi themes like organic agriculture and animal rights while articulating novel right-wing interpretations of concepts like biodiversity, decentralism, deep ecology, bioregionalism, anti-capitalism, Indigenism, and anarchism. It will explore the core themes and political actors within the milieu, as well as how ideological cross-pollination has resulted in left-right resonance and, at times, political collaboration.

The chapter concludes by discussing the potential political role of alt-right ecology in the present historical conjuncture, drawing on theoretical frameworks developed by Fraser (2017) and Brown (2006).

How the alt-right has shifted conservative environmental discourse

Environmentalism remains a contested topic within right-wing politics. While a reflexive pro-business anti-environmentalism remains dominant within mainstream conservatism, the insurgent character of the alt-right has made it more receptive to environmental ideas. As an article on altright.com proclaims,

> The anti-environmentalism of the present 'conservative' movement is nothing but a modernist, capitalist, classically Liberal, materialist Jewish conception which has found fertile ground in the hands of billionaire plutocrats wiping out nature all over the Earth for the sake of higher profit margins'.
>
> *Ahab (2017)*

Concern for the environment has therefore become an important issue for distinguishing the alt-right from mainstream conservatism: 'The Alt Right was born as an "alternative" to Bush-Cheney era neoconservatism, when this strain of vile anti-environmentalism became entrenched within the Republican Party' (ibid.).

Why is the alt-right more receptive to ecological ideas? On the one hand, it reflects the growing acceptance of the unavoidable reality of climate change and ecological degradation, especially among young people. Another factor is the alt right's reclamation of previously taboo political traditions like fascism. This has facilitated a rediscovery of the historical connection between far-right and ecological ideologies and ecological ideologies, from Ernst Haeckel, the racialist who coined the term ecology, to Nazi environmentalism (Biehl and Staudenmaier 1995). Greg Johnson (2018), editor of the prominent alt-right site Counter-Currents, describes himself as a "very pro-ecology person" and actively seeks to reclaim this legacy:

> Although today ecology is considered a preserve of the Left, the truth of the matter is that if you go back far enough ecology was actually something that was pioneered by a lot of figures that today would be considered figures on the Right.
>
> *Greg Johnson (2018)*

Yet another factor is the intellectual bent of the alt-right. As its political vision centres on the allegedly 'natural' condition of human inequality, there has been a resurgence of interest in right-wing nature philosophy. Various strains of esoteric, mystical, and quasi-anarchist fascism are important currents within the movement, especially its ecological wing. Alt-right media outlets like Counter-Currents and Arktos prominently feature European right-wing ecologists like Savitri Devi, Troy Southgate and Pentti Linkola alongside mystical fascists like Julius Evola.

Lastly, the predominantly white demographic makeup of the environmental movement also makes it highly attractive to the alt-right. '[T]he modern environmentalist movement is overwhelmingly dominated by white people, to the point where one might be tempted to say environmentalism (…) is the last bastion of implicit whiteness' (Ahab 2017). Since contemporary environmental activism is overwhelmingly liberal or left in orientation, alt-right ecology is primarily intellectual and 'metapolitical' in nature rather than activist, following the neo-Gramscian strategy of the French *Nouvelle Droit* that seeks to shift the cultural and intellectual landscape towards receptivity to right wing politics.

Ecology is an important front in this war of ideas, and its articulation has shifted with the changing times. The classic right ecological emphasis on population and immigration is still there but takes new ideological forms. It is now often part of an expanded political terrain of alt-right ecology that includes new themes like deep ecology, decentralism, anti-modernism, anarchism, Indigenism, and veganism.

Overpopulation: Immigration and white 'habitat loss'

Overpopulation has remained a central theme for right ecology because it shifts responsibility for environmental problems away from questions of overconsumption or capitalism and focuses instead on either the sins of an undifferentiated 'humanity' or racist anti-immigration arguments about the profligacy of non-western cultures. The first coincides with the alt-right's elitist, declinist and clash of civilizations narratives, while the second overlaps with white nationalism's fixation on 'white genocide'. Stern (2005) has shown how early conservationists in California viewed Mexicans as both a cultural and ecological threat, while Bhatia (2004) has described the broad network of right-wing groups that try to seduce progressives by channelling environmental concern into support for anti-immigration policies. Groups like Progressives for Immigration Reform, The Pioneer Group and Apply the Brakes have recruited prominent ecologists like David Foreman and Garret Hardin to give anti-immigration politics an environmental veneer. Many of these groups are funded by John Tanton, who the Southern Poverty Law Center calls 'the racist architect of the modern anti-immigration movement' (SPLC 2010). White nationalists have worked to establish an academic presence; white separatist Virginia Abernethy was editor of the scholarly journal *Population and Environment* from 1989 to 1999, followed by Kevin MacDonald, an evolutionary psychologist popular with neo-Nazis for his quasi-academic updates of classical antisemitic motifs.

In addition to pseudo-scientific anti-immigration discourse, the mystical and esoteric orientation of the alt-right has affinities with deep ecology, attractive because of a shared antimodernist politics and preference for changing culture over statist solutions. Writing for the alt-right website Amerika.org, author Brett Stevens states 'the Alt Right movement is a philosophical descendant of Deep Ecology' (Stevens 2017). Deep ecology's emphasis on protecting an abstract 'wilderness' or 'nature' has been a consistent discursive terrain where right-wing positions have surfaced in the radical environmental movement. Earth First! was criticized for espousing racist and sexist anti-immigration and pro-HIV positions in the late 1980s, leading to a split where self-professed 'redneck for wilderness' Dave Foreman and others advocating a narrow focus on wilderness protection left the group while lamenting the growing influence of humanist ideologies like feminism and anarchism (1987; Bookchin and Foreman 1999). Foreman is now a spokesperson for the anti-immigration group Apply the Brakes, and his most recent book *Man Swarm* (2011) was glowingly reviewed by anti-immigration groups (Kolankiewicz 2015). Opposition to immigration in the name of overpopulation has long been a theme for many deep ecological thinkers and groups (Olsen 1999), from Edward Abbey's warnings against a 'mass influx of (…) culturally-morally-genetically impoverished people' (Abbey 1988: 43) to Devall and Sessions (1985) neo-Malthusian fixation on third-world population growth.

Finnish deep ecologist Linkola (2006) has become a darling of the alt-right for statements like, '[t]here is no use counting the immigrants at the border: one should wait a while and look in their nurseries' (2006: 130). His books are published by alt-right press Arktos and praised on Counter-Currents website, which notes that, 'Linkola's sympathies lie squarely with fascism' (Hawthorne 2011). A growing number of alt-right actors, like whitebiocentrism.com and amerika.org, fuse deep ecology with explicit white supremacy, translating deep ecology's concern for habitat loss and species extinction into fears about 'white genocide' and the displacement of 'indigenous' white people by 'invasive species'. Greg Johnson states, 'what is now happening to the European peoples [is] habitat loss' due to competition with 'similar creatures' (Minkowitz 2017). White supremacist Harold Covington frames his call for a whites-only ethno-state in terms of habitat loss, 'The wolves have to have a habitat, and the white man has to have a habitat' (Francey 2013), as does Tom Metzger of White Aryan Resistance: 'being only about 10 per cent of the population, we [whites] begin to sympathize, empathize more, with the wolves and other animals' (quoted in Bhatia 2004: 201).

Ethnopluralism and right-wing bioregionalism

The flip side of alt-right ecology's discourse of overpopulation, racialized environmental disruption, and white genocide is the creation of a white homeland, a place to realize David Lane's '14 words' slogan ('We must secure the existence of our people and a future for white children'). The alt-right' has resurrected a blood and soil politics which echoes the 'love of land and militant racist nationalism' articulated by Nazi ecofascism (Biehl and Staudenmaier 1995: 6). Yet many contemporary far-right groups have traded in the language of overt white supremacy for ethnopluralism, a vision wherein distinct groups live separately but allegedly equal, free to pursue their ethnic interests. Ethnopluralism is often embedded in the discourse of diversity or biodiversity, capitalizing on the progressive antiracist and environmental associations (Forchtner 2019). The white nationalist group Identity Evropa proclaims:

> [W]e are ethno-pluralists: We believe that all ethnic and racial groups should have somewhere in the world to call home – a place wherein they can fully express themselves and enjoy self-determination.
>
> *Identity Evropa (2017)*

Many strains of the alt-right reject traditional nationalism and the state in favour of regionalism, decentralism, or tribalism. Bioregionalism has become an increasingly attractive political vehicle to articulate these views. Growing out of the ecology movement, bioregionalism advocates living in decentralized communities that correspond to bioregions rather than artificial borders. Although bioregionalism has mostly been articulated in a progressive environmental idiom, it is also open to nationalist interpretations contain seeds for exclusion which have made it attractive to the right (Olsen 2000; Park 2013; Lokting 2018).

The Pacific Northwest, which has among the whitest populations in the United States, has long been targeted by white supremacists as the natural place to create a white homeland, a strategy called the Northwest Territorial Imperative (Durham 2007). This same geographic area is also known by environmentalists as the Cascadia bioregion, which spans from Northern California to Southern British Columbia. The notion of Cascadia was first popularized in the pages of Ernest Callenbach's 1975 utopian novel *Ecotopia*, which depicts life in a breakaway eco-community spanning Northern California, Oregon and Washington. Informed by the emergent ecology movement, New Left, counterculture and back to the land movements, the book describes a largely progressive environmental utopia. Yet it also hints at problems of nationalism and identity; Ecotopia's borders are guarded from outsiders and voluntary ethnic separatism the norm.

Cascadian bioregionalists have sought to make this fictional ecotopia into reality; there are a number of different organizations pursuing this goal united by their own Cascadian flag. While the majority of Cascadian bioregionalists are liberal or progressive environmentalists, the open parameters of place-based politics in an overwhelmingly white area has created an opportunity for participation by right actors.[1] One such group is True Cascadia, which fuses Cascadian bioregionalist discourse with overt white separatism to promote 'a White ethnic consciousness in the Pacific Northwest and prevent, as well as reverse, the increasingly discriminatory policies enacted in opposition to Whites in our own homelands' (Kavanaugh 2018). The group was invited to speak at the 2016 Northwest Forum in Seattle, the premier intellectual gathering for the far right, and their hats emblazoned with the Cascadia flag were a visible presence at the 2017 Unite the Right march in Charlottesville, Virginia (Stern 2019, Pacific Northwest Antifascist Workers Collective 2018).

The Northwest Front is another white bioregionalist group led by long-time white supremacist Harold Covington. The group deploys a quasi-ecological discourse which seeks to 'preserv[e] our race from biological and cultural extinction' by creating 'a sovereign and independent nation on the continent of North America for White people only' (Northwest Front 2010a, 2010b). Northwest Front's website depicts happy white families frolicking in the natural beauty of the Pacific Northwest against the backdrop of the Cascadian bioregional flag, the design for which they offer a white supremacist interpretation: 'The sky is the blue, and the land is the green. The white is for the people in between' (Northwest Front 2010c).

Some actors are attracted to bioregionalism precisely because it seems to transcend left and right. Casey Brian Corcoran (2014), a prominent activist in the Cascadia movement, states:

> I believe Place is more powerful and alive than Race or Ideology. The 'far Right' is not my enemy, nor is the 'far Left' (…) I am hopeful that Bioregionalism can be formulated in a way that opposes both toxic ideologies.
> *Casey Brian Corcoran (2014)*

Corcoran's background is in the radical ecology and green anarchist movements, but he has come to abandon both, stating 'I am not an environmentalist (…) Call me a European Indigenist' (quoted in Jacob 2013). Although insisting he is not personally racist, the broader goal of creating a Cascadia comprised of diverse 'tribes' requires working with them: '[r]efusing to scapegoat even white "racialists" is, I feel, both morally courageous and *highly intelligent*' (Corcoran 2014).

Indigenism, decolonization, tribalism

As Cascadia overlaps many sovereign indigenous nations, there is signficant interaction between bioregionalism and Indigenism, a political perspective which asserts the rights and sovereignty of indigenous people. Chinook tribal member Robert Izatt states, 'I totally see overlap with bioregionalism and current indigenous resistance (…) People with European ancestry need to reinhabit the land (…) taking only what you need, not just taking it because you can, you want it, or want to sell it' (Sears 2016). Despite this ecological anti-capitalist sentiment, a politics tightly bound to place, culture, and ancestry resonates with core themes of right-wing politics. In recent years the discourse of 'indigeneity' and 'decolonization' have been taken up by the alt-right, especially its ecological wing, as they are open to blood-and-soil, essentialist, and traditionalist interpretations. Despite radically different political motivations, indigenism's valorization of place and identity, suspicion of modernity, valorization of traditional lifeways, critiques of the state, and transcendence of right/left echo central themes of the alt-right.

Corcoran's close associate Vince Rinehart is a prominent figure in the convergence of indigenist, bioregionalist, and far-right movements. A member of the Tlingit tribe, he advocates a 'tribal anarchism' wherein different distinct tribes – not necessarily but potentially racial in nature – unite to bring down the United States government and corporations. Although Native Americans have actual tribes, white settlers can create bioregional tribes which then ally together 'against colonialism and imperialism' (Rinehart 2013). Although Rinehart's pleas are garbed in the language of the left, he is closely allied with cryptofascist groups please are mostly articulated in with ties to white nationalism like Attack the System, where he is an editor. Rinehart has written sympathetic like the website Attack articles defending the armed occupation of the Malheur National Wildlife Refuge, despite many occupiers' ties to white supremacist and anti-Indian movements (Rinehart 2016a and b; See also Boggs 2019 in this volume). Tulalip tribal member Eric Lee Flores, active in the Patriot movement, was also among those arrested at the Malheur Occupation (Rural Organizing Project 2017).

The related discourse of decolonization has become another arena for left-right crossover, often intertwining with ecological politics. The journal *Autonomy Cascadia: A Journal of Bioregional Decolonization*, for example, explores this intersection, seeking 'a bioregional practice rooted in the repatriation of Indigenous land' (Autonomy Cascadia 2013). Yet in other quarters, the 'repatriation of indigenous land' takes on a very different meaning. Far-right populists across Europe have argued against recent immigration in terms of defending indigenous Europeans from colonization by alien cultures, deploying the language and tropes of the postcolonial and indigenist lefts, often in combination with ecological themes. Marine LePen of the far-right French National Rally party combines these elements, attacking "nomadic" people like immigrants because they "do not care about the environment" and "have no homeland," while her advisor Hervé Juvin speaks of the need to reclaim Europe for its "indigenous people, on our land, in our countries, with our traditions" (Aronoff 2019). Right-wing actors from Eastern Europe to Africa have also utilized post-colonial discourse to argue against the imposition of 'western' concepts such as gay rights, feminism, or free speech on indigenous populations (Tax 2012; Taylor 2017). While decolonization and indigeneity mostly associated with the antiracist left, they are not immune to ecofascist and other right-wing interpretations.

Right decentralism: Secessionism, tribalism, Green anarchism

At the other corner of the United States is The Second Vermont Republic (SVR), a self-described 'peaceful secessionist' movement. Founded in 2003 by Thomas Naylor, the group speaks in the progressive idiom of sustainability, community, and independence from Wall Street and corporate America. Pioneering bioregionalist and ecologist Kirkpatrick Sale is a leading voice in

SVR. Despite this progressive orientation, SVR worked closely with the neo-Confederate secessionist group League of the South, designated a hate group by the Southern Poverty Law Center, until finally being forced to cut ties in 2008 after increased public scrutiny and pressure. But the relationship seems to have significantly affected Sale's political and intellectual interests; he now writes almost exclusively about secessionism. His last book, *Emancipation Hell* (2012), blames President Abraham Lincoln and the Emancipation Proclamation for the Civil War and contemporary race relations. SVR continues to collaborate with League of the South via the Abbeville Institute, devoted to 'what is true and valuable in the Southern tradition'. Sale wrote the introduction to Naylor's 2008 book *How Vermont and all the Other States Can Save Themselves from the Empire*, published by Feral House Press, which features a wide variety of fringe thinkers including white supremacists (Reid Ross 2017). In 2004, Sale and others from SVR founded the Middlebury Institute, a think tank dedicated to 'the study of separatism, secession, and self-determination' (Sale 2016). Its website features links to the website Attack the System, another 'secessionist' group that combines ecological and far-right themes.

Attack the System (ATS) is a 'pan-secessionist' website that grew out of the National Revolutionary Vanguard organization. Both were founded by Keith Preston, a former left-wing anarchist who now identifies as an 'anarcho-pluralist' and 'pan-secessionist'. Pan-secessionism seeks to build 'an international coalition against the global plutocratic super-class' comprised of 'regional federations of autonomous and self-determined communities reflecting an infinite variety of themes' (Preston 2015a). Although Preston does not identify as a white nationalist and several ATS associates are people of colour, these 'communities' include white nationalist separatists. Like most crypto-fascist ecologists, Preston (2015b) insists '[s]uch a project necessarily involves transcending ordinary divisions of the kind that normally define the conventional Left and Right'.

Preston and Attack the System are closely related to the National Anarchist and Third Positionist movements. National Anarchism advocates a decentralized, racialized socialism inspired by figures from the left and ecological wings of the Nazi Party like Walther Darré and the Strasser brothers (Reid Ross 2017; Lyons 2018). Despite their affinities with fascism, they reject the label due to its statist connotations. National Anarchists are active participants on the white supremacist website Stormfront. Third Positionism is a similar tendency which seeks to create a red-brown alliance of communists and fascists, also drawing from green Nazism. Along with other classically left positions like anti-imperialism and anti-capitalism, Third Positionists and National Anarchists actively embrace ecology and veganism (Sunshine 2008).

British activist Troy Southgate has been a key figure in developing a synthesis of fascist and green anarchist ideas within the Third Positionist and National Anarchist milieus. A onetime British National Front member, Southgate founded the National Revolutionary Faction in 1996, which became the main

National Anarchist group in the anglophone world. Southgate was influenced by the British ecofascist Richard Hunt, who edited the journal *Green Anarchist* before being expelled and founding *Green Alternative*, which Southgate later edited. Southgate has articulated an anti-egalitarian right-wing anarchism that draws on anarcho-primitivist John Zerzan as well as mystical anarchist Hakim Bey (Macklin 2005).

National Anarchists have sought to infiltrate groups and movements in the United States with some success, targeting the anarchist, environmental, anti-capitalist, indigenous, and animal rights movements (Griffin 2005). The main US group was the Bay Area National Anarchist Movement, founded in 2007. According to the Southern Poverty Law Center, 'BANA and other likeminded national anarchists cloak their bigotry in the language of radical environmentalism and mystical tribalism, pulling recruits from both the extreme right and the far left' (Sanchez 2009).

When their neo-fascist commitments were discovered, BANA was expelled from the anarchist movement. Yet they insisted that 'National Anarchists are genuinely sympathetic with many green anarchist ideas [and] deeply allied with Green Anarchist ideals' (Bay Area National Anarchists 2007). Their response also highlighted their ecofascist orientation: '[t]hose who misuse or destroy the environment deserve a swift and merciless response that such behavior is unacceptable' (ibid.). Although BANA disbanded in 2011, National Anarchist ideas continue to circulate in the left-right crossover milieu.

The discourse of racialized ecological anarchism that characterizes much alt-right ecology also resonates with the argument of Timothy Snyder's 2015 book, *Black Earth: The Holocaust as History and Warning*, which argues Hitler was less a nationalist than a 'racial anarchist' fuelled by an ecological worldview. This took the form of an 'ecological antisemitism' wherein Jews were *unnatur*, the embodiment of unnaturalness, and had to be eradicated to restore the world to its natural state (Snyder 2015: 5–10). Nazism sought to do this by attacking the liberal state form which protected Jews, thus returning to a 'natural' order characterized by the struggle of racialized groups for resources and dominance. Synder's reinterpretation of Hitler's political vision closely resembles that of National Anarchism, Third Positionism, and other antisemitic, conspiracist, and white nationalist groups concerned with ecology.

The Wolves of Vinland (WoV), described as the 'environmentalist component' of the alt-right (Caldwell 2016), are a right decentralist group founded in 2006 which combine quasi-ecological tribalism with masculinist authoritarianism. 'Vinland', which refers to the section of eastern North America reached by Vikings before Columbus, is a cultural reference for a variety of white nationalist groups. Although not primarily focused on race, one member was convicted of attempted arson on a black church and various members openly praise white nationalism (Woodruff 2015; Amend and Piggot 2017). A large part of the Wolves' allure is their distinct aesthetic, which draws from the neo-folk and black metal music scenes, biker subculture, weightlifting and martial arts communities, and neo-paganism. Members engage in pagan rituals involving blood, runes and animal skulls wearing black-metal inspired face paint. Group membership is denoted by patches on biker vests or 'cuts' and prominent tattoos; many are heavily muscled from training for the fight clubs that characterize their gatherings.

Ideologically, the group draws on European paganism, a cult of masculinity, self-help ideology, egoism, and other romantic themes to envision a political imaginary of tight-knit Viking warrior tribes subordinated in a meritocratic hierarchy bound together through ritual and martial training. It aligns closely with the mystical fascism of Julius Evola, sharing his rejection of nationalism as an overly inclusive, statist, and modernist ideology and social form. Their ecological antimodernism targets the 'Empire of Nothing', described as unfulfilling capitalist materialism, unnatural and emasculating egalitarianism, and the corrupt state which protects it. Nature represents a timeless hierarchical order and the alternative to modernity, whose multiculturalism, feminism, wealth, urbanism, and statist bureaucracy breeds decadence, weakness and inferiority.

The Wolves prepare for an inevitable social and ecological collapse that will create a lawless world where real men can 'restart the world'. This requires building tribal communities that are 'are immersed in nature, in spirituality, free and independent of the materialistic hell of capitalism' (Waggener quoted in Wallace 2015). These tribes will take many forms, 'whether it be (…) religious tribalism or racial tribalism or whatever' (Southern Poverty Law Center 2017). It also requires reasserting traditional gender roles, especially the hierarchical 'gang masculinity' advocated by prominent member Jack Donovan, a popular author in the alt-right and manosphere movements. Although Donovan is homosexual, he rejects gay identity in favour of a hyper-masculine 'androphila'. This process of becoming 'barbarians' is also a spiritual battle that rejects the 'dead gods' while resurrecting 'old gods' through a combination of pagan spirituality, marital masculinity, and nature worship (Donovan 2014). Pagan and other pre-Christian spiritual practices have become an important ideological vector for combining white nationalism and ecology, including overtly racist variants like Wotanism, but also others like Ásatrúar and Odinism, which have become battlegrounds between racist and non-racist interpretations (Weber 2018).

In recent years eco-fascist and alt-right groups have also embraced green, primitivist and anti-civilization anarchist thinkers. These tendencies are united by a palingenetic impulse coupled with romantic longing for a prelapsarian universe of authenticity and ecological harmony unsullied by the corruption of modernity. The ultra-violent Atomwaffen Division considers Ted Kaczynski, imprisoned Unabomber and green anarchist icon, as one of their 'holy trinity' alongside Timothy McVeigh and Anders Breivik (Anti-Defamation League ND). For Kaczynski, National Socialism had redeemable aspects because it was 'partly a revolution against civilization' (Kaczynski 2017). Anarcho-primitivist theorist John Zerzan, a friend of Kaczynski's, rejects not only the state and capitalism but civilization, language and abstract thought. Sunshine (2009) has demonstrated how this critique echoes Nazi rhetoric discourse which identifies Jews with the ills of abstraction and modernity, while drawing from shared intellectual sources like Martin Heidegger, Oswald Spengler, and Mircea Eliade. Indeed, Zerzan has been praised by far-right figures including Russian fascist Aleksandr Dugin and on the white supremacist site Stormfront (Reid Ross 2017). Similarly, Deep Green Resistance (DGR), a group founded by green anarchist author Derrick Jensen, has been scorned by the left and embraced by the right for anti-trans and anti-immigration statements (Reid Ross 2017). At least three green anarchists have embraced fascism while imprisoned for environmental activism (NYCABF 2008; Valdinoci 2014). Green and primitivist anarchism share significant philosophical and political terrain with the ecofascist right, including ecological antimodernism, civilizational decline narratives, blood and soil sympathies, and hostility towards the left.

While most right ecological discourse romanticizes a premodern imaginary, it also takes urban neoliberal forms. At the opposite end of the aesthetic and temporal spectrum exists a hypermodern ecofascism. McFarlane (2018) argues that Silicon Valley's unique combination of ecology, neoliberalism, and techno-elitism has made it 'a significant incubator of "neoreaction" (NRx) and neo-fascist thought'. Drawing on accelerationist thinkers like Nick Land and the Italian Futurists, this tendency sees heroic individuals using technology to solve ecological problems created by the unenlightened masses, either via technological fixes or escape – walling themselves off from the teeming throngs or abandoning Earth altogether for the greener pastures of space.

Veganism and animal rights

While far from dominant in alt-right circles, veganism has become an increasingly visible theme within white nationalist and far-right groups (Forchtner and Tominc 2017). It is often advocated as natural and pure, but also an elitist gesture that dissociates one from the masses and, for some, aligns oneself with the diet of Adolf Hitler. Aryanism.net has a veganism subsection which states that 'a lifestyle that generates no demand for animal products, has always been a hallmark of an authentic National Socialist' (Aryanism.net). It continues, 'veganism is a direct consequence of the Aryan instinct of universal compassion. Far from its misassociation with the pacifistic hippy caricatures presented to us by the Jewish media, its most accurate spiritual association is with the archetypical heroic warrior' (ibid.).

Jayme Louis Liardi, founder of a popular YouTube channel called Simply Vegan, evolved into a white nationalist advocating an anti-globalist, racial separatist stance. In a video titled *My Awakening/Globalism vs Nationalism*, he outlined his transformation from a liberal, vegan, feminist multiculturalist into a pagan European nationalist, describing veganism as a search for personal truth informed by his European heritage (Liardi 2016). Capitalist consumerism – coded as Jewish – fills the vacuum of meaning for those without ethno-racial identity, thus 'If you have no identity, one will be installed for you by the kosher forces of the state' (ibid.). He advocates rediscovery of 'the Native European way' in order to create 'a future for our children free of this consumerism, this insanity, this anti-nature world' (ibid.) and is a regular guest on various far right media outlets (de Coning 2017).

One reason for veganism's newfound popularity is the rediscovery of the ecofascist vegetarian Savitri Devi, widely read in the alt-right. Prominent Northwest neo-Nazi Matthew Stafford lists her as an inspiration for his fascist veganism (Reitman 2018). Andrew Anglin, founder of the white supremacist Daily Stormer website, is a former vegan. Troy Southgate (2011) emphasizes the role of veganism in Third Positionism, and fascists in England have been actively involved in animal rights activism. By contrast, for the US right veganism remains mostly an individual lifestyle choice and intellectual interest; there is no evidence of participation in the wider animal rights movement. Nonetheless, the prominence of veganism in contemporary alt-right discourse cannot be reduced to purely strategic considerations. Instead, it shows these issues have no stable or neutral political valence, but rather are available for authoritarian and anti-egalitarian articulation.

Conclusion

The crossover between left and right politics is a confusing mixture of cynical co-optation alongside sincerely-held political convictions. While some groups strategically use ecological discourse solely to recruit new members from the left or public at large, many others are genuine in their advocacy of a far-right ecological perspective, especially tribalist, quasi-anarchist, and esoteric actors. Alexander Reid Ross suggests that this 'shared ideological space cannot be tidily blamed on co-optation', but instead constitutes a 'unique response to the same material conditions' (2017: 1).

Somewhat surprisingly, climate change does not seem to be a core theme in the groups and traditions under consideration here. This is likely due in part to the conflicting views found within the alt-right, where some dispute climate change as a 'cultural Marxist' lie while others accept its reality but direct blame onto immigration, the third world, or 'globalism'. What results is a big-tent ecumenicism that can accommodate both, wherein '[p]ersonal beliefs on whether or not climate change is anthropogenic do not matter since Alt-Right policies would be positive for the environment' (Evolalinkola 2017).

The current political moment has seen existing political institutions, parties, and ideologies lose legitimacy or collapse, opening new discursive terrain to actors on the left and right alike to offer new political visions and interpretations of events. The global surge of far-right populism across Europe, in the Philippines, Brazil, and the election of Trump in the United States has revealed the depth of dissatisfaction with the status quo and normalized previously far-right views. The Trump administration's ties to the alt-right have been extensively documented, including discursive as well as personnel overlap (Hawley 2017; Main 2017; Nagle 2017). The enemy identified by the ecological right closely approximates what Fraser calls 'progressive neoliberalism', an 'alliance of mainstream currents of new social movements (feminism, anti-racism, multiculturalism, and LGBTQ rights), on the one side, and high-end "symbolic" and service-based business sectors (Wall Street, Silicon Valley, and Hollywood), on the other' (2017). This political formation aptly describes the Clinton-Obama paradigm, which formally emancipated a variety of subaltern groups while simultaneously liberating the destructive forces of global capitalism. Progressive neoliberalism therefore contributes to a social dynamic ripe for the scapegoating political discourse of right populism. According to Fraser (2017), by 'Rejecting globalization, Trump voters also repudiated the liberal cosmopolitanism identified with it. For some (though by no means all), it was a short step to blaming their worsening conditions on political correctness, people of color, immigrants, and Muslims'.

Alt-right ecology adds an ecological dimension to this critique of progressive neoliberalism, blaming environmental problems on a left-identified form of global capitalism and the multiculturalism and immigration it ostensibly champions (Mix 2009).

Brown's (2006) essay 'American Nightmare' also illuminates recent developments on the right. She argues that by subordinating democratic values to a fundamentally amoral market logic, neoliberalism produced 'de-democratized' subjectivities confronting a social world emptied of normative meaning. This in turn sets the stage for neoconservatism to fill this ethical void, also bypassing democratic values, but this time in favour of an authoritarian form of Christian militarism. Contemporary right ecology performs a similar function, channelling the same sentiment into a different ideological container which reflects new political realities. They confront the same 'empire of Nothing' produced by neoliberalism but reject the war, centralized state power, and Judeo-Christian values which characterized neoconservatism, an ideology further tainted by the fact that many neocons are both Jewish and former leftists. They are not alone on the right in viewing neoconservatism as a suspiciously leftist species of conservatism: overly urban, intellectual, Jewish, and ideological. Alt-right ecology addresses the pervasive sense of social and ecological crisis in a precarious world facing climate change, offering new ideologies – anarchism, tribalism, paganism – which reject the state and Christianity as viable sources for social, ecological, and spiritual renewal.

On this reading, capitalism and fascism are closely intertwined. Of course, Frankfurt School Critical Theory has long insisted on centering this relationship, with Horkheimer's famous quip that 'whoever is not prepared to talk about capitalism should also remain silent about fascism' (quoted in Bronner and Kellner 1989: 78). Yet whereas the connection between the racialized scapegoating of traditional fascism and the dislocations of capital were rather opaque, mediated, and channelled into the state, alt-right ecology and ecofascism make this connection clearly and directly; capitalism causes ecological degradation and the state cannot save us (Bookchin 1982; Moore 2003). As social and ecological dystopia feature ever more prominently in popular culture, the vision of stateless gangs living off the land in a brutal social Darwinian universe is already a familiar one. Alt-right ecology warns of this dystopian future while also embracing its potentiality, giving it a veneer of inevitability and desirability. As social and ecological crises continue to deepen, and with few emancipatory political alternatives in sight, alt-right ecology is likely to keep growing.

Indeed, in March 2019 a self-professed ecofascist killed 49 worshippers at a mosque in Christchurch, New Zealand. In a manifesto published online before the attack he declared: "I am an Ethno-nationalist Eco-fascist. Ethnic autonomy for all peoples with a focus on the preservation of nature, and the natural order" (Pothast 2019). Yet at roughly the same time, as the 2020 US presidential election approaches a host of Democratic candidates have embraced the Green New Deal as a bold political vision for solving climate and economic problems at the same time. While debate continues about the ecological limits to "green growth" or structural constraints on national political action within global capitalism (Bernes 2019, Riofrancos 2019), the embrace of an idea only recently dismissed as too radical signals a promising shift in public discourse, one that reflects growing acceptance of the fact that avoiding or at least mitigating the looming climate crisis will require a more fundamental transformation of not only consciousness but political and economic institutions.

Note

1 Anti-racist and left-wing bioregionalists, like Cascadia flag designer Alexander Baretich, actively counter right-wing attempts to co-opt the term Cascadia.

References

Abbey, Edward (1988): *One Life at a Time Please*. New York: Henry Holt and Co.
Ahab (2017): Environmentalism and the alt right, *Altright.com*, 6 July. https://altright.com/2017/06/06/environmentalism-and-the-alt-right/ [10 July 2018].
Amend, Alex and Piggott, Stephen (2017): The Daily Caller has a white nationalist problem, *Southern Poverty Law Center*, 16 August. www.splcenter.org/hatewatch/2017/08/16/daily-caller-has-white-nationalist-problem [14 July 2018].
Anti-Defamation League (ND): Atomwaffen Division, *Adl.com*, undated. www.adl.org/resources/backgrounders/atomwaffen-division-awd [14 July 2018].
Aryanism.net (2018): Veganism, *Aryanism.net*. http://aryanism.net/culture/veganism/ [13 July 2018].
Aronoff, Kate (2019): The European Far-Right's Environmental Turn, *Dissent*, 31 May. www.dissentmagazine.org/online_articles/the-european-far-rights-environmental-turn [14 June 2019].
Autonomy Cascadia (2013): Autonomy Cascadia: A journal of bioregional decolonization, *Unsettling America*, 30 September. https://unsettlingamerica.wordpress.com/2013/09/30/autonomy-cascadia-a-journal-of-bioregional-decolonization/ [12 July 2018].

Bay Area National Anarchists (2007): A smear salad, *Bayareanationalanarchists.com*, 7 October. http://bayareanationalanarchists.com/blog/2007/10/a-smear-salad-nick-griffin-gre.ht [16 July 2018].

Bhatia, Rajani (2004): Green or brown? White nativist environmental movements. In: Abby L. Ferber (Ed). *Home-Grown Hate: Gender and Organized Racism*. New York: Routledge, pp. 205–225.

Bernes, Jasper (2019): Between the devil and the Green New Deal, *Commune*, 2. https://communemag.com/between-the-devil-and-the-green-new-deal/ [June 14 2019].

Biehl, Janet and Staudenmaier, Peter (1995): *Ecofascism: Lessons from the German Experience*. Oakland: AK Press.

Boggs, Kyle (2019): The rhetorical landscapes of the 'Alt Right' and the Patriot Movements: Settler entitlement to native land. In: Bernhard Forchtner (Ed). *The Far Right and the Environment: Politics, Discourse and Communication*. London: Routledge.

Bookchin, Murray (1982): *The Ecology of Freedom*. Palo Alto: Cheshire Books.

Bookchin, Murray and Foreman, Dave (1999): *Defending the Earth: A Dialogue between Murray Bookchin and Dave Foreman*. Boston: South End Press.

Bronner, Stephen Eric and Kellner, Douglas (1989): *Critical Theory and Society: A Reader*. New York: Routledge.

Brown, Wendy (2006): American nightmare: Neoliberalism, neoconservatism, and de-Democratization, *Political Theory*, 34(6): 690–714.

Caldwell, Christopher (2016): What the Alt-Right really means, *New York Times*, 2 December. www.nytimes.com/2016/12/02/opinion/sunday/what-the-alt-right-really-means.html?_r=0 [3 July 2018].

Coning, Alexis de (2017): Why so many white supremacists are into veganism, *Vice.com*, 22 October. www.vice.com/en_us/article/evb4zw/why-so-many-white-supremacists-are-into-veganism [18 July 2018].

Corcoran, Casey (2014): Slackwater rising: An open letter to Rose City Antifa. April 8. https://milegaiscioch.wordpress.com/2014/04/08/slackwater-rising-an-open-letter-to-rose-city-antifa/ [9 July 2018].

Devall, Bill and Sessions, George (1985): *Deep Ecology: Living As if Nature Mattered*. Salt Lake City: Gibbs Smith.

Donovan, Jack (2014): A time for wolves, *Jackdonovan.com*, 14 June. www.jack-donovan.com/axis/2014/06/a-time-for-wolves/&num=1&strip=1&vwsrc=0 [2 July 2018].

Durham, Martin (2007): *White Rage: The Extreme Right and American Politics*. London: Routledge.

Evolalinkola (2017): The alt-right is Green: Not a Pepe meme, *Altright.com*, 26 July. https://altright.com/2017/07/26/the-alt-right-is-green-not-a-pepe-meme/

Forchtner, Bernhard (2019): Nation, nature, purity: Extreme-right biodiversity in Germany. In: Christoffer Kølvraa and Bernhard Forchtner (Eds). *Cultural Imaginaries of the Extreme Right*. Special issue of *Patterns of Prejudice*, 53(2).

Forchtner, Bernhard and Tominc, Ana (2017): Kalashnikov and cooking-spoon: Neo-Nazism, veganism and a lifestyle cooking show on YouTube, *Food, Culture & Society*, 20(3): 415–441.

Foreman, Dave (1987): Interview with Bill Devall. *Simply Living* 12.

Francey, Matthew (2013): This guy wants to start his own Aryan country, *Vice.com*, 12 February. www.vice.com/en_us/article/4wqe33/this-guy-wants-to-start-his-own-aryan-country [16 August 2018].

Fraser, Nancy (2017): The end of progressive neoliberalism, *Dissent*, 2 January. www.dissentmagazine.org/online_articles/progressive-neoliberalism-reactionary-populism-nancy-fraser [16 August 2018].

Griffin, Nick (2005): Nick Griffin, 'National-Anarchism: Trojan Horse for white nationalism', *Green Anarchy* 19. http://web.archive.org/web/20090825115331/; www.greenanarchy.org:80/index.php?action=viewwritingdetail&writingId=150 [15 July 2018].

Hawley, George (2017): *Making Sense of the Alt-Right*. New York: Columbia University Press.

Hawthorne, Derek (2011): In praise of Pentti Linkola, *Countercurrents.com*. www.counter-currents.com/2011/06/in-praise-of-pentti-linkola/ [16 August 2018].

Identity Evropa (2017): About us, *Identityevropa.com*. www.identityevropa.com/about-us [16 August 2018].

Jacob, R.J. (2013): Cascadia: Interview with Casey Corcoran, *Attack the System*, September 1. https://attackthesystem.com/2013/09/01/cascadia-interview-with-casey-corcoran/ [9 July 2018].

Johnson, Greg (2018): Greg stark interviews Greg Johnson on eco-fascism, *Countercurrents*, April 14. www.counter-currents.com/2018/04/robert-stark-interviews-greg-johnson-on-eco-fascism-2/ [2 July 2018].

Kaczynski, Ted (2017): Ted Kaczynski on individualists tending toward savagery, *The Wild Will Project*, November 28. www.wildwill.net/blog/2017/11/28/ted-kaczynski-individualists-tending-toward-savagery/ [20 July 2018].

Kavanaugh, Shane Dixon (2018): Oregon Wells Fargo mortgage consultant out amid Charlottesville, white nationalist allegations, *The Oregonian*, March 19. www.oregonlive.com/pacific-northwest-news/index.ssf/2018/03/oregon_wells_fargo_mortgage_co.html [14 July 2018].

Kolankiewicz, Leon (2015): Get the new editions of man swarm by CAPS advisory board member Dave Foreman, *Californians for Population Stabilization Website*. www.capsweb.org/blog/get-new-edition-man-swarm-caps-advisory-board-member-dave-foreman-laura-carroll [13 July 2018].

Liardi, Jayme Louis (2016): My awakening: Globalism vs. nationalism, *Youtube.com*. www.youtube.com/watch?v=TAH8on_kS1s

Linkola, Pentti (2006): *Can Life Prevail? A Radical Approach to the Environmental Crisis*. Budapest: Arktos Press.

Lokting, Britta (2018): Fear and loathing in Cascadia, *The Baffler*, February 27. https://thebaffler.com/latest/cascadia-lokting [15 July 2018].

Lyons, Matthew (2018): *Insurgent Supremacists: The U.S. Far Right's Challenge to State and Empire*. Montreal: Kersplebedeb.

Macklin, Graham D. (2005): Co-opting the counter culture: Troy Southgate and the National Revolutionary Faction, *Patterns of Prejudice*, 39(3): 301–326.

Main, Thomas J. (2017): *The Rise of the Alt-Right*. Washington, DC: Brookings Institution.

McFarlane, Key (2018): The greenhouse effect, *Metamute*, February. www.metamute.org/editorial/articles/greenhouse-effect [2 July 2018].

Minkowitz, Donna (2017): Hiding in plain sight: An American renaissance of white nationalism, *The Public Eye*, Fall issue. www.politicalresearch.org/2017/10/26/hiding-in-plain-sight-an-american-renaissance-of-white-nationalism/ [16 August 2018].

Mix, Tamara L. (2009): The greening of white separatism: Use of environmental themes to elaborate and legitimize extremist discourse, *Nature and Culture*, 4(2): 138–166.

Moore, Jason W. (2003): The modern world-system as environmental history? Ecology and the rise of capitalism, *Theory and Society*, 32(3): 307–377.

Nagle, Angela (2017): *Kill All Normies: Online Culture Wars From 4Chan And Tumblr To Trump and The Alt-Right*. Portland: Zero Books.

New York City Anarchist Black Cross Federation (2008): ELP withdraws support for Christopher 'Dirt' McIntosh, *Nycindymedia.org*, 18 February. https://nyc.indymedia.org/en/2008/02/94825.html [21 July 2018].
Northwest Front (2010a): Northwest Front homepage. http://northwestfront.org/ [12 July 2018].
Northwest Front (2010b): Fundamental principles of Northwest migration, *Northwestfront.org*. http://northwestfront.org/about/principles-of-migration/ [13 July 2018].
Northwest Front (2010c): About the Tricolor flag, *Northwestfront.org*. http://northwestfront.org/about/tricolor-flag/ [13 July 2018].
Olsen, Jonathan (1999): *Nature and Nationalism: Right-Wing Ecology and the Politics of Identity*. New York: St. Martin's Press.
Olsen, Jonathan (2000): The perils of rootedness: On bioregionalism and right wing ecology in Germany, *Landscape Journal*, 19(1/2): 73–83.
Pacific Northwest Antifascist Workers Collective (2018): Rebranding Fascism and Refinancing Mortgages: Andrew Murphy Harkins, Portland's Nazi Banker. 8 March. http://pnwawc.com/tag/true-cascadia/
Park, Mi (2013): The trouble with eco-politics of localism: Too close to the far right? Debates on ecology and globalization, *Interface*, 5(2): 318–343.
Pothast, Emily (2019): What the New Zealand Killer's manifesto tells us about the radicalization of white men, *Medium*, 15 March. https://medium.com/@emilypothast/what-the-christchurch-killers-manifesto-tells-us-about-the-radicalization-of-white-men-c55857149b33 [June 14, 2019].
Preston, Keith (2015a): Upcoming appearances by Keith Preston, *Attack the System*, 22 October. https://attackthesystem.com/2015/10/22/upcoming-appearances-by-keith-preston-and-some-humor-from-the-critics/ [17 July 2018].
Preston, Keith (2015b): More anarchistic than thou, *Attack the System*, 16 November. https://attackthesystem.com/2015/11/16/more-anarchistic-than-thou/ [17 July 2018].
Reid Ross, Alexander (2017): *Against the Fascist Creep*. Oakland: AK Press.
Reitman, Janet (2018): All American Nazis, *Rollingstone.com*, 2 May. www.rollingstone.com/politics/politics-news/all-american-nazis-628023/ [23 July 2018].
Rinehart, Vince (2013): PIELC panel: Cascadian bioregionalism and the indigenous future, *Lingit Latseen*, 4 March. https://lingitlatseen.wordpress.com/2013/03/04/pielc-panel-cascadian-bioregionalism-and-the-indigenous-future/ [16 July 2018].
Rinehart, Vince (2016a): A native American white nationalist? *Attack the System*, 18 October. https://attackthesystem.com/2016/10/18/a-native-american-white-nationalist/ [17 July 2018].
Rinehart, Vince (2016b). LaVoy Finicum's message to Native Americans, *Attack the System*, 29 January. https://attackthesystem.com/2016/01/29/lavoy-finicums-message-to-native-americans/ [17 July 2018].
Rural Organizing Project (2017): Racism & identity in the Patriot Movement. www.rop.org/up-in-arms/up-in-arms-section-i/patriot-movement-historically-nationally/racism-identity-movement/ [20 July 2018].
Riofrancos, Thea (2019): Plan, mood, battlefield – reflections on the Green New Deal. *Viewpoint*, 16 May. www.viewpointmag.com/2019/05/16/plan-mood-battlefield-reflections-on-the-green-new-deal/ [June 14, 2019].
Sale, Kirkpatrick (2016): Is secession a right? *Middlebury Institute* webpage, 26 March. https://kirkpatricksale.wordpress.com/ [13 July 2018].
Sanchez, Casey (2009): California racists claim they're Anarchists, *Southern Poverty Law Center Intelligence Report*, 29 May. www.splcenter.org/fighting-hate/intelligence-report/2009/california-racists-claim-they%E2%80%99re-anarchists [22 July 2018].

Sears, Kelton (2016): A bioregional declaration of interdependence, *Seattle Weekly*, 16 November. www.seattleweekly.com/arts/a-bioregional-declaration-of-interdependence/ [12 July 2018].

Snyder, Timothy (2015): *Black Earth: The Holocaust as History and Warning*. New York: Tim Duggan Books.

Southern Poverty Law Center (2017): A chorus of violence, *Southern Poverty Law Center*, 27 March. www.splcenter.org/hatewatch/2017/03/27/chorus-violence-jack-donovan-and-organizing-power-male-supremacy [17 July 2018].

Southgate, Troy (2011): Troy Southgate Interview from 2001, *Herjafather*, 29 March. https://herjafather.wordpress.com/2011/03/29/troy-southgate-interview-from-2001/

SPLC (2010): Greenwash: Nativists, Environmentalism, and the Hypocrisy of Hate, Southern Poverty Law Center, 1 June. www.splcenter.org/20100630/greenwash-nativists-environmentalism-and-hypocrisy-hate [17 July 2018].

Stern, Alexandra (2005): *Eugenic Nation: Faults and Frontiers of Better Breeding in Modern America*. Berkeley: University of California Press.

Stern, Alexandra Minna (2019): *Proud Boys and the White Ethnostate: How the Alt-Right Is Warping the American Imagination*. Boston: Beacon Press.

Stevens, Brett (2011): Nationalists unite, *Amerika.org*, 7 September. www.amerika.org/politics/nationalists-unite/?doing_wp_cron=1340368118.8628630638122558593750&replytocom=7785 [21 July 2018].

Stevens, Brett (2017): Deep Ecology and the Alt Right, *Amerika.org*, 7 February. www.amerika.org/politics/deep-ecology-and-the-alt-right/ [3 July 2018].

Sunshine, Spencer (2008): Rebranding fascism: National anarchism and third positionism, *Political Research Associates*, 28 January. www.politicalresearch.org/2008/01/28/rebranding-fascism-national-anarchists/ [24 July 2018].

Sunshine, Spencer (2009): John Zerzan's twilight of the machines, *Fifth Estate*. www.fifthestate.org/archive/381-summer-fall-2009/john-zerzans-twilight-machines/ [6 July 2018].

Tax, Meredith (2012): *Double Bind: The Muslim Right, the Anglo-American Left, and Universal Human Rights*. New York: Centre for Secular Space.

Taylor, Blair (2017): Ruthless critique or selective apologia: The postcolonial left in theory and practice, *Amerikastudien/American Studies*, 62(4): 649–661.

Valdinoci (2014): Former ELF/Green scare prisoner 'Exile' now a fascist, *NYCAntifa*, 5 August. https://nycantifa.wordpress.com/2014/08/05/exile-is-a-fascist/ [2 July 2018].

Wallace, Eric (2015): Eco-Punks, *Blue Ridge Outdoors*, 5 May. https://web.archive.org/web/20151119022642/; www.blueridgeoutdoors.com/go-outside/eco-punks-the-wolves-of-vinland-badasses-dare-you-to-re-wild-yourself/ [2 July 2018].

Weber, Shannon (2018): White supremacy's old Gods: The far right and neopaganism, *Political Research Associates*, 1 February. www.politicalresearch.org/2018/02/01/white-supremacys-old-gods-the-far-right-and-neopaganism/ [7 July 2018].

Woodruff, Betsy (2015): Inside Virginia's church-burning werewolf white supremacist cult, *The Daily Beast*, 11 November. www.thedailybeast.com/inside-virginias-church-burning-werewolf-white-supremacist-cult [9 July 2018].

17
THE RHETORICAL LANDSCAPES OF THE 'ALT RIGHT' AND THE PATRIOT MOVEMENTS

Settler entitlement to native land

Kyle Boggs

> What is an ideology without a space to which it refers, a space which it describes, whose vocabulary and links it makes use of, and whose code it embodies?
>
> Henri Lefebvre, *The Production of Space (2011: 44)*

Introduction

This chapter forges new connections between white settler colonialism and environmental communication at the confluence of the 'Alt-Right'[1] movement and western contingencies of the 'Patriot' movements, which are both subgroups of the far right in the United States. I pluralize 'movements' to emphasize that these actors are not monolithic or ideologically coherent, but do share beliefs. For example, they distain the federal government, which they consider over-reaching, particularly as it relates to control over public lands in the western United States, and belief in the importance and vulnerability of the Second Amendment, which grants citizens the right to bare arms. The far-right patriot movements are a growing faction of white rural conservative nationalists and land-use militants, who include militia groups like the Oath Keepers (made up largely by former and current members of the military and law enforcement) and Three-Percenters (a name derived from a mistaken claim that only 3% of colonists fought in the American Revolution[2]), as well as so-called 'sovereign citizens', who draw a distinction between themselves as 'original citizens' of the states, and US citizens or 'Fourteenth Amendment citizens' (Abanes 1996; Sunshine et al. 2016). Therefore, they do not recognize the legitimacy of the federal government, and do not believe they are obligated to follow most federal laws (Potok 2012; Pogue 2015). Their anti-government rhetoric obscures their investment in white settler colonialism, as it hinges on a reading of history that emphasizes the end of settlement, when the federal government took possession of unclaimed land, selectively ignoring the stages leading up to it, when the government systematically and violently seized Native land for white settlement (Gallaher 2016; Inwood and Bonds 2017). Both the rhetorical erasure of Native presence and the disavowal of the violence that underpins white settlement produces an understanding of the environment as an empty slate in which white settler belongings are inscribed through various mythologies. These mythologies accommodate an environmental aesthetic which positions the land in service of white settlers. In an effort to better understand the communicative practices of far right groups in the United States that aim to form alliances across shared ideologies, connecting identity to land, this chapter explores the convergence of far right white nationalism and the frontier narratives that animate the mythic west – where the social imaginary coheres to the physical landscape.

The problem for activists and academics who write about the far right is that it's in a constant state of flux – coalescing in some areas, fracturing in others – moment by moment fractured ends connect in unexpected ways. This chapter focuses on and attempts to contextualize one of these moments of coalescence, which is the growing synchronicity and cooperation between the overt alt-right white nationalism as seen in the 2017 'Unite the Right' rally in Charlottesville, Virginia, and elsewhere and far-right patriot movements as seen in the western United States, specifically the 2014 standoff at Cliven Bundy's ranch and the 2016 armed occupation of the Malheur National Wildlife Refuge in Oregon (for a more general look at the alt right and the natural environment, see Taylor 2019 in this volume). Like Alexander Reid Ross (2017), I do not suggest that groups like those that constitute the Patriot movements are 'becoming fascist', but increased hybridization suggests they are 'part of a larger process that facilitates fascism's creep into power' (16). Also, the Patriot movements indeed have 'fascist roots', originally 'created by fascists out of the broader settler-colonial orientation for the purposes of advancing fascism' (Ross 2017: 118). However it is necessary to analyse them 'in light of their present form, rather than simply their genealogy' (Ross 2017: 118). Fused with the militia movement of the 1980s, they 'found their early support base not in overt neo-Nazi organizations, but in the Wise Use movement', a group of western ranchers, loggers and local law enforcement galvanized over how public lands are managed, and whom valued the environment for its utilitarian and extractive uses rather than aesthetic or ecological values (Ross 2017: 23). In the interest of those communities seeking to understand and learn how best to respond to these groups, however, it is crucial that they are contextualized. In short, because fascism once foregrounded the Patriot movements, there lies much potential for it to happen again in new ways. Activists on the ground during the armed occupation of the Malheur National Wildlife Refuge have observed this connection. 'Their core ideas – that the federal government lacks authority to own public land, and that county sheriffs are the highest form of law enforcement – are rooted in hard-right racist ideologies created to sidestep federal civil rights laws by elevating local authority' (McKinnon 2016). This analysis, therefore, is also part of the large spectrum of anti-racist work that monitors the relationships white racists seek to build across the far right political landscape, but also aims to show how the physical landscape itself has played an active role in shaping far-right political discourse.

This rhetorical analysis deploys a material-discursive understanding of language, that is, that language is shaped by an affective relationship between physical or material, reality and discourse, which can include historical and cultural articulations of place often rooted in deeply entrenched power dynamics (Barad 2008). This approach is important in shedding light on how these two factions are finding common ground, and to decode the settler logic that foregrounds how they articulate their relationship and entitlement to stolen Native land. A material-discursive lens is dependent on an understanding of rhetoric as an inquiry into the ways in which discourse affects and is affected by the material environments in which we engage, shedding light on how the production and performance of discourse reflect those very effects.[3] In addition to the historical background and application of critical, cultural and rhetorical theory necessary for this analysis, I draw much of my examples from the work of journalists who have covered the far right and Patriot movements, reporting on the ground from courtrooms, protests, press conferences, and elsewhere.

In the sections that follow, I will provide a brief summary of what constitutes the so-called alt-right movement in the United States, and how recent collaborations with the Patriot movements signal a renewed understanding that white supremacy is necessitated by control over land, not just land as material space, but as cultural space. Next, I will emphasize the significance of the mythic west, valorized stories of the 'frontier', a metaphor for white expansion and entitlement to Native lands. Articulations of the land within the Patriot movements are inevitably woven into these mythologies and play a key role in the formation of settler identities in the United States, but particularly among the western contingencies of the Patriot movements. Finally, I hone in on the Nazi propaganda slogan 'blood and soil', which was chanted by white nationalists in Charlottesville, Virginia, as a point of rhetorical cohesion between the two groups that at once connects discourses of ethnic identity (blood) and the materiality of stolen Native land (soil) within the settler colonial context of the United States. In short, I argue that alt-right white nationalists see their 'blood and soil' ideology at work in the actions and tactics of the Patriot movements, and those points of cohesion tie identity to land.

Alt-right white nationalism and patriot collaborations

First articulated in 2008 as 'alternative right' by professor emeritus Gottfried (2018),[4] and popularized as 'alt right' in 2016 by white nationalist Richard Spencer, the full scope of the term 'alt right' is still in flux. Though a spectrum of radical far-right groups and subgroups exist, the Alt-Right is less a group itself than it is a set of far-right ideologies – sometimes competing ideologies – united in their rejection of mainstream conservatism and in favour of white nationalism and white supremacy. They are Nazis, neo-Nazis, Ku Klux Klan members, neo-Confederates, and others who are invested in transforming the country into a 'white ethnostate'. Richard Spencer's white nationalism has helped rebrand fascist ideologies for a new generation under the guise of 'free speech'. As this 'relatively new and amorphous group' has evolved, its appeals to intellectualism and political correctness have made 'space for overtly pro-white rhetoric in mainstream US American public discourse' (Hartzell 2018: 7). This branding has allowed them to recruit others, but also to 'construct rhetorical distance between white nationalism and white supremacy' as they infiltrate the mainstream public discourse (ibid.: 9).

'Alt right' is not just a movement or an ideology, but an argument, one deployed specifically to articulate uniformity and cohesion across the far right where it does not actually exist. This argument was most clearly articulated in the language chosen for the name of the now infamous and tragic August 2017 rally in Charlottesville, which was 'Unite the Right'. The call to unite, which is as much an admission of discontinuity, sought to bring in as many far-right groups together under the same banner in reaction to the city's planned removal of Jim Crow-era Confederate monuments, specifically Confederate General Robert E. Lee. For these groups, the 'alt right' identification softens their otherwise neo-fascist image, and helps them to recruit others.

Also present at the Charlottesville rally were heavily armed militiamen in military-style camouflage standing between alt-right demonstrators and anti-fascist counter protesters. At least at the organizational level – these groups have distanced themselves from the label, primarily so as not to appear racist. Upon publication of photographs of the event, the national militia group, the 'three percenters', told supporters to 'stand down' (The Three Percenters n.d.). The statement said, 'we cannot have this organization tainted by news outlets as they will almost certainly report that we have aligned ourselves with white supremacists and Nazis' (The Three Percenters n.d.). Despite statements such as this, and other expressions of neutrality, there is growing evidence that militia groups have moved toward protecting white supremacists, acting as 'security' or 'private police' during street demonstrations (Michel 2017).

This observation follows the historical trajectory of militia groups in the United States, which Dunbar-Ortiz (2017) has traced to pre-Civil War era slave patrols and even further back to their role in killing off Native people for white settlement. More recently, anti-immigrant militia groups like the 'Minutemen' have been involved in patrolling the US/Mexico border, and others such as the 'Oath Keepers', have served as security for an anti-Muslim hate group (Morlin 2017). They did the same thing for white property owners in Ferguson, Missouri, during protests on the anniversary of the death of a black teen who was killed a year earlier by a white police officer (Larimer and Phillip 2015). Spencer Sunshine with Rural Organizing Project and Political Research Associates have more specifically traced some individual members of militia groups that constitute the growing Patriot movements in the western United States to white nationalist organizations (2016: 28–33).

In the fallout after the Charlottesville rally, the white nationalist group, Vanguard America strategically rebranded the group 'Patriot Front', according to Southern Poverty Law Center (2017). 'Patriot Front' rhetorically signals both 'Patriot' groups, and neo-Nazi groups that have utilized names like 'Storm Front', among others. The name and others like it allow white supremacists to access mainstream culture, as journalist Shane Burley (2018) has observed. 'To gain a connection to the mass culture, white supremacists require a crossover point, a group of people close enough to their ideas and built into the larger conservative social framework to help move them onto a public stage'. This observation is supported by rhetorical theorists like Burke (2013: 55) who have written about how affinities can be developed through the process of 'identification', that is, identifying points of cohesion and shared motives and values by speaking the other's language 'by speech, gesture, tonality, order, image, attitude, idea'. Burke also frames the use of language 'as a symbolic means of inducing cooperation in beings that by nature respond to symbols' (ibid.: 43). The material environment, rich with cultural myths, and competing histories also constitute discursive symbols that affect language and communication. The landscapes in which we engage also induce particular rhetorical responses: ways of knowing and being in such spaces. Applying Burke's (1984: 49) observations on exclusion, that 'a way of seeing is also a way of not seeing', the rhetorical response also produces assumptions about who belongs, and how they belong, but also cultural assumptions about those who do not.

Last year, the Patriot Front received national media attention for posting black and white flyers that tout different white nationalist messages. One of them featured a map of the country divided by regions that mark the dates of white expansion moving from east to west with text that reads, 'Not Stolen Conquered'. Here, 'conquest' is not just a legacy – but one that must be maintained today to preclude Native peoples from access and control over their own lands and cultural resources. On all the flyers was a link to the group's website, 'bloodandsoil.org', a reference to Nazi propaganda that was chanted by alt-right white nationalists on the streets of Charlottesville.

Alt-right white nationalism seeks to build a white 'ethnostate', a country inhabited only by white people, which reflects their inherently violent anti-immigrant views, but is also predicated on a hyper-nationalist racism reminiscent of the Nazi's historic concept of the *Volk*, 'an almost mystical faith in the goodness of the ethnically pure "'common German'"' (Brüggemeier et al. 2005: 5). The *Volk* 'colored all aspects of the German political, social, and intellectual life after World War I – the nature protection movement most of all' (ibid.: 5). In general, German conservation in the 1920s and 1930s echoed Nazi ideology, positioning Germany's landscape as the 'foundation for an allegedly superior race of people' (Closmann in Brüggemeier et al. 2005: 26). Clearly alt-right white nationalists see an ideological opportunity in the largely white Patriot movements in the western United States, whose adherents exploit a romanticized image of the rural rancher – a similar image to the German *Volk* – to defend and claim land for white people through paramilitary action. This environmental ethos is predicated on constructions of purity – racial and ecological – and performances of conquest situated in place. The space of the western United States, which is at once the physical landscape and the mythologies that produce it, actively make these connections possible.

The material-discursive frontier

Foucault (1986) has observed that space is discursive, that it has a history. Informed by cultural geography and critical cultural studies, theories of rhetoric that articulate the materiality of cultural spaces as integral to the construction of discourse,[5] is particularly useful to analyse how these group's politics are informed by the relationships they draw between the mythic west – which exists in a hyper-romanticized social imaginary where white settlers are written from the centre – and the actual lived material environments in which they engage.

Particularly in the rural western areas of the United States – eastern Oregon, Idaho, Wyoming, Montana, Nevada, Utah, Arizona, and western Colorado – entitlement to Native land is tied to identity. Space that is discursive is also persuasive, therefore, producing a way of understanding and being in space; accommodating a material-discursive view of the landscape, Clark (2004) has observed that Americans in particular have always experienced the landscapes of the United States rhetorically. He (ibid.: 3) notes that the 'relational encounter' that includes the physical material landscape tied to discourses about it 'constitute a person's social and cultural experience'. A key figure in nationalism studies, Anderson (2016: 6) observes that 'fellow members of even the smallest nation will never know most of their fellow members (…) yet in the minds of each lives the image of the communion (…) the style in which they are imagined'. Certainly, the material landscape constitutes part of this 'image' and 'style' that play a significant role in the formation of and performance of a national identity, where belongings are tied to the landscape in highly gendered and racialized ways.

This rural landscape is steeped in cultural myths that sustain white supremacy – the settler, the frontiersman, the cowboy. The mythic American West tells the story of a self-stylized rugged masculinity forged through a sense of independence and self-reliance – where white men are virtuous and Native people, if present at all in the story, must be destroyed, and that land was justly and heroically acquired. This is a 'region where the federal administration of millions of acres of public lands has long sat uneasily next to a pervasive and mythologized ethos of bootstraps individualism and white property rights' (Inwood and Bonds 2017: 255). The mythic American West, or the 'frontier' is less a geographical reference than it is ideological. The concept of the frontier was popularized by historian Fredrick Jackson Turner in 1893 (2014), when he used the term as a metaphor to understand

American identity and politics. Turner understood the frontier not only as a male-dominated space, but 'the meeting point between savagery and civilization' (ibid.: 10). He also drew a distinction between the American frontier as lying on the 'edge of free land' (or stolen Native land) and the frontiers of Europe as being more sharply defined outside of dense populations (ibid.: 11). Today, historians understand the frontier as a myth (Slotkin 1998), defined as 'a set of narratives that acquire through specifiable historical action a significant ideological charge' (ibid.: 31). The myth of the frontier has been persistent and pervasive, evolving from ideals of America as the 'land of opportunity' where the strongest are valorized through conquest, from exploitation of land through industrialization, to the twentieth century and the present where it is seemingly cemented as cultural ideology, in popular culture, and embedded in law and politics.

This myth is reflected more broadly and popularly in the construction of wilderness itself, a fusion of material space and white supremacy that DeLuca and Demo (2001: 550–555) refer to as 'white wilderness'. The latter is reflected in literature, art, and the canonization of white environmental thinkers like John Muir. Finney (2014: 3) argues, whiteness as a way of knowing, becomes *the* way of understanding our environment, and through representation and rhetoric becomes entrenched in our national psyche. For white rural communities that embrace the confluence of ideologies that constitute the patriot movement, it is also an ethos that must be constantly revisited, justified, and performed to legitimize their perceived dominion over stolen Native land. Indeed, colonization did not just happen, but *is happening* (Wolfe 2006; Veracini 2015). The mythic west ultimately tells one story, however, obscuring Native genocide and displacement that precipitated white settlement, as well as the resilience of Native communities who resist colonization through the present, and maintain claims to their ancestral lands. 'The settlers of the West took the view that the land was there to be taken, and that the rules and regulations of the government did not change their natural rights as citizens' (Malcolm Rohrbough quoted in Limerick 1987: 61). The anti-government rhetoric of armed land-use militants described in this chapter embody this same sense of entitlement today. Ironically, they believe it is the federal government who is exerting power over their land, and in so doing, reinforce the same settler colonial power structures that originally fostered settlement of stolen Native lands, animating Wolfe's (2006: 388) observation that settler colonialism 'destroys to replace'. The mythic West has redefined the material space of the landscape to sustain white settler belongings.

The settler colonial situation

In the context of the United States, it is impossible to interrogate far-right appeals to nationalism on stolen Native[6] land without situating these ideologies as firmly rooted in white settler colonialism. Settler Colonialism is transnational and functions through what Wolfe (2006: 387) describes as a 'logic of elimination', a process by which indigenous populations are replaced by invasive settlers.

I agree with Morgensen (2014) who uses the term 'white settler colonialism' as a 'tactic for drawing white people to address white-supremacist settler colonialism multidimensionally'. This multidimensionality builds on Wolfe's (2006: 388) understanding of settler colonialism as a 'structure, not an event', opening up new possibilities to analyse colonialism in terms of discourse and performativity. Further, white settler colonialism is a process that is 'regenerative' and 'situational' (Veracini 2010: 3). It is also, therefore, rhetorical.

Rhetorical discourse is not merely called into existence by the situation, as Bitzer (1968) argues. 'Situations' themselves 'are rhetorical', and 'obtain their character from the rhetoric which surrounds them or creates them' (Vatz 1973: 159). Debauer's (2005: 9) recontextualization of the rhetorical situation as 'rhetorical ecologies' is useful as it is based on an 'amalgamation of processes and encounters', which frame rhetoric within the context of 'temporal, historical, and lived fluxes' that accounts for 'a circulating ecology of effects, enactments, and events'. Debauer's 'rhetorical ecologies' is, therefore, suited to address the full range of material, historical, and cultural elements of environmental communication specific to settler colonialism as a 'situation' 'premised on the traumatic, that is, *violent*, replacement and/or displacement of Indigenous others' (Veracini 2008: 364).

Veracini (2008: 364) describes 'settler colonialism's need to disavow its violence' as the 'settler colonial situation'. This rhetorical situation, according to Veracini, produces 'lingering anxieties over settler legitimacy and belonging' (ibid.). When it comes to issues of belonging, it is often more revealing to look at *how* one belongs. For the Patriot movements in western United States, the anxiety produced by the settler colonial situation has resulted in a near constant performance of frontier masculinity that is inherently anti-indigenous, where cultural ties to the land are legitimated by white men with guns wearing cowboy hats. The importance of such cultural symbols was emphasized in 2016 during the trial of Ammon Bundy, one of the men who engaged in the armed occupation of the Malheur National Wildlife Refuge in Oregon to protest government ownership of land. His lawyer argued for his client's right to wear cowboy boots at his trial (Bernstein 2016). 'These men are cowboys, and given that the jury will be assessing their authenticity and credibility, they should be able to present themselves to the jury in that manner' (ibid.). The subtext of such statements implies that authentic cultural ties to land are legitimated and delegitimated in certain aesthetic ways. For white racists who link culture to biology, such performances that seemingly cement ethnic identity to land have proven to be extremely attractive to form alliances across the far-right political spectrum.

Blut und Boden from the Nazis to Charlottesville

The rally to 'Unite the Right' in Charlottesville was one of the largest white nationalist rallies in at least a decade, described by the city's mayor as a 'display of visual intimidation' (Ruiz and Mccallister 2017). Alongside the swastikas,

Confederate flags, and burning torches reminiscent of a lynch mob, there were also forms of verbal intimidation: 'Jews will not replace us', 'white lives matter', and 'blood and soil'.

Scores of articles immediately started popping up attempting to contextualize 'blood and soil' (Epstein 2017). Devised by nineteenth century German nationalists, *Blut und Boden!* was popularized as Nazi propaganda to 'invoke patriotic identification with native national identity' (Neiwert 2017). The phrase is meant to unite ethnic identity (blood) to land (soil), idealizing the rural farm life of the German-Nordic peasant – viewed as noble and sedentary – against the racist and anti-Semitic view that Jewish people corrupted urban spaces, and were inherently nomadic (Kiernan 2007: 2). It was also regarded as a kind of environmentalism, important for National Socialist ideologues and 'leading conservationists' 'like Alfred Rosenberg, Richard Darré, and other[s], [who] were coming to the conclusion that the mystical connection of "blood and soil" existed between the people and the land' (Closmann in Brüggemeier et al. 2005: 26). Darré in particular was instrumental in bringing the idea of 'blood and soil' to his agrarian ideology, who wanted '"blood" to be understood as "race"', referring to the German peasantry as the 'life source of the Nordic race' (Gerhard in Brüggemeier et al. 2005: 132). The phrase was embraced by the Third Reich as exemplifying the German concept of *Lebensraum*, an ideological principal that sought to reclaim and expand what they believed to be historically German areas of Eastern Europe, comprising practices of settler colonialism not radically different from the guiding principles and policies of 'Manifest Destiny' that continues to drive settler entitlement to stolen Native land in the United States today (Friedrichsmeyer et al. 1998; Wolfe 2006; Kiernam 2007).

'Blood and soil' is also a material-discursive argument for highly racialized place-based belongings. The two words, individually signal a relationship between discourse and materiality, as well as together. The slogan evokes 'blood' as a means of articulating ethnicity and nationality – relying on the false assumption that biology is connected to culture and ethnicity. In her analysis of the way in which Native American DNA is problematically used to prove or disprove tribal relations, Tallbear (2013) extends Haraway's notion of the 'material semiotic' to describe DNA itself as a 'material-semiotic object of knowledge' that obscures 'far more complex political histories of relations and power' (70–71). 'Soil', is similarly used to denote the materiality of physical land and national space. 'Blood and soil' at once flags ethnicity and nationality as inextricably linked to a social space that is ever-expanding, affirming Massey's (1994: 5) view of the particularity of social space as a 'vast complexity of interlocking and articulating social relations'. At the same time as it articulates belonging, these discursive 'processes of exclusion' also 'limit what can be said and what can be counted as knowledge' (Mills 1997: 57). The slogan obscures any discussion regarding who inhabited the land before, how they lived there, and how historically German areas were mapped out. 'Blood and soil' begins from the assumption that there was no 'before'.

Bundy standoffs and the patriot appeal to 'blood and soil' ideology

The 2014 'Bundy standoff', an armed confrontation in rural Nevada that attracted roughly 400 supporters – many of them heavily armed militia members from those groups that constitute the Patriot movements – ensued over a legal dispute spanning two decades in which rancher Cliven Bundy refused to pay the government the nominal fee it charges all ranchers to graze their cattle on public lands under the jurisdiction of the US Bureau of Land Management (BLM) (Montero 2017). Bundy owes over $1 million in back fees, which include fines accrued for allowing his cattle to graze on federal land protected to preserve crucial habitat for a rare desert tortoise. Two years later, his sons galvanized 25–40 members of these same militia groups to seize and occupy the federally operated Malheur National Wildlife Refuge in Oregon (Levin 2016). Both standoffs were carried out to demand that federal lands across the western United States be ceded to the states, counties, and private land owners. Both events are also part of a movement dating back to the mid-1970s in the American West aimed at turning over federally owned lands to states and counties to manage, known broadly as the Sagebrush Rebellion (Cawley 1993, 2016), the cause of which has been taken up by the Patriot movements.

The Bundy's claims to the land, and the ideologies that support those claims, are ultimately dependent on three false premises: that Native people are no longer present, the land has 'always' belonged to them, and that their family's occupation of that land is not predicated on government-assisted theft of Native lands. The Patriot movements have not embraced 'blood and soil' as a slogan, however, in addition to the fact that white supremacists and patriots have been showing up to support each other, the ideological positioning of 'blood and soil' is in line with the both groups sense of entitlement to stolen Native land that begins from the assumption that they are the true inheritors of the land.

A number of examples indicate that the Bundys and others in the Patriot movement embrace 'the myth of the vanishing Indian' (Tuck and Wayne Yang 2012; Dunbar-Ortiz and Gilio-Whitaker 2016: 9). Dunbar-Ortiz and Gilio-Whitaker (ibid.) discuss how the myth was used to 'advance dubious – even nefarious – political agendas aimed at the continual seizure of Indian lands and resources'. The myth is perpetuated by Ryan Bundy (Keeler 2017: 3), who said, 'we also recognize that the Native Americans had the claim to the land, but they lost that claim (…). There are things to learn from cultures of the past, but the current culture is the most important'. By positioning Native peoples as 'of the past', he reinforces the myth that Native people no longer exist. Further, Bundy's entitlement to stolen Native land, belongings predicated on a win/lose framework that positions him as an innocent beneficiary, accommodates white nationalists who also clearly believe the 'current culture' is one characterized by white people, as evidenced by prominent white nationalist and alt-right leader,

Richard Spencer (Cobler and Watkins 2016), who said: 'White men conquered this continent (...) we won, and we got to define what America means. America, at the end of the day, belongs to white men'.

The Nazi propaganda film *Blut und Boden* (1933) provides striking aesthetic and ideological parallels between the alt-right white nationalists who evoke the phrase and the western land-use militants, like the Bundys, who appear to them to embody it. The film begins with soft music and images of a peaceful rural farm landscape, in which the fields are diligently worked, and inside women prepare food for a struggling family. The narrator chimes in, 'A farmer is a man who, in his spirited deep-rootedness, tends to his land. And sees his labour as a duty to his family and folk'. The film then establishes the fact that the family had been working that land for centuries and, like their neighbours, had fallen on hard times as the German market is flooded by foreign goods, and their debt increases every year. 'The liberal economic policies of the state means the farmer can't afford anything he needs', says the narrator. 'One German farm after another goes out of business'. Ammon Bundy (Mock 2016) also casts himself as a humble rural yeoman, the voice for other victims of an over-reaching government and outside influences that threaten their way of life. 'Their land and resources have been taken from them to the point where it's putting them literally in poverty' (ibid.).

Like the Bundys and their supporters, the film establishes a clear cause-and-effect relationship between 'liberal land use policy' and devastation on white rural family life. Fearing a government takeover of land, Ryan Payne (Johnson and Healy 2016), one of the occupiers of the wildlife refuge in Oregon, similarly proclaimed 'land is power'. Referring to the occupation, Ryan Bundy (Mock 2016) said, 'the best possible outcome' would be that the ranchers get to 'reclaim their land, and the wildlife refuge will be shut down forever and the federal government will relinquish such control'. In both the film and the actual expressions of patriot groups, there is an implied ownership of land – 'his land' and 'their land' – a premise that must be accepted in order for their logic as victims of land theft to work.

Among the Patriot movement occupiers there is an anxiety at work, that the federal government will do to them what was done, and continues to be done, to Native people. This is a logic that attempts to co-opt indigeneity, another central component of settler colonialism (Wolfe 2006: 399). At one point, the occupiers in Oregon asked that all federally owned land be handed over to the descendants of the previous owners, by which they meant ranchers, selectively ignoring the fact that the Burns Paiute Tribe are the original inhabitants of the land, many of whom still live in the area and maintain rights to use the refuge for tribal purposes (Sunshine et al. 2016; Keeler 2017: 4). The chairwoman of the tribe, Charlotte Rodrique, held a news conference during the occupation in which she reiterated this point (Allen 2016). 'This land belonged to Paiute people as wintering grounds long before the first settlers, ranchers and trappers ever arrived here. We haven't given up our rights to the land' (ibid.). The fantasy that North

America was peacefully settled and not colonized (and still being colonized, and resisted) is the cornerstone of the settler state, where European conquest and colonization are strategically denied so that white ownership is not questioned (Razack 2002: 2).

The Homestead Acts of the mid-nineteenth century encouraged easterners to move west by guaranteeing ownership of land at no cost. This resulted in more than 270 million acres of public land (i.e. Native land claimed by the federal government) given away, overwhelmingly to white settlers west of the Mississippi River, displacing Native people from already limited reservation land (Limerick 1987). 'Over time and through law and policy … land is recast as property and as a resource' where value is based on production and indigenous people are cast as 'in the way' (Tuck and Wayne Yang 2012: 6). Putting stolen "unimproved land" in the hands of white settlers was viewed as an improvement to the land itself' (Loman and Barker 2015: 60). The descendants of those early settlers not only benefit from this legacy today, but also in the way western discourses always already accommodate them based on the kind of relationship they have with the land in terms of productivity.

The film *Blut und Boden* (1933) concludes with Nazi youth marching and taking control over rural land, reclaiming it for Germany, singing and celebrating in abundance of food and prosperity. 'Take note! There is something of lasting good; the homeland is yours. Keep it pure'. 'German soil, German blood should always be sacred to you'. When alt-right white nationalists in Charlottesville chanted, 'blood and soil', the irony was obvious. How could a group of people, whose ancestors came here from another country, claim that the soil of the United States was somehow connected to their ethnicity? More to the point, Rifkin (2009: 8) asks, 'What happens when the "homeland" discursively and institutionally is constituted in ways that deny that the colonizers are in fact "foreign"'? A self-identified fascist, from the alt-right rally in Charlottesville, clad in a black Nazi-style helmet, told a journalist the reason why he was there: 'We're defending our heritage', he said (Thompson 2017). Similarly, when asked in court why he decided to initiate the occupation at the wildlife refuge in Oregon, Ammon Bundy replied that families like his are 'in a situation (…) where we're losing our heritage' (Flaccus 2017).

Conclusion

The natural environment affects the process by which far-right appeals to national identity become narrated, and in turn, these narrations impact the landscape, how it is understood, who belongs there, and how they belong there. The formation of the mythic frontier was an important part of the colonization process. 'The territory over which the United States claimed authority', or stolen Native land, 'had to be narrated as part of the nation' (Rifkin 2009: 8). Settler colonial theory, however, serves as an important framework for understanding this process as a present phenomenon, that land *has to* be narrated as part of the nation, and is necessary to maintain and accommodate white supremacy. The confluence of white supremacy and far-right appeals to nationalism are rooted in the settler colonial context of the United States itself and, indeed, makes these relationships not just possible, but inevitable. What follows is a view of the natural environment that cannot be separated from notions of conquest, dominance and entitlement to land.

The far-right Patriot movements of the western United States – illustrated most strikingly by the Bundys and their supporters – embody the ideologies associated with 'blood and soil' white nationalism. Their performance of the white rural rancher whose 'heritage' and 'way of life' is under attack by outside influences, has proven to be a point of rhetorical cohesion, where white nationalists see an ethos that fits their ideology. While 'blood and soil' was not uttered during the Patriot standoffs, nor is Nazi ideology explicitly embraced as part of the larger Patriot movements, the slogan is profoundly applicable in the context of both group's inherent investments in white settler colonialism. Inquiry into the resurgence of this phrase within alt-right white nationalism can shed light on how far-right politics shape environmental discourses, and how these discourses sustain white settler belongings in the United States.

Notes

1 I deliberately place 'Alt Right' in quotes upon first reference in this way to draw attention to it as a re-branding deployed as a rhetorical gesture by white nationalists. I do not wish to validate white nationalist's effort toward mainstreaming their ideologies by accommodating them based on their preferred terminology. However, the term is used throughout this chapter because of the confluence of discursive formations the term currently signals, which is a necessary component of this analysis. See Hartzell (2018) for a full discussion on the use of or rejection of this term.
2 www.npr.org/2017/08/23/545509627/armed-militias-face-off-with-the-antifa-in-the-new-landscape-of-political-protes [8 July 2018].
3 This definition of rhetoric was scaffolded by Sid Dobrin's (2002: 18) definition of rhetoric from his chapter, 'Writing Takes Place', in *Natural Discourse: Toward Ecocomposition* (18). It was modified to more directly accommodate the material as an active agent in the production of discourse.
4 Gottfried (2018), a Jewish-American scholar and self-identified 'paleoconservative', has claimed that he is a 'major source' of the alt right's 'ideas and attitudes', and has worked with Spencer before, but notes 'strategic differences' with the prominent white nationalist.
5 For a more in-depth discussion on the affective relationship between discourse and materiality, see Barad 2008. Of tertiary interest is Haraway's concept of the 'material-semiotic' (2008), and Soja's articulation of 'realandimagined' space (1996).
6 In this chapter, I deploy the term 'Native' to denote people of a variety of tribes and nations who may identify as such, or Indigenous, Native American, or American Indian. Some reject all these terms and simply prefer to go by their recognized or unrecognized tribal affiliation.

References

Abanes, Richard (1996): *American Militias: Rebellion, Racism & Religion*. Downers Grove: InterVarsity Press.

Allen, Jonathan (2016, January 6): Oregon native tribe uneasy with armed standoff over land rights, *Reuters*. www.reuters.com/article/us-oregon-militia-tribe/oregon-native-tribe-uncomfortable-with-armed-standoff-over-land-rights-idUSKBN0UK1FS20160106 [8 July 2018].

Anderson, Benedict (2016): *Imagined Communities: Reflections on the Origin and Spread of Nationalism*. London: Verso.

Barad, Karen (2008): Posthumanist performativity: Toward and understanding of how matter comes to matter. In: Stacy Alaimo and Susan Hekman (Eds). *Material Feminisms*. Bloomington: University of Indiana Press, pp. 120–156.

Bernstein, Maxine (2016, September 7): Ammon Bundy's lawyer argues for his client's right to wear cowboy boots at trial. www.oregonlive.com/oregon-standoff/2016/09/ammon_bundys_lawyer_argues_for.html [8 July 2018].

Bitzer, Lloyd (1968): The rhetorical situation, *Philosophy and Rhetoric*, 1: 1–14.

Brüggemeier, Franz-Josef, Cioc, Mark and Zeller, Thomas (Eds) (2005): *How Green Were the Nazis? Nature, Environment, and Nation in the Third Reich*. Athens: Ohio University Press.

Burke, Kenneth (1984): *Permanence and Change: An Anatomy of Purpose*. Berkeley: University of California Press.

Burke, Kenneth (2013): *A Rhetoric of Motives*. Berkeley: University of California Press.

Burley, Shane (2018, June 19): How patriot prayer is building a violent far-right movement in Portland. https://truthout.org/articles/how-patriot-prayer-is-building-a-violent-far-right-movement-in-portland/ [8 July 2018].

Cawley, McGreggor R. (1993): *Federal Land, Western Anger: The Sagebrush Rebellion and Environmental Politics*. Lawrence: University of Kansas Press.

Cawley, McGreggor R. (2017): Behind the Oregon standoff, you'll find big questions about Democracy, *The New York Times*, 21 December. www.nytimes.com/2016/01/08/magazine/behind-the-oregon-standoff-youll-find-big-questions-about-democracy.html [8 July 2018].

Clark, Gregory (2004): *Rhetorical Landscapes in America: Variations on a Theme from Kenneth Burke*. Columbia: University of South Carolina Press.

Cobler, Nicole and Watkins, Matthew (2016, December 7): At A&M, protests, big crowds and tension over white nationalist's speech. www.texastribune.org/2016/12/06/m-protests-big-crowds-and-tension-over-white-natio/ [8 July 2018].

Debauer, Jenny (2005): Unframing models of public distribution: From rhetorical situation to rhetorical ecologies, *Rhetoric Society Quarterly*, 35(4): 5–24.

DeLuca, Kevin and Demo, Anne (2001): Imagining nature and erasing class and race: Carleton Watkins, John Muir, and the construction of wilderness, *Environmental History*, 6(4): 541–560.

Dobrin, Sidney (2002): Writing takes place. In: Sidney Dobrin (Ed). *Natural Discourse: Toward Ecocomposition*. Albany: University of New York Press, pp. 11–26.

Dunbar-Ortiz, Roxanne (2017): *Loaded: A Disarming History of the Second Amendment*. San Francisco: City Lights Books.

Dunbar-Ortiz, Roxanne, and Dina Gilio-Whitaker (2016): *"All the Real Indians Died Off" and 20 Other Myths about Native Americans*. Boston: Beacon Press.

Empire HD (1933): *Blut Und Boden*. film. www.youtube.com/watch?v=M7LwaSquBW0 [8 July 2018].

Epstein, Adam (2017, August 13): 'Blood and soil': The meaning of the Nazi slogan chanted by white nationalists in Charlottesville. https://qz.com/1052725/the-definition-of-the-nazi-slogan-chanted-by-white-nationalists-in-charlottesville/ [8 July 2018].

Finney, Carolyn (2014): *Black Faces, White Spaces: Reimagining the Relationship of African Americans to the Great Outdoors*. Chapel Hill: The University of North Carolina Press.

Flaccus, Gillian (2017): Ammon Bundy testifies in second Oregon standoff trial. www.usnews.com/news/best-states/oregon/articles/2017-02-28/ammon-bundy-on-stand-in-second-oregon-standoff-trial [8 July 2018].

Foucault, Michel (1986): Of other spaces, *Diacritics*, 16(1): 22–27.

Friedrichsmeyer, Sara, Lennox, Sara, and Zantop, Suzanne (Eds) (1998): *The Imperialist Imagination: German Colonialism and Its Legacy*. Ann Arbor: University of Michigan Press.

Gallaher, Carolyn (2016): Placing the militia occupation of the Malheur National Wildlife Refuge in Harney County, Oregon, *ACME: An International Journal for Critical Geographies*, 15(2): 293–308.

Gottfried, Paul (2018): Paul Gottfried: Don't call me the 'godfather' of those alt-right neo-Nazis. I'm Jewish, *National Post*, 17 April. https://nationalpost.com/opinion/paul-gottfried-dont-call-me-the-godfather-of-those-alt-right-neo-nazis-im-jewish [8 July 2018].

Haraway, Donna (2008): Otherworldly conversations, terran topics, local terms. In: Stacy Alaimo and Susan Hekman (Eds). *Material Feminisms*. Indiana University Press.

Hartzell, Stephanie L. (2018): Alt-white: Conceptualizing the 'alt-right' as a rhetorical bridge between white nationalism and mainstream public discourse, *Journal of Contemporary Rhetoric*, 8(1/2): 6–25.

Inwood, Joshua F.J. and Anne Bonds (2017): Property and whiteness: The Oregon standoff and the contradictions of the U.S. settler state, *Space and Polity*, 21(3): 253–268.

Johnson, Kirk, and Healy, Jack (2017, December 21): Protesters in Oregon seek to end policy that shaped West, *The New York Times*, 20 December. www.nytimes.com/2016/01/06/us/protesters-seek-to-end-policy-that-shaped-west.html [8 July 2018].

Keeler, Jacqueline (2017): *Native Voices: At the Edge of Morning*. Salt Lake City: Torrey House Press.

Kiernan, Ben (2007): *Blood and Soil: A World History of Genocide and Extermination from Sparta to Darfur*. New Haven: Yale University Press.

Larimer, Sara, and Phillip, Abby (2015, August 11): Who are the Oath Keepers, and why has the armed group returned to Ferguson? www.washingtonpost.com/news/morning-mix/wp/2015/08/11/who-are-the-oath-keepers-and-why-has-the-armed-group-returned-to-ferguson/ [8 July 2018].

Lefebvre, Henri (2011). *The Production of Space*. Malden: Blackwell.

Levin, Sam (2016): Oregon standoff tension mounts as so-called '3%' groups refuse to leave, *The Guardian*, 10 January. www.theguardian.com/us-news/2016/jan/10/oregon-standoff-three-percenter-groups [8 July 2018].

Limerick, Patricia (1987 [2006]): *The Legacy of Conquest: The Unbroken Past of the American West*. New York: W.W. Norton.

Loman, Emma Battell, and Adam J. Barker (2015): *Settler: Identity and Colonialism in 21st Century Canada*. Halifax & Winnipeg: Fernwood Publishing.

Massey, Doreen B. (1994): *Space, Place, and Gender*. Minneapolis: University of Minnesota Press.

McKinnon, Taylor (2016, October 31): Why Oregon standoff verdicts set dangerous and far-reaching precedent. www.commondreams.org/views/2016/10/31/why-oregon-standoff-verdicts-set-dangerous-and-far-reaching-precedent [8 July 2018].

Michel, Casey (2017): How militias became the private police for white supremacists. www.politico.com/magazine/story/2017/08/17/white-supremacists-militias-private-police-215498/ [8 July 2018].

Mills, Sara (1997): *Discourse: The New Critical Idiom*. 2nd edition. New York. Routledge.

Mock, Brentin (2016, January 4): The Oregon standoff: Race, land use, and environmental protection. www.citylab.com/weather/2016/01/the-oregon-standoff-race-land-use-and-environmental-protection/422526/ [8 July 2018].

Montero, David (2017, November 14): Prosecutor describes tense, armed standoff at Cliven Bundy ranch: 'We were outgunned'. www.latimes.com/nation/la-na-bundy-opening-statements-20171114-story.html [8 July 2018].

Morgensen, Scott L. (2014, May 26): White settlers and indigenous solidarity: Confronting white supremacy, answering decolonial alliances. https://decolonization.wordpress.com/2014/05/26/white-settlers-and-indigenous-solidarity-confronting-white-supremacy-answering-decolonial-alliances/ [8 July 2018].

Morlin, Bill (2017, June 12): ACT's anti-Muslim message fertile ground for Oath Keepers. www.splcenter.org/hatewatch/2017/06/12/act%E2%80%99s-anti-muslim-message-fertile-ground-oath-keepers [8 July 2018].

Neiwert, David (2017): When white nationalists chant their weird slogans, what do they mean? www.splcenter.org/hatewatch/2017/10/10/when-white-nationalists-chant-their-weird-slogans-what-do-they-mean/ [8 July 2018].

NPR – National Public Radio (2017, August 23): Armed militias face off with the 'Antifa' in the new landscape of political protest, *Fresh Air*. www.npr.org/2017/08/23/545509627/armed-militias-face-off-with-the-antifa-in-the-new-landscape-of-political-protest [8 July 2018].

Pogue, James (2015, September 13): The Oath Keepers are ready for war with the Federal Government. www.vice.com/en_us/article/exq8en/miner-threat-0000747-v22n9 [8 July 2018].

Potok, Mark (2012, March 1): The 'patriot' movement explodes. www.splcenter.org/fighting-hate/intelligence-report/2012/patriot-movement-explodes [8 July 2018].

Razack, Sherene (Ed) (2002): *Race, Space, and the Law: Unmapping a White Settler Society*. Toronto: Between the Lines.

Rifkin, Mark (2009): *Manifesting America: The Imperial Construction of U.S. National Space*. Oxford: Oxford University Press.

Ross, Alexander R. (2017): *Against the Fascist Creep*. Oakland: AK Press.

Ruiz, Joe, and Mccallister, Doreen (2017, August 12): Events surrounding white nationalist rally in Virginia turn fatal. www.npr.org/sections/thetwo-way/2017/08/12/542982015/home-to-university-of-virginia-prepares-for-violence-at-white-nationalist-rally [8 July 2018].

Slotkin, Richard (1998): *The Fatal Environment: The Myth of the Frontier in the Age of Industrialization, 1800–1890*. Norman: University of Oklahoma Press.

Soja, Edward (1996): *Thirdspace: Journeys to Los Angeles and Other Real-and-Imagined Places*. Malden, MA: Blackwell Publishers.

SPLC – Southern Poverty and Law Center (2017): Meet 'Patriot Front': Neo-Nazi network aims to blur lines with militiamen, the alt-right. www.splcenter.org/hatewatch/2017/12/11/meet-patriot-front-neo-nazi-network-aims-blur-lines-militiamen-alt-right [8 July 2018].

Sunshine, Spencer, Campbell, Jessica, HoSang, Daniel, Beda, Steven and Berlet, Chip (2016): *Up in Arms: A Guide to Oregon's Patriot Movement*. Somerville: Political Research Associates.

TallBear, Kim (2013): *Native American DNA: Tribal Belonging and the False Promise of Genetic Science.* Minneapolis: University of Minnesota Press.

Taylor, Blair (2019): Alt-right ecology: Ecofascism and far-right environmentalism in the United States. In: Bernhard Forchtner (Ed). *The Far Right and the Environment: Politics, Discourse and Communication.* London: Routledge.

The Three Percenters (n.d.): The Three Percenters official statement regarding the violent protests in Charlottesville. www.thethreepercenters.org/single-post/2017/08/12/The-Three-Percenters-Official-Statement-Regarding-the-Violent-Protests-in-Charlottesville [8 July 2018].

Thompson, A. C. (2017, August 12): Police stood by as mayhem mounted in Charlottesville. www.propublica.org/article/police-stood-by-as-mayhem-mounted-in-charlottesville [8 July 2018].

Tuck, Eve and Wayne Yang, K. (2012): Decolonization is not a metaphor, *Decolonization: Indigeneity, Education & Society,* 1(1): 1–40.

Turner, Fredrick J. (2014): *The Significance of the Frontier in American History.* Mansfield Centre: Martino Publishing.

Vatz, Richard E. (1973): The myth of the rhetorical situation, *Philosophy and Rhetoric,* 6: 154–161.

Veracini, Lorenzo (2008): Settler collective, founding violence and disavowal: The settler colonial situation. *Journal of Intercultural Studies,* 29(4): 363–379.

Veracini, Lorenzo (2010): *Settler Colonialism: A Theoretical Overview.* Basingstoke: Palgrave Macmillan.

Veracini, Lorenzo (2015): *The Settler Colonial Present.* Basingstoke: Palgrave Macmillan.

Wolfe, Patrick (2006): Settler colonialism and the elimination of the native, *Journal of Genocide Research,* 8(4): 387–409.

18
LOOKING BACK, LOOKING FORWARD

Some preliminary conclusions on the far right and its natural environment(s)

Bernhard Forchtner

Introduction

Looking at the wide variety of actors and issues considered in the previous chapters on far-right environmental communication, this concluding chapter seeks to connect some of the points raised. As mentioned in Chapter 1 of this volume (Forchtner 2019a), the objective underlying this publication has been to provide a broad overview of how far-right actors make sense of the natural environment, how they, in fact, construct different natural environments. As such, this volume examines (a) differences and similarities within the far-right spectrum, ranging from radical-right party and non-party actors who oppose elements of *liberal* democracy to extreme-right, anti-democratic ones (Mudde 2007), as well as diverging historical backgrounds, and diverse discursive and political contexts. Furthermore, it explores (b) tensions within individual actors, for example due to competing ideological elements. In revisiting these objectives at the end of the volume, I hope to offer a foundation for future research on the contemporary far right and the ways in which it engages with the natural environment.

Consequently, the following remarks are organised in two dimensions: first, I consider the patterns and puzzles which emerge from the contributions to this volume. Key themes I highlight are, on the one hand and more abstract, the significance of 'ideological stance' in relation to environmental protection. That is, to what extent do more extreme-right, organic ethnonationalist actors show greater affinity towards environmental protection? On the other hand and more concrete, I point to how this relation is realised in articulations of what is maybe the most pressing environmental issue today: anthropogenic climate change. Second, I highlight a series of directions to explore in the future, directions which emerge against the background of the findings presented in this book. These include a selection of specific topics, but also, more generally, calls to look closely at contexts, at genres in which the far right performs, at how to conduct comparative research and at the usage of a broader array of methods of data collection and analysis. In doing so, this final chapter of *The Far Right and the Environment: Politics, Discourse and Communication* hopes to inspire subsequent work on the intersection of political changes and environmental issues at the beginning of the twenty-first century.

Looking back: Existing patterns?

Amongst the factors influencing far-right stances vis-à-vis the natural environment pointed to is the ideological position of actors (where actors are positioned on the far-right continuum ranging from anti-liberal democracy to anti-democracy). That is, are those closer to the extreme end of this far-right continuum more likely to share a world view prone to ecological positions? Indeed, there are both empirical and theoretical grounds which support such an assumption. After all, such actors in particular tend to see an organic connection between land and people.

This argument is forcefully supported by Voss (2019), in his chapter on the Freedom Party of Austria (FPÖ). The FPÖ is historically rooted in ethnic nationalist thought and, as Voss argues, its organicism leads the party to push for nature protection, making it one of the FPÖ's comprehensive and fundamental concerns. Voss further supports this observation by showing that the ideologically less far right Alliance for the Future of Austria has been less serious in this area. Some of the findings introduced in the chapter on Germany by Forchtner and Özvatan (2019) indicate the same (see, for example, the differences between the Alternative for Germany and the extreme-right ecological magazine *Umwelt & Aktiv* [Environment & Active]). The assumption is furthermore supported by the Hungarian case, though this time based on differences within an actor: while Kyriazi (2019) argues that the environment has long been a prominent feature in Jobbik's programme due to its organicism, she points to Jobbik's recent moderation which has pushed their organic nationalism to the back. When looking specifically at anti-liberal, (nominally) democratic far-right actors whose organic ethnonationalist outlook might well be less pronounced, the aforementioned assumption seems to be proven, too. For example, Kølvraa (2019), in his chapter on the Danish People's Party (DPP), convincingly separates their anthropocentric 'environmental imaginary' from an organic ethnonationalist, rather ecological one. As he illlustrates in his analysis of the 'wolf debate' in the spring of 2018, the DPP's call to eliminate reimmigrating wolves from the Danish countryside, based on what he calls 'popular lupophobia', illustrates the centrality of anthropocentrism in the party's environmental imaginary and connects well with its 'poujadist populism' – not organic ethnonationalism.

Thus, while *organic* ethnonationalism, which is arguably more strongly present in extreme-right actors, appears to lead to greater affinity for environmental protection than is the case in 'mainstream' radical-right actors, the question remains how strong is this pattern actually today? That is, if organic ethnonationalism strengthens the environmental component of actors, is this a condition unchecked by other ideological elements? In fact, competition between ideological elements is not new in this area but is visible already in, for example, National Socialist stances vis-à-vis environmental protection (see Chapter 13 in this volume by Forchtner and Özvatan 2019). While putting forward a few pro-environment positions (and realising them via legislation), massive damage was ultimately inflicted under National Socialism in its pursuit of modernisation and war preparations. Indeed, this is certainly more than an exception and questions any automatic link between more 'extreme' organic ethnonationalism and environmental protection. For example, Kyriazi (2019) points out that other ideological concerns might override environmental ones when saying that 'autarchy and economic preponderance of the nation tends to come before the protection of the environment' (see also Forchtner and Kølvraa 2015; Forchtner et al. 2018). Similarly, Tarant (2019), in his chapter on the far right in the Czech Republic, argues that while parts of 'the Czech far-right tended to head in the direction of "nativist environmentalism"', others do not make eco-friendly suggestions, but cast doubt on, for example, global environmental issues due to what he calls "nativist anti-environmentalism"' (see also Schaller and Carius 2019).

Such ideologically motivated tensions are perhaps best illustrated by considering the concrete case of climate change. While contributions to this volume regularly agree that countryside and landscape are important to the far right (explicitly, for example, in chapters by Tarant [2019]; Turner-Graham [2019] on the United Kingdom; Boukala and Tountasaki [2019] on France; and Boggs [2019] on the United States), climate change is repeatedly viewed rather sceptically, at least in comparison to mainstream attitudes in Europe. Indeed, it has been noted that opinions on climate change (and, more specifically, its anthropogenic origins) are often divided (Forchtner forthcoming). Climate-change scepticism is reported in chapters by Turner-Graham (2019), Kølvraa (2019), Hultman et al. (2019) on Sweden, Forchtner and Özvatan (2019), Bennett and Kwiatkowski (2019) on Poland and Kaiser (2019) on German-US networks. Against this background, the question emerges why actors which are (more or less) ideologically motivated to protect the natural environment, and which seem to be so concerning a variety of other issues, doubt this key threat to 'the land'?

According to existing literature and contributions to this volume, climate-change sceptic voices cover the entire range of the far-right spectrum, i.e. both those driven by organicism and those not. Anthropogenic climate change might be doubted and climate-change policies rejected by the far right as activities combatting climate change become linked to 'cosmopolitanism', 'globalism', a 'liberal world government' and a loss of sovereignty, themes strongly opposed by these actors. Furthermore, opposition to climate-change policies enables them to present themselves as defenders of economic interest of 'the little guy' vis-à-vis 'the elite'. Even in the case of the FPÖ, a party with a rather *völkisch* world view and which 'aims to combat anthropogenic climate change by limiting fossil fuels and industrial agriculture' (Voss 2019), the situation seems to be ambivalent. After all, Austrian media de facto affiliated with the party (but not being official party publications) predominantly articulate climate-change scepticism while the long-term FPÖ-chairman Heinz-Christian Strache (2005-2019) and Susanne Winter, back then environmental spokesperson of the FPÖ, openly questioned anthropogenic climate change (Forchtner 2019b).

Underlying the various types of climate-change scepticism, a number of related forms of thinking and communicating can be identified. Especially when it comes to climate change, the far right's propensity to conspiracy theories is visible (see, for example, Kaiser 2019). Indeed, even those accepting the thesis of anthropogenic climate change, and who thus do not perform evidence scepticism, doubt established processes and responses to climate change by questioning, among other things, the credibility of scientists, the rationality in public debates and the (material and ideological) goals behind climate protection. As such, here too the far right provokes and scandalises (the '*right-wing populist perpetuum mobile*', Wodak 2013). In fact, it is in the area of climate change that the far right both warns against 'fearmongering' (by the mainstream and the left, the hysteria they supposedly push) and pushes it (*they* screw *us*; on fearmongering as a discursive strategy, see Wodak 2019 in her chapter on the far right). Here, strategic considerations are often involved. For example, concerning the Hungarian far-right party Jobbik, which is not denying climate change, Kyriazi (2019) argues that 'principled-ideological and instrumental-strategic components' are combined in their 'anti-establishment critique'. Relatedly, Hatakka and Välimäki (2019) describe the Finns Party, a party which has not questioned the validity of climate change, as articulating a 'reasonable, down-to-earth and virtuous people (…) patronized by out-of-touch elites' – but being 'nearly void of all structurally motivated ideological content except for the thin centre or core of populism'. Indeed, climate change is frequently mentioned as a vehicle for performing populism (see also Hultman et al. 2019; Kølvraa 2019, for performing a populist style, see Moffitt 2016).

Whether or not similar observations concerning climate change (and beyond) can be made in the United States, where, for example, the so-called 'alt right' stands in opposition to a climate-sceptic mainstream right, remains to be seen. Given its opposition to traditional pro-business conservatism, it stands to reason that eco-friendly stances are both ideologically and strategically present. And yet, Taylor (2019), in his chapter on the alt right, reports conflicting views on climate change. However, fairly little is known about alt-right stances in this area and, as such, this is one area for further research – something that points me to the next section in this chapter.

Looking forward: Further avenues for research?

In addition to the above, there remains a wide range of issues to be explored. Following a brief note on a few specific topics which I deem particularly beneficial to the study of the far right and the field of environmental communication,

I turn to the issues of context, genre, comparison and methods to indicate a range of broader areas which invite further exploration.

There are numerous topics that this volume has not been able to touch upon and/or could not be sufficiently discussed here. Indeed, I do not claim to provide a complete and exhaustive list of such topics. Yet, climate change, given the wider relevance of the phenomenon that touches on so many aspects of life on our planet, most clearly calls for further investigation. Here, I find the interplay between, on one hand, evidence scepticism (scepticism concerning the trend, causes and/or impact of climate change) and, on the other hand, process and response scepticism specifically interesting. van Rensburg (2015) has called the latter two 'concomitant objects of scepticism' which strengthen scepticism (see also Capstick and Pidgeon [2014] for a separation of epistemic from response scepticism). As neither process nor response scepticism do necessarily deny the existence of anthropogenic climate change, though possibly having similar effects as evidence/epistemic scepticism, this complexity calls for differentiated analysis. Indeed, exploring the presence of and link between these types of scepticism will furthermore benefit research on 'mainstream' conservative climate-change scepticism. Another aspect of climate-change scepticism is indicated by Hultman et al. (2019), who argue that the Swedish far-right's climate-change denial is also due to an 'industrial masculinity' (on climate change and gender more generally, see Cohen 2017; on 'ecological masculinities', see Hultman and Pulé 2018). The particularities of 'gender' in the world of the far right have been explored (see Köttig et al. 2017); but the role of 'women' and masculinity/femininity in relevant discourses warrants further research on the specific gendered relationships between them and (protection of) the natural environment.

Energy security (the role of nuclear energy, fossil fuels and renewables) is another area which warrants further exploration. After all, the topic is likely to remain a significant policy issue, linked to discussions about a just energy transition (Fraune and Knodt 2018), and with a strong presence among the wider public. Moreover, energy self-sufficiency, aesthetic and populist concerns have led to presentations of renewables by the far right in both positive and negative terms.

Furthermore, the connection between lifestyle and environmental topics remains underexplored. This includes, for example, food consumption (e.g. a meat-free diet) and actual 'eco-activism' – both areas commonly connected to the left, but also present in far-right politics (for indications, see Forchtner and Tominc 2017; Forchtner and Özvatan 2019; Tarant 2019; Taylor 2019). Knowing more about communication and practices surrounding these issues could potentially tell us a lot about 'cultural imaginaries' of the far right (Kølvraa and Forchtner 2019).

However, let me go beyond listing specific topics for further research and, instead, suggest broader areas for research. First, and going beyond ideology, context, both in terms of, for example, media landscape and history, is an important object for further research. After all, context influences, for example, 'micro-politics', i.e. discursive strategies and performative elements, thus being key for creating a plethora of political imaginaries and identity narratives (Wodak 2019). Concerning historical context, Hansen (2019) notes in his piece on environmental communication research that environmental themes 'draw from and rework historically and culturally deep-seated views of nature and the environment', and Wodak (1996) has long emphasised the historical character of 'discourse' as being 'always connected to other discourses which were produced earlier'. That is, there is path dependency, but no path determinacy. The way actors construct their past (here in relation to the natural environment) thus affects how both the wider public and specific collectives make sense of the natural environment. Sometimes, this is strikingly visible, for example, when juxtaposing Italy (Chapter 6) and Germany (Chapter 13). While the far right in Germany is prone to talk about the environment due to historical reasons (environmental themes have long been present) and the wider societal prominence of the topic, the environment has not been a salient electoral issue in Italy. It is this weakness, Bulli (2019) argues, which made the environment attractive to non-conventional forces on the far right during the 1970s – though it is also this weakness which constrains the appeal of CasaPound Italia's utilisation of the environment today. While the German context is well researched, the historical evolution of far-right contributions to the discourse about the environment in many other countries appears to be less often sufficiently reconstructed.

The contributions to this volume have, furthermore, covered a wide range of genres, from legislative and social media texts to newspapers and magazines, to name but a few. A genre is, in the words of Fairclough (1992), a 'relatively stable set of conventions that is associated with, and partly enacts, a socially ratified type of activity'. Indeed, communication cannot be disconnected from the genre in which it occurs, and even differences in the content of messages might be due to genre. Hansen (2019) is one of those in this volume who points to the significance of investigating differences between genres. This resonates with, for example, Voss' (2019) investigation of legislative motions, which finds the FPÖ putting forward policies combating climate change – though the non-affiliated, but nevertheless FPÖ-aligned, press exhibits climate-change scepticism (Forchtner 2019b). It is in this light that changes in the media and communications landscape, and the ways in which claim-makers make use of the affordances of digital media (see Turner-Graham 2013, 2019; Bennett and Kwiatkowski 2019; Hatakka and Välimäki 2019; Kaiser 2019), have to be considered in great detail as the relative freedom of these channels allows for voices which might represent far-right sentiment, but which do not have access to established far-right parties, blogs and/or print publications. The question of whether there is a systematic difference due to 'genre' offers considerable scope for further research.

A related issue is the need for more comparative research – in terms of both country-to-country and country-internal analyses, comparing both far-right actors and far-right ones with centre/left-wing ones. In this volume,

Turner-Graham (2019), for example, points to similarities between the far right and more mainstream sources in the United Kingdom, and Bennett and Kwiatkowski (2019) speak of far-right discursive frames and strategies which migrate into mainstream discussions in Poland. Closely related to the investigation by Turner-Graham (2019) is a point raised by Hansen (2019) on the relation between far right and mainstream themes known from advertising, film and other pop-cultural representations (on far-right bricolage, see Forchtner and Kølvraa 2017). Bulli, for example, looks at the presence of J.R.R. Tolkien's middle-Earth in Italy's far right, pointing to the Hobbit Camps organised in 1977, 1978 and 1980. In such wider societal sites, including particularly in advertising (Hansen 2015), nature is often romanticised and/or viewed nostalgically, as authentic and genuine – something manifestly not at odds with the views of the far right. So, how are signs known to the wider public utilised, what kind of bricolage exists in the area of 'the environment', both in terms of actual content and in the how (the form) of talking about nature?

Research on central and eastern European far-right actors, in particular, requires comparative research. On the one hand, these actors – partly due to the predominance of ethnic models of nation tend to be more aligned with organic ethnonationalism, thus taking environmental questions potentially rather seriously (see the assumption formulated at the beginning of the previous section). On the other hand, and although central and eastern European countries have brought together political and environmental activism, for example in Latvia, environmental concern appears to be less salient in this region (Marquart-Pyatt 2012). Chapters on the Czech Republic, Poland and Hungary provide a range of indications concerning how this tension might play out, but as so often: more research is required.

Next, analyses of transnational links in and through which stories about the environment circulate within the far right are worth conducting. Kaiser's (2019) analysis of hyperlink connections between Germany and the United States is a case in point. Talking about the United States, a timely area of research on transnational links concerns the so-called alt right in the United States and its connections and overlaps with the European far right. Such comparative research appears to be, across all issues, in its infancy and there is certainly a need for a conversation on both (a) form and technology as well as (b) actual positions. Sure, such transatlantic connections are not new, but it is in the changing (media) context of the twenty-first century that this exchange is picking up again. In fact, the European and the US far right are already having this discussion and are aware of the need to further intensify it, something one of the key members of the Identitarian movement, Austrian Martin Sellner (2017), has made clear when pointing to his exchange with the alt right on online and offline activism. The significance of such connections is touched upon by Bulli (2019), who talks about the French *Nouvelle Droite* and its influence on young cadres in Italy during the 1970s and 1980s. Similarly, central and eastern European far-right actors might draw on thoughts and repertoires practised in western Europe, combining them with their own mythologies (see Tarant 2019 in this volume).

The final area for further work I want to mention concerns the methods of data collection and analysis utilised. First, while most contributions in this volume are qualitative in nature, network analysis, corpus linguistics and other quantitative approaches might help to identify patterns. And yet, there is clearly also more space for qualitative research; for example, research on far-right environmental communication could benefit from existing work in the wider field of environmental communication to do with the analysis of visual representations. Second, the reception of environmental communication by far-right actors is a relevant area to be explored. How, for example, is climate-change scepticism – in Europe within a societal context where scepticism is far less legitimate than in the United States – perceived by followers and the wider public? Is, for example, relevant communication noticed at all? Is it shrugged off or embraced as a major element in the fight against 'the elite'? How exactly is analogous thinking, for example in relation to biodiversity and 'invasive species', happening? Are relevant texts read at all – and by whom? How are they decoded? While this would illuminate the reception of environmental communication by far-right audiences, it would also speak to arguments about the strategic use of environmental communication, that is, the far right talks 'green' so to 'appeal to' and 'sneak' far-right thought into the mainstream. While there is research on large populations (e.g. analysis of data from the Eurobarometer survey offers insights into how climate change is viewed, including by those who position themselves at the very right, see McCright et al. 2016), qualitative research which can look more specifically at issues, such as through interviews, focus groups and ethnographic observations, might prove more beneficial at this point. Although there are clearly ethical questions involved in such steps (Massanari 2018), members of the far right, even at the 'extreme' end of the spectrum, have been open towards such research.

The above is not meant to constitute a once-and-for-all, complete list of patterns and possible further tasks. But it is meant to point to the fact that left-wing thinking so commonly associated with present-day talk about 'the environment' is not the only way in which the latter has been addressed. Indeed, while one might question the environmental merits of far-right proposals, far-right ways of engaging with the natural environment can seriously question contemporary societies' ways of prolonging and exaggerating the environmental crises of our times. Yet, how humans and their relationships are defined through far-right articulations of what 'the environment' is, and how it should be interacted with, will never be neutral, but will have implications for social life (and beyond). Such articulations will continue to change depending on wider societal developments, requiring new explanations so as to better understand far-right actors. The authors contributing to this volume hope that they have added their bit to this project.

References

Bennett, Samuel and Kwiatkowski, Cezary (2019): The environment as an emerging discourse in Polish far-right politics. In: Bernhard Forchtner (Ed). *The Far Right and the Environment: Politics, Discourse and Communication*. London: Routledge.

Boggs, Kyle (2019): The rhetorical landscapes of the 'alt right' and the Patriot Movements: Settler entitlement to native land. In: Bernhard Forchtner (Ed). *The Far Right and the Environment: Politics, Discourse and Communication*. London: Routledge.

Boukala, Salomi and Tountasaki, Eirini (2019): From Black to Green: Analysing *Le Front National*'s 'Patriotic Ecology'. In: Bernhard Forchtner (Ed). *The Far Right and the Environment: Politics, Discourse and Communication*. London: Routledge.

Bulli, Giorgia (2019): Environmental politics on the Italian far right: Not a party issue? In: Bernhard Forchtner (Ed). *The Far Right and the Environment: Politics, Discourse and Communication*. London: Routledge.

Capstick, Stuart Bryce and Pidgeon, Nicholas Frank (2014): What is climate change scepticism? Examination of the concept using a mixed methods study of the UK public, *Global Environmental Change*, 24: 389–401.

Cohen, Marjorie (Ed) (2017): *Climate Change and Gender in Rich Countries*. London: Routledge.

Fairclough, Norman (1992): *Discourse and Social Change*. Cambridge: Polity Press.

Forchtner, Bernhard (forthcoming): Climate change and the far right, *Wires Climate Change*.

Forchtner, Bernhard (2019a): Far-right articulations of the natural environment: An introduction. In: Bernhard Forchtner (Ed). *The Far Right and the Environment: Politics, Discourse and Communication*. London: Routledge.

Forchtner, Bernhard (2019b): Articulations of climate change by the Austrian far right: A discourse-historical perspective on what is 'allegedly manmade'. In: Ruth Wodak and Pieter Bevelander (Eds). *'Europe at the Cross-road': Confronting Populist, Nationalist and Global Challenges*. Lund: Nordic Academic Press, pp. 159–179.

Forchtner, Bernhard and Kølvraa, Christoffer (2015): The nature of nationalism: Populist radical right parties on countryside and climate, *Nature and Culture*, 10(2): 199–224.

Forchtner, Bernhard and Kølvraa, Christoffer (2017): Images of radical authenticity: The aesthetics of the extreme right in social media, *European Journal of Cultural and Political Sociology*, 4(3): 252–281.

Forchtner, Bernhard, Kroneder, Andreas and Wetzel, David (2018): Being skeptical? Exploring far-right climate change communication in Germany, *Environmental Communication*, 12(5): 589–604.

Forchtner, Bernhard and Özvatan, Özgür (2019): Beyond the 'German forest': Environmental communication by the far right in Germany. In: Bernhard Forchtner (Ed). *The Far Right and the Environment: Politics, Discourse and Communication*. London: Routledge.

Forchtner, Bernhard and Tominc, Ana (2017): Kalashnikov and cooking-spoon: Neo-nazism, veganism and a lifestyle cooking show on YouTube, *Food, Culture and Society*, 20(3): 415–441.

Fraune, Cornelia and Knodt, Michèle (2018): Sustainable energy transformations in an age of populism, post-truth politics, and local resistance, *Energy Research & Social Science*, 43, 1–7.

Hansen, Anders (2015): Nature, environment and commercial advertising. In: Anders Hansen and Robert Cox (Eds). *The Routledge Handbook of Environmental Communication*. Oxon: Routledge, pp. 270–280.

Hansen, Anders (2019): Environmental communication research: Origins, development and new directions. In: Bernhard Forchtner (Ed). *The Far Right and the Environment: Politics, Discourse and Communication*. London: Routledge.

Hatakka, Niko and Välimäki, Matti (2019): The allure of exploding bats: The Finns Party's populist environmental communication and the media. In: Bernhard Forchtner (Ed). *The Far Right and the Environment: Politics, Discourse and Communication*. London: Routledge.

Hultman, Martin, Björk, Anna and Viinikka, Tanya (2019): Far-right and climate change denial. Denouncing environmental challenges via anti-establishment rhetoric, marketing of doubts, industrial/breadwinner masculinities enactments and ethno-nationalism. In: Bernhard Forchtner (Ed). *The Far Right and the Environment: Politics, Discourse and Communication*. London: Routledge.

Hultman, Martin and Pulé, Paul (2018): *Ecological Masculinities. Theoretical Foundations and Practical Guidance*. Oxon: Routledge.

Kaiser, Jonas (2019): In the heartland of climate scepticism: A hyperlink network analysis of German climate sceptics and the US right-wing. In: Bernhard Forchtner (Ed). *The Far Right and the Environment: Politics, Discourse and Communication*. London: Routledge.

Kølvraa, Christoffer (2019): Wolves in sheep's clothing? The Danish far right and 'wild nature'. In: Bernhard Forchtner (Ed). *The Far Right and the Environment: Politics, Discourse and Communication*. London: Routledge.

Kølvraa, Christoffer and Forchtner, Bernhard (2019): Cultural imaginaries of the extreme right: an introduction, *Cultural Imaginaries of the Extreme Right. Special Issue of Patterns of Prejudice*, 53(3): 227–235.

Köttig, Michaela, Bitzan, Renate and Petö, Andrea (2017): *Gender and Far Right Politics in Europe*. Cham: Palgrave.

Kyriazi, Anna (2019): The environmental communication of Jobbik: Between strategy and ideology. In: Bernhard Forchtner (Ed). *The Far Right and the Environment: Politics, Discourse and Communication*. London: Routledge.

Marquart-Pyatt, Sandra (2012): Concern for the environment among general publics: How do mass publics in Central and Eastern Europe compare with other regions of the world, *Czech Sociological Review*, 48: 641–666.

Massanari, Adrienne (2018): Rethinking research ethics, power, and the risk of visibility in the era of the 'Alt-Right' gaze, *Social Media + Society*, 4(2): 1–9.

McCright, Aaron M., Dunlap, Riley E. and Marquart-Pyatt, Sandra T. (2016): Political ideology and views about climate change in the European Union, *Environmental Politics*, 25(2): 338–358.

Mudde, Cas (2007): *Populist Radical Right Parties in Europe*. Cambridge: Cambridge University Press.

Moffitt, Benjamin (2016): *The Global Rise of Populism. Performance, Political Style, and Representation*. Redwood City: Stanford University Press.

Schaller, Stella and Carius, Alexander (2019): *Convenient Truths: Mapping Climate Agendas of Right-Wing Populist Parties in Europe*. Berlin: adelphi.

Sellner, Martin (2017): Arcadi, Boulevard und Kulturrevolution – Yannick Noé im Gespräch. www.youtube.com/watch?v=aDLuakpfvM4 [13 September 2018].

Tarant, Zbyněk (2019): Is brown the new green? The environmental discourse of the Czech far right. In: Bernhard Forchtner (Ed). *The Far Right and the Environment: Politics, Discourse and Communication*. London: Routledge.

Taylor, Blair (2019): Alt-right ecology: Ecofascism and far-right environmentalism in the United States. In: Bernhard Forchtner (Ed). *The Far Right and the Environment: Politics, Discourse and Communication*. London: Routledge.

Turner-Graham, Emily (2013): 'An intact environment is our foundation of life': The Junge Nationaldemokraten, the Ring Freiheitlicher Jugend and the cyber-construction of nationalist landscapes. In: Andrea Mammone, Emmanuel Godin and Brian Jenkins (Eds). *Varieties of Right-Wing Extremism in Europe*. London: Routledge, pp. 233–248.

Turner-Graham, Emily (2019): 'Protecting our green and pleasant land': UKIP, the BNP and a history of green ideology on Britain's far right. In: Bernhard Forchtner (Ed). *The Far Right and the Environment: Politics, Discourse and Communication*. London: Routledge.

van Rensburg, William (2015): Climate change scepticism: A conceptual re-evaluation, *SAGE Open*, April–June, pp. 1–13.

Voss, Kristian (2019): The ecological component of the ideology and legislative activity of the Freedom Party of Austria. In: Bernhard Forchtner (Ed). *The Far Right and the Environment: Politics, Discourse and Communication*. London: Routledge.

Wodak, Ruth (1996): *Disorders of Discourse*. London: Longman.

Wodak, Ruth (2013): 'Anything Goes!' – The Haiderization of Europe. In: Ruth Wodak, Majid KhosraviNik and Brigitte Mral (Eds). *Right-Wing Populism in Europe. Politics and Discourse*. London: Bloomsbury, pp. 23–37.

Wodak, Ruth (2019): The trajectory of far-right populism – A discourse-analytical perspective. In: Bernhard Forchtner (Ed). *The Far Right and the Environment: Politics, Discourse and Communication*. London: Routledge.

INDEX

Note: Page numbers in italic and bold refer to figures and tables, respectively.

Abbey, Edward 278
Abernethy, Virginia 277
Adelsteen, Pia 114
ad hominem arguments 28
advertising 45–7
aesthetic dimension 8
Age of Ecology (Radkau) 4
Aktion für Deutschland 29
Alfsson, Thoralf 124
Alleanza Nazionale (AN) 93
All-Polish Youth (Młodzież Wszechpolska, MW) 239
Alternative für Deutschland (AfD, Alternative for Germany) 23, 217, 225
alt-right, defined 295–6
alt-right ecology 275–92; Arktos 277; Atomwaffen Division 284, Attack the System (ATS) 282; Bay Area National Anarchist Movement 283; Cascadia 279–80; Counter-Currents 277; decolonization 281; Deep Green Resistance (DGR) 285; ecological anarchism 282–5; ethnopluralism 278–80; Green anarchism 282–5; immigration 277–8; indigenism 280–1; National Anarchists 282–3; National Revolutionary Vanguard 282; neoreaction 285; Northwest Front 280; overpopulation 277–8; Patriot movement 281; right decentralism 281–5; right-wing bioregionalism 278–80; secessionism 281–2; the Second Vermont Republic (SVR) 281–2; shifting conservative environmental discourse 276–7; Third Positionists 282; tribal anarchism 281; True Cascadia 279; veganism and animal rights 285–6; white habitat loss 277–8; Wolves of Vinland (WoV) 283–4
alt-right nationalism and patriot collaborations 295–8
alt-right rhetorical landscapes 293–309; alt-right defined 295–6; alt-right nationalism and patriot collaborations 295–8; blood and soil 300–4; Bundy standoffs 302–4; Charlottesville rally 296–7, 300–1; colonization as a continuing process 299; entitlement to Native land 298; Homestead Acts 303; Malheur National Wildlife Refuge occupation 294–5, 300, 302; material-discursive frontier 298–9; material-discursive understanding of language 295; mythic American West 298–9; Oath Keepers 293, 296; Patriot Front 297; Patriot movement 294, 300, 302; settler colonialism 299–300; Three-Percenters 293; Vanguard America 297; and the *Volk* 297–8; white wilderness 299

American Thinker 264, 269, 271
Anderson, Benedict 82, 109, 258, 298
Anglin, Andrew 286
animal protection, rights and welfare 170–1
anthropogenic climate change 109–10, 174, 225, 230–1, 286, 312–14
anti-anthropocentrism and organicism 177–8
anti-elitism 26
anti-establishment rhetoric 125–6
Anti-Federalist League 64
anti-feminism community 264
anti-pluralism 26
apologies 28
appeal to ignorance 79
Apply the Brakes 277–8
Arktos 277
Arndt, Ernst Moritz 7
Atomwaffen Division 284
Attack the System (ATS) 282
Austria: Austrian Freedom Party 23; and climate change 7–8; ecological ideology in 153–83; Freedom Party 7–8; nature protection in FPÖ documents and legislative motions **157–69**; salience of nature protection in legislature **156**; VDU 23
Austrian Freedom Party 23
authenticity 31
authoritarianism 26, 79
Autonomous Nationalists 205–7
Avatar 209
Azione Giovani 93

bad manners 29
Barthes, Roland 108, 115–16
Batten, Gerard 66
Bay Area National Anarchist Movement 283
beauty, concept of 95
Bednarski, Wolfram 220
Belgium: and climate change 7; Flemish Interest 7
Benoist, Alain de 75
Bergström, Hans 128
Berlusconi, Silvio 93
Berlusconization 31
Bern, Lars 128
Bey, Hakim 283
biodiversity 226–7
Black Earth: The Holocaust as History and Warning (Snyder) 283

Blake, William 58
blood and soil 300–4
blood-and-soil arguments 63
Blut und Boden (Nazi propaganda film) 303–4
Boggs, Kyle 12
Bolsonaro, Jair 2
Bookchin, Murray 6, 278, 288
BP (British Petroleum) Deepwater Horizon oil spill of 2010 42
Brandberg, Tomas 127
Brazil 2
Breitbart News 65
Breivik, Anders 284
Brexit 2
Britain: blood-and-soil arguments 63; British country life 58–9; British National Party 63, 67–9; British Union of Fascists 6; cities as a unhealthy living space 61, 63; and climate change 7; country life 58–9; fascists and environmental issues 6; green ideology in 57–71; jingoistic view of Britain's past 59–60; National Party 7; organic farming 62–3; Soil Association 62; United Kingdom Independence Party 23, 63–6; visions of the 'real' 58–60
Britain First 67
British National Party (BNP) 30, 67–9; anti-immigration stance 67–9; environmental policies of 58–9, 67–9; overview of 67; *völkisch* movement 68
British Union of Fascists 6, 62–3
Bund Heimatschutz 7
Bündnis Zukunft Österreich (BZÖ) 155, 178, 180
Bundy, Ammon 300, 304
Bundy, Cliven 302
Bundy standoffs 302–4

calculated ambivalence 29
Callenbach, Ernest 279
Captus 127
Carson, Rachel 65
CasaPound Italia 94–8; concept of beauty 95; ecology and 95; electoral program 95–6; enemies and 94–5; and environmentalism 94–8; Italian biodiversity 96; la Foresta che Avanza 97; la Muvra 97; la Salamandra 97; terra redenta (redeemed soil) 96

Cascadia 279–80
Castoriadis, Cornelius 81–2
Cato Institute 264, 268, 271
charisma 30–2
Charlottesville rally 296–7, 300–1
Chávez, Hugo 23
Children of the Sun (Schaefer) 57
cities as a unhealthy living space 61, 63
Civic Platform (Platforma Obywatelska, PO) 239
climate change 7–9
climate change denial 121–35; anti-establishment rhetoric 125–6; climate friendly Swede 129–30; forms of 125–30; Green Fatwá 128–9; historical and political context 122–4; industrial/breadwinner masculinities 128–9; marketing of doubts 126–7; nationalism 129–30; overview of 121–2; Stockholm Initiative (SI) 122; Sweden Democrats 121–35; and Swedish parliament politics 124–5; watermelon metaphor 125
climate change denialism 122–4, 128–31
climate change doubts, marketing of 126–7
climate change network 264
climate network 261–2, *265*
climate scepticism 257–74, 312–13; American Thinker 264, 269, 271; anti-feminism community 264; Cato Institute 264, 268, 271; climate change network 264; climate network 261–2, *265*; collection of links 260–1; communities identified 262–7, **263**; community topics of relevance 262, **263**; connection between German and US sites 267–8, **268**; Counter Currents 271; critical and left-wing blogs community 266; Daily Caller 264, 269, 271; economics, religion, and conspiracy community 264; far-right and conspiracy community 265–6; general conspiracy community 265; German climate sceptics and the US far right 268–70; Heartland Institute 264, 268–9, 271; mono-thematic sceptics *vs* political climate sceptics 271; networked public sphere 258–9; networks identified 261; prominence of US sites in climate change community 267; research questions 260; (far-)right news and conspiracy community 265;

right-wing news sites 269; structuralist approach to 8; types of climate sceptic communities 270
Colli, Paolo 92–3
colonization as a continuing process 299
communications strategies 42
comparative research 315–16
connection between German and US sites 267–8, **268**
conservatism 26
conspiracy theories 30
context as area of future research 314–15
continuous campaigning 31
Corcoran, Casey Brian 280
Counter Currents 271, 277
Country Life 58–9
Covington, Harold 278
critical and left-wing blogs community 266
Czech far right 201–15; aesthetic role of nature 201–2; Autonomous Nationalists 205–7; and climate change denial 208–9; decline of forests 209; eco-action 206, 209; eco-activism as advertisement 206; eco-pioneers of 203–4; environmental issues 204–5; environmental topics as marginalia 208; Freedom and Direct Democracy 210–11; lack of environmental agenda 210; *Patriot Front* (*Vlastenecká Fronta*, VF) 203–4; Republican Party 204–5; Republican Party (Sdružení pro republiku – Republikánská strana Československa) 202; shifting discourse 207–8; source material 202–3; White-Power Skinheads 202, 204, 206; Workers' Party 207–10; Workers' Party (Dělnická strana) 202; Workers' Youth (Dělnická mládež) 202, 208–9
Czech Republic 201–15

Dahl, Kristian Thulesen 109–10
Daily Caller 264, 269, 271
Danish Peoples Party (DPP) 107–20; academic study of 109; and actual wolves in Denmark 111–12; and anthropogenic climate change 110; classical myth of the wolf 112–13; cultural semiotics of the wolf 112–13; ecological imaginary of Nazism 112–13; and emotionality 115–16; environmental imaginaries 111, 116–17; history of 109; legitimacy

Danish Peoples Party (DPP) (*Continued*) of fear of wolves 114–15; nature of 109–11; pivoting away from factual space 114; politics of lupophobia 113–17; popular lupophobia 114–16; rejection of the wolf 113–14; response skepticism 110; wild nature 112; wolf-debate 107–8, 111–12; wolves and fairy tales, mythology, and religion 112; zoo-semiotics of the wolf 112
Dansk Folkeparti 33
Darré, Richard Walther 219, 282, 301
Davies, Peter 72
decolonization 281
dédiabolisation 76
deep ecology 4, 6, 275, 277–8
Deep Green Resistance (DGR) 285
Defence Union of Shopkeepers and Craftsmen (UDCA) 74
degeneration 62
Delingpole, James 65
demos 26
Denmark 7, 11, 22, 66, 107, 110–11, 113–14, 117, 178, 247
Devi, Savitri 277, 286
DHA 73, 77–8, 222; and Aristotelian tradition 77–8; endoxon 77; fallacies 77–8; topoi 77–8
Di Pietro, Alessandro 90
Discourse Historical Approach (DHA) to Critical Discourse Studies *see* DHA
Discourses of Global Climate Change 123
disease metaphors and national identity 244
Dohrmann, Jørn 110
Donovan, Jack 284
The Doomsday Clock (Bergström) 128
Due, Karina 115
Dugin, Aleksandr 285

Earth in the Balance (Gore) 205
eco-action 206, 209
eco-activism as advertisement 206
eco-authoritarianism 8
ecofascism 14, 65, 139, 278, 281, 283–8
ecological anarchism 282–5
ecological ideology 153–83; animal protection, rights and welfare 170–1; anti-anthropocentrism and organicism 177–8; conservation of nature 171; economics and 171–2; energy and 172; fish and marine policy 172–3; human health and bio-ethics 173; immigration and nationalism 173; individual protection and participation 174; international relations 174; nature protection 155, 170–7; salience of nature protection in Austrian legislature **156**; science and technology 174–5; spatial planning 175; traditional national culture and nature 175; transportation 175–6; waste management 176–7
ecological imaginary 108, 112–13, 231
ecologism 5, 12–13, 96, 99
ecology 95
eco-naturalism 8
economics, religion, and conspiracy community 264
eco-organicism 8
Ecotopia (Callenbach) 279
Eibel-Eibesfeldt, Irenäus 220
Eichberg, Henning 220
electoral program 95–6
Eliade, Mircea 285
elite 3, 8, 12, 21–2, 25–8, 34, 45, 63, 74, 109–10, 117, 122, 125–6, 142–6, 185–6, 226, 228, 238, 312, 314, 317
Emancipation Hell (Sale) 282
emotionality 115–16
Endeavour (TV series) 57
enemies 94–5
energy 227–9
energy security 314
English Defence League 67
environment: care of as a threat to the natural order 244; and post-fascism 89–93; ways to care for 243
Environmental Communication 39
environmental communication 184–200, 216–36; audiovisual media 192–5; eco-social national economy 187–8; environmentalism in Hungary 185; Jobbik and the broader media environment 192; Jobbik in the Hungarian political space 185–7; Jobbik's manifestos 188–91; Jobbik's priorities 191–2; social media 195–8
environmental communication research 38–53; advertising 45–7; communications strategies 42; constructions of nature and power of environmental communication 46; drivers of news coverage 41; *Environmental Communication* 39; environmental journalism 40–2; environmental news, implications of 42–5; environmental pressure groups 42; film 45; framing theory 43–5; front groups 42; International

Environmental Communication Association 39; management of public communication 41; mediated communication 39; nature as a key rhetorical component 45–6; news coverage 40–1; non-news media representation of environment 45–6; *Palgrave Studies in Media and Environmental Communication* 39; political persuasion 46–7; popular culture representations 45; public perception of risk 39–40; reframing 43–4; right-wing populist parties 44–5; rise of 39–40; rural imaginary 45; social constructionist perspective 40; source influence 41–2; uncertainty-framing 43; visual environmental communication 46–7
environmental discourses 3–4
environmental imaginary 11, 107–8, 110–11, 113, 116–17, 231, 311
environmental journalism 40–2
environmental news, implications of 42–5
environmental politics 88–103; Alleanza Nazionale (AN) 93; Azione Giovani 93; CasaPound Italia and environmentalism 94–8; concept of beauty 95; ecology and 95; electoral program 95–6; enemies and 94–5; environment and post-fascism 89–93; Evolian doctrine 90; Fare Verde (Go Green) 92; Federazione dei Verdi (Federation of the Greens) 93; Forza Nuova 93; Fronte della Gioventù (FdG) 90, 93; Gruppi di Azione Ecologica 92; Gruppi di ricerca ecologica (GRE) 90–1; Hobbit Camps 91–2; ideas from Nouvelle Droite 91; Italian biodiversity 96; la Foresta che Avanza 97; la Muvra 97; la Salamandra 97; L'Ulivo (the Olive tree) 93; Movimento Sociale Italiano (MSI) 90, 92–4; MSI Fiamma Tricolore (Tricolour Flame) 93; regeneration of unity of people and nature 89–90; resistance of mainstream parties 88–9; Sinistra Ecologia Libertà (Left, Ecology Freedom) 89; terra redenta (redeemed soil) 96
environmental position of contemporary far right 63–9
environmental pressure groups 42
environmental protection 226
environmental protection *vs* economic growth 141

Espersen, Søren 111, 114, 116
ethnonationalism and authoritarianism 3
ethnopluralism 3, 5–6, 227–80
ethnos 26
Europe, far-right populism in 23–4
evidence scepticism 314
Evola, Julius 277, 284
Evolian doctrine 90
exploding bats campaign 143

Fagerström, Jonny 128
fallacy of authority 80
Farage, Nigel 63, 65–6
Fare Verde (Go Green) 92
far-right and conspiracy community 3, 265–6
far right and the environment: aesthetic dimension 8; alt-right ecology 275–92; alt-right rhetorical landscapes 293–309; in Austria 153–83; Brexit 2; in Britain 57–71; British Union of Fascists 6; climate change 7–9; climate change denial 121–35; climate scepticism 257–74, 312–13; comparative research 315–16; concern for the natural environment 4; context 314–15; crisis of liberal democracy 2; Czech far right 201–15; in Denmark 107–20; Donald Trump 2; eco-authoritarianism 8; ecological ideology 153–83; eco-naturalism 8; eco-organicism 8; energy security 314; environmental communication 184–200; environmental communication research 38–53; environmental discourse 3–4, 237–53; environmental politics 88–103; ethnonationalism and authoritarianism 3; evidence scepticism 314; existing patterns 311–13; far-right populism 21–37; fascists 6; in Finland 136–50; in France 72–87; future research 313–17; German far right 216–36; in Germany 6–7, 216–36, 257–74; green ideology in Britain 57–71; green radicalism 4; in Hungary 184–200; introduction 1–17; intruders 5; in Italy 88–103; Jair Bolsonaro 2; *landschap* 5; lifestyle and environmental topics 314; limits and survival 4; the local and the particular 5; material dimension 9; nationalism's nature 4–5; nation's

far right and the environment (*Continued*) landscape 4–5; natural right and ideology 1–2; *Nouvelle Droite* 5–6; organic ethnonationalism 311–12; patriotic ecology 72–87; in Poland 237–53; populist environmental communication 136–50; problem-solving 3–4; process and response scepticism 314; Soil Association 6; structuralist approach to climate-change scepticism 8; sustainability 4; in Sweden 121–35; symbolic dimension 8–9; transnational links 316–17; in the US 257–74, 275–92, 293–309

far-right mind-set *29*

far-right populism 21–37; ad hominem arguments 28; Aktion für Deutschland 29; Alternative für Deutschland 23; anti-elitism 26; anti-pluralism 26; apologies 28; Austrian Freedom Party 23; authenticity 31; authoritarianism 26; bad manners 29; Berlusconization 31; British National Party 30; calculated ambivalence 29; charisma 30–2; conservatism 26; conspiracy theories 30; continuous campaigning 31; Dansk Folkeparti 33; defined 24–7; demos 26; the elites 25–6, 28; ethnos 26; in Europe 23–4; far-right mind-set *29*; fascists and Nazis 21; fear-mongering 27; Fidesz 23–4; financial crises 24; form of the performance 30–1; Forza Italia 23; FPÖ 24; Front National 24; front-stage *vs* back-stage activities 30; Golden Dawn 30; Greenback Party 23; group cohesion 27; Haiderization 33; hasty generalization fallacy 28; historical revisionism 26; history and foundations of 22–4; homogenous ethnos 26; identity narratives 21; illiberal democracy 23–4; instrumentalizing the media 31; in Latin America 23; leadership 30–2; left-wing populist parties 22; mediatization 30–2; micropolitics of 27–32; mixing of content and form 25; Nationaldemokratische Partei Deutschlands 30; nationalism 26; National Socialism 23; nativism 26; normalization of tabooed ideologies 22–3; the others 28; the people 25–6, 28; perceived danger to national identity 22; perpetuum mobile 32, *32*; political correctness 29; political imaginaries 21; Populist Party 23; populum 26; Prawo i Sprawiedliwość 24; Progressive Party 23; refugee crisis 24; savior 27; scapegoats 27–8; Schweizerische Volkspartei 33; shamelessness 27; shared narrative of the past 26; Share Our Wealth movement 23; straw-man fallacy 28; Tea-Parties 23; as thin ideology 25–6; threat scenarios 26; tipping points 32–2; Tory Party 24; traditional Christian conservative-reactionary agenda 22; United Kingdom Independence Party 23–4; in the US 23; VDU 23; *volonté générale* 26; welfare chauvinism 26; Yobbik 30

fascists 6; British Union of Fascists 6; and environmental issues 6; and Nazis 21

fear-mongering 27

Federazione dei Verdi (Federation of the Greens) 93

Fidesz 23–4

film 45

financial crises 24

Fini, Gianfranco 93

Finland: Academic Karelia Society (AKS) 138; far-right and environmental politics in 137–9; Finnish Resistance Movement (Suomen Vastarintaliike) 138; Green League (Vihreä liitto) 139; Green movement 139; Lapua movement 138; Patriotic People's Movement (Isänmaallinen kansanliike, IKL) 138; populist environmental communication in 136–50; preservation of countryside 140; Suomen Sisu 138

Finnish energy policy 141

Finnish Rural Party (Suomen Maaseudun Puolue, SMP) 139

Finns Party 136–50; campaign against wind power 142–4; environmental policy of 139–42; environmental protection *vs* economic growth 141; explanations of communication strategies 145–6; exploding bats campaign 143; factions of 140; Finnish energy policy 141; Finnish Rural Party (Suomen Maaseudun Puolue, SMP) 139; Green policies as hysteria 142; health hazards

Index **327**

of wind turbines 143–4; ideological heartland 140; opposition to wind power 141–4; preservation of Finnish countryside 140
fish and marine policy 172–3
Flemish Interest 7
Flores, Eric Lee 281
Foreman, Dave 278
Foreman, David 277
form of the performance 30–1
Forza Italia 23
Forza Nuova 93
Foundation for Free Enterprise 127
framing theory 43–5
France: appeal to ignorance 79; authoritarianism 79; dédiabolisation 76; Discourse Historical Approach (DHA) to Critical Discourse Studies 73; fallacy of authority 80; Front National (FN, National Front) 76–7; imagined community 82; Marine Le Pen's green agenda 78–84; national identity 82–3; nativism 79; new ecology movement 73, 82–3; New Right (la Nouvelle Droite) 75–6; patriotic ecology in 72–87; populism 79; Poujadisme 74–5; topos of analogue consequence 80; topos of induction 81, 83; topos of responsibility 80; topos of the consequential 80; topos of the FN'S vision of France 81, 83; topos of threat 80; Vichy regime 74
Fransson, Josef 125–6, 128
Freedom and Direct Democracy (Svoboda a přímá demokracie) 210–11
Freedom Party 7–8; in Netherlands 210
Freiheitliche Partei Österreichs (FPÖ) 24, 153–83; animal protection, rights and welfare 170–1; anti-anthropocentrism and organicism 177–8; conservation of nature 171; economics and 171–2; energy and 172; fish and marine policy 172–3; human health and bio-ethics 173; immigration and nationalism 173; individual protection and participation 174; international relations 174; legislative activity compared to other parties 178, **179**, 180; nature protection 153–4; nature protection in FPÖ documents and legislative motions **157–69**; science and technology 174–5; spatial planning 175; traditional national culture and nature 175; transportation 175–6; waste management 176–7
Fronte della Gioventù (FdG) 90, 93
front groups 42
Front National (FN, National Front) 24, 72–87; appeal to ignorance 79; authoritarianism 79; contradictory arguments of 79; ecology of 83–4; nativism 79; new ecology movement 73; nouvelle écologie 78–89; overview of 76–7; populism 79; topos of the FN'S vision of France 81
front-stage vs back-stage activities 30
Fujimori, Alberto 23
future research 313–17

Gardiner, Rolf 61
general conspiracy community 265
German climate sceptics and the US far right 268–70
German far right: Alternative für Deutschland (AfD, Alternative for Germany) 217, 225; biodiversity and 226–7; Christlich Demokratische Union Deutschlands (CDU, Christian Democratic Union of Germany) 220; climate change and 224–5; conservationists and the Nazi party 219; content studied 217, 221–2; discourse analyzed 222, 224–30; Discourse-Historical Approach 217; energy and 227–9; environmental protection 226; environmental topics **223**; German forest and 216, 218, 224; GMOs 227; *Heidelberger Manifest* 220; homeland and nature 218; homeland protection 218–19; material pollution 226; National Socialism 219; nature and nation 217–20; nature protection 219–20; Neue Rechte (New Right) 220; Ökologisch-Demokratische Partei (ÖDP, Ecological-Democratic Party) 220; overpopulation and 224; topos of alienation 229–30; topos of autarky 228; topos of (quasi-religious) irrationalism 225; topos of naturalness 227; topos of scientific untrustworthiness 225; topos of sustainability 226; topos of 'we' first 225; Unabhängigen Ökologen Deutschlands (Independent Ecologists of Germany) 220; *völkisch* ideas 218; (wo)man-nature relations 229–30

Germany 6–7, 60–1; Aktion für Deutschland 29; Alternative für Deutschland 23; *Bund Heimatschutz* 7; and climate change 7–8; climate scepticism in 257–74; far right and the natural environment 6–7; German climate sceptics and the US far right 257–74; German far right *see* German far right; *Heimatschutz* movement 7; National Democratic Party 7–8; National Socialism 23; *Reichsnaturschutzgesetz* (Reich Nature Protection Law) 7
global climate change as a national issue 129
GMOs 227
Golden Dawn 30
Gore, Al 205
Göring, Hermann 219
Gramsci, Antonio 75
Greece: and climate change 7; Greek Popular Orthodox Rally 7; *Syriza* party 22
Greek Popular Orthodox Rally 7
Green anarchism 282–5
Green Answer (Zöld Válasz) 192
Greenback Party 23
Green Fatwá 128–9
green ideology in Britain 57–71; blood-and-soil arguments 63; British country life 58–9; British National Party 63, 67–9; British Union of Fascists 62–3; cities as a unhealthy living space 61, 63; degeneration 62; environmental position of contemporary far right 63–9; historical context 60–3; interwar back-to-the-landers 61–2; Jewish people as unhealthy city dwellers 61; jingoistic view of Britain's past 59–60; mainstream visions 58–60; modernity as antithesis of healthy living space 61; organic farming 62–3; rural revival 61–2; Soil Association 62; United Kingdom Independence Party 63–6; visions of the 'real' Britain 58–60; *völkisch* movement 60–1
green radicalism 4
Griffin, Nick 57, 63, 67
group cohesion 27
Gruhl, Herbert 220
Gruppi di Azione Ecologica 92
Gruppi di ricerca ecologica (GRE) 90–1

Guz, Tadeusz 244
Gyurcsány, Ferenc 185–6

Haeckel, Ernst 218
Haiderization 33
Handbook of the Radical Right (Rydgren) 25
Hardin, Garret 277
hasty generalization fallacy 28
Haverbeck, Werner 220
Heartland Institute 264, 268–9, 271
Hedges, Mark 59
Heidegger, Martin 285
Heimatschutz (Homeland Protection) 60
Heimatschutz movement 7
Hess, Rudolf 219
Himmler, Heinrich 60
historical and political context 122–4
historical revisionism 26
Hitler, Adolf 65
Hobbit Camps 91–2
Holík, Jaroslav 210
Homestead Acts 303
homogenous ethnos 26
How Vermont and All the Other States Can Save Themselves from the Empire (Sale) 282
Hulot, Nicolas 79
human health and bio-ethics 173
Hungary: environmentalism in 185; Fidesz 23–4; Politics Can Be Different (Lehet Más a Politika, LMP) 185
Hunt, Richard 283

Iannone, Gianluca 94
Identity 68
Identity Evropa 279
identity narratives 21
ideological heartland 140
illiberal democracy 23–4
imaginaries: ecological imaginary 108, 112–13, 231; environmental imaginaries 111, 116–17; political imaginaries 21; rural imaginary 45
imagined community 82
immigration 277–8
immigration and nationalism 173
An Inconvenient Truth 122
indigenism 280–1
individual protection and participation 174
industrial/breadwinner masculinities 128–9
International Environmental Communication Association 39

international relations 174
interwar back-to-the-landers 61–2
intruders 5
Italian biodiversity 96
Italy: Alleanza Nazionale (AN) 93; Azione Giovani 93; CasaPound Italia and environmentalism 94–8; and climate change 7; concept of beauty 95; ecology and 95; enemies and 94–5; environment and post-fascism 89–93; environmental politics in 88–103; Evolian doctrine 90; Fare Verde (Go Green) 92; fascists and environmental issues 6; Federazione dei Verdi (Federation of the Greens) 93; Forza Italia 23; Forza Nuova 93; Fronte della Gioventù (FdG) 90, 93; Gruppi di Azione Ecologica 92; Gruppi di ricerca ecologica (GRE) 90–1; Hobbit Camps 91–2; ideas from Nouvelle Droite 91; Italian biodiversity 96; la Foresta che Avanza 97; la Muvra 97; la Salamandra 97; L'Ulivo (the Olive tree) 93; Movimento Sociale Italiano (MSI) 90, 92–4; MSI Fiamma Tricolore (Tricolour Flame) 93; Northern League 7; regeneration of unity of people and nature 89–90; resistance of mainstream parties 88–9; Sinistra Ecologia Libertà (Left, Ecology Freedom) 89; terra redenta (redeemed soil) 96
Izatt, Robert 280

Jenks, Jorian 6, 62
Jensen, Derrick 285
Jewish people as unhealthy city dwellers 61
Jirásek, Alois 201
Jobbik 184–200; audiovisual media 192–5; and the broader media environment 192; eco-social national economy 187–8; electoral results **186**; environmental agenda of 187–98; in the Hungarian political space 185–7; major themes on Facebook page *197*; manifestos of 188–91; opinions related to the environment **186**; priorities of 191–2; social media 195–8
Johnson, Greg 207, 276, 278

Kaczynski, Ted 284
Kepli, Lajos 191, 192–5, *194*
The Killing of the Countryside 69

Kinnunnen, Martin 127
Kjærsgaard, Pia 109–10
Klaus, Václav 124, 208
Klose, Hans 219
Korwin-Mikke, Janusz 239
Kurz, Sebastian 153
Kyoto Protocol 124

la Foresta che Avanza 97
la Muvra 97
landschap 5
Lane, David 278
Lange, Alex 92–3
la Salamandra 95–7
Latin America 23
La voce della fogna (The voice of the sewer) 91
Law and Justice party (Prawo i Sprawiedliwość, PiS) 238
leadership 30–2
League of Polish Families (Liga Polskich Rodzin, LPR) 238–9
Lebensraum 65, 301
Lebensreform (Life Reform) Movement 60
left-wing populist parties 22
Le Pen, Jean-Marie 74–5, 79
Le Pen, Marine 281; environmental agenda of 72–3; green agenda of 78–84; national identity 82–3; political revival of Front National 76–7; threats to France 80; use of first-person instead of third person 81; vision for patriotic ecology 79–80
Liardi, Jayme Louis 285–6
liberal democracy, crisis of 2
Liberty (Wolność) 239
lifestyle and environmental topics 314
limits and survival 4
Lindzen, Richard 126
Linkola, Pentti 277
The Little Green Book of Eco-fascism: The Plan to Frighten Your Kids, Drive Up Energy Costs and Hike Your Taxes (Delingpole) 65
local and the particular 5
Lockwood, Matthew 145
logging 246–7
Löns, Hermann 218
Lorenz, Konrad 220
Louvain algorithm 262
L'Ulivo (the Olive tree) 93
lupophobia 113–17

MacDonald, Kevin 277
Mácha, Karel Hynek 201
mainstream parties, resistance of 88–9
Malheur National Wildlife Refuge occupation 294–5, 300, 302
Man Swarm (Foreman) 278
Massingham, Harold John 61
material dimension 9
material-discursive frontier 298–9
material-discursive understanding of language 295
material pollution 226
McVeigh, Timothy 284
media, instrumentalizing the 31
mediated communication 39
mediatization 30–2
Messerschmidt, Morten 109
Mészáros, Lőrinc 194
Metzger, Tom 278
Minutemen 296
mixing of content and form 25
modernity as antithesis of healthy living space 61
Monist League 218
mono-thematic sceptics *vs* political climate sceptics 271
Morales, Evo 23
Mosley, Diana 57
Mosley, Oswald 6
Movimento Sociale Italiano (MSI) 90, 92–4
MSI Fiamma Tricolore (Tricolour Flame) 93
My Awakening/Globalism vs Nationalism (Liardi) 285–6
mythic American West 298–9
Mythologies (Barthes) 108

National Anarchists 282–3
national culture and nature 175
National Democratic Party 7–8
Nationaldemokratische Partei Deutschlands 30
national identity 22, 82–3
nationalism 10–11, 13–14, 26, 73–6, 108–9, 121–2, 124–31, 186–90, 205–7, 217–18
nationalism's nature 4–5
National Movement (Ruch Narodowy) 240
National Party 7
National Radical Camp (Obóz Narodowo-Radykalny, ONR) 239
National Revolutionary Vanguard 282
National Socialism 3, 7, 23, 34, 112, 153–4, 177–8, 203, 207, 219–20, 284, 312
nation's landscape 4–5

Native land, entitlement to 298
nativism 26, 79
Nativist environmentalism 205, 212, 312
NATO and environmentalism 205
natural environment, concern for 4
natural right and ideology 1–2
nature and national identity 242–3
nature as a key rhetorical component 45–6
nature conservation 171
nature protection 153–4, **157–69**
Naylor, Thomas 281–2
Nemcová, Božena 201
neoreaction 285
networked public sphere 258–9
new ecology movement 73, 82–3
New Right (la Nouvelle Droite) 75–6
New Right (Nowa Prawica) 239
news coverage 40–1
non-news media representation of environment 45–6
normalization of tabooed ideologies 22–3
Northern League 7
Northwest Front 280
Nouvelle Droite 5–6, 11, 73, 75–6, 89, 91–2, 96, 99, 277, 316
nouvelle écologie *see* new ecology movement
nuclear energy 9, 43, 89, 95, 139, **162**, 172, 177–8, 180, 205, 209, 211, 221, 228, 314
nuclear power 58, 78–9, 88, 93, 100, 171–2, 177, 191, 195, 209, 220, 227–8

Oath Keepers 293, 296
Okamura, Tomio 210
opposition to wind power 141–4
Orbán, Viktor 30, 194
organic ethnonationalism 311–12
organic farming 62–3
Österreichische Volkspartei (ÖVP) 153, 178
others 28
overpopulation 277–8

Palgrave Studies in Media and Environmental Communication 39
Paris Agreement 124, 128
Patriot Front 297
Patriot Front (Vlastenecká Fronta, VF) 203–4
patriotic ecology 72–87; appeal to ignorance 79; authoritarianism 79; dédiabolisation 76; Discourse Historical Approach (DHA) to Critical Discourse Studies 73; fallacy of authority 80; Front National (FN, National Front) 76–7; imagined

community 82; Marine Le Pen's green agenda 78–84; national identity 82–3; nativism 79; new ecology movement 73, 82–3; New Right (la Nouvelle Droite) 75–6; populism 79; Poujadisme 74–5; topos of analogue consequence 80; topos of induction 81, 83; topos of responsibility 80; topos of the consequential 80; topos of the FN'S vision of France 81, 83; topos of threat 80; Vichy regime 74
Patriot movement 281, 294, 300, 302
the people 25–6, 28
Perón, Juan Domingo 23
perpetuum mobile 32, *32*
Pétain, Phillippe 74
Pieta, Stanislaw 240
The Pioneer Group 277
pivoting away from factual space 114
Polish far-right politics 237–53; All-Polish Youth (Młodziez Wszechpolska, MW) 239; analysis tools 241–2; Civic Platform (Platforma Obywatelska, PO) 239; content analysis 242–8; disease metaphors and national identity 244; division in Polish politics 248; environment as a topic of concern 237; environmental care as a threat to the natural order 244; historical context 238–40; issue-based discourse 238; lack of cohesion in 248; Law and Justice party (Prawo i Sprawiedliwość, PiS) 238, 249; League of Polish Families (Liga Polskich Rodzin, LPR) 238–9; Liberty (Wolność) 239; logging 246–7; methodology and data 240–2; National Movement (Ruch Narodowy) 240; National Radical Camp (Obóz Narodowo-Radykalny, ONR) 239; nature and national identity 242–3; New Right (Nowa Prawica) 239; Polish politics after 1989 238–9; post-2015 shift to the right 239–40; Ruch Narodowy 240; smog 246; social media data 240–1; symbolic actions 243; ways to care for environment 243
political correctness 29
political imaginaries 21
political persuasion 46–7
Politics Can Be Different (Lehet Más a Politika, LMP) 185
popular culture representations 45

populism 79
populist environmental communication 136–50; anti-intellectual approach 144; appreciation of Finnish nature 140; explanations of communication strategies 145–6; exploding bats campaign 143; health hazards of wind turbines 143–4; performance and 137, 142–4; sovereignty of the people 140–1; stylistic elements of 143
Populist Party 23
populum 26
Poujade, Pierre 74
Poujadisme 74–5
Powell, Enoch 66
Prawo i Sprawiedliwość 24
Preston, Keith 282
process and response scepticism 314
Prodi, Romano 93
Progressive Party 23
Progressives for Immigration Reform 277
prominence of US sites in climate change community 267
public communication, management of 41
Putkonen, Matti 144

Rauti, Pino 90
Red Mud Disaster 193–4, 196
reframing 43–4
refugee crisis 24
regeneration of unity of people and nature 89–90
Reichsnaturschutzgesetz (Reich Nature Protection Law) 7
Reid, Julia 64
Republican Party (Sdružení pro republiku – Republikánská strana Československa, SPR-RSČ) 202, 204–5
Research and Study Group for European Civilisation (GRECE) 75
response scepticism 110
right decentralism 281–5
right-wing bioregionalism 278–80
right-wing news sites 269
right-wing populist parties 44–5
Rinehart, Vince 281
risk, public perception of 39–40
Rodrique, Charlotte 303
Ronchi, Edoardo 93
Rosenberg, Alfred 301
Ross, Alexander Reid 286, 294
Ruch Narodowy 240
rural imaginary 45
rural revival 61–2
Rydzyk, Tadeusz 244

Sale, Kirkpatrick 281–2
Sarkozy, Nikola 79
Saunders, Robert 61
savior 27
Scanio, Alfonso Pecoraro 93
scapegoats 27, 28
Schaefer, Max 57
Schoenichen, Walther 219
Schultze-Naumburg, Paul 219
Schwab, Günther 220
Schweizerische Volkspartei 33
science and technology 174–5
secessionism 281–2
Second Vermont Republic (SVR) 281–2
Seifert, Alwin 219
settler colonialism 299–300
shamelessness 27
shared narrative of the past 26
Share Our Wealth movement 23
shifting conservative environmental discourse 276–7
Sinistra Ecologia Libertà (Left, Ecology Freedom) 89
Sládek, Miroslav 204
Smetana, Bedřich 201
Smith, Anthony 4, 82, 216
smog 246
Sneider, Tamás 186
Snyder, Timothy 283
social constructionist perspective 40
social media 195–8
social media data 240–1
Soil Association 6, 62
Soini, Timo 144
source influence 41–2
Southgate, Troy 277, 282–3, 286
Sozialdemokratische Partei Österreichs (SPÖ) 153, 178
Spain 22
spatial planning 175
Spencer, Richard 295–6, 303
Spengler, Oswald 61, 285
Springmann, Baldur 220
Stafford, Matthew 286
Stapledon, Reginald George 62–3
Štepánek, Jiří 209
Stilbs, Peter 128
Stockholm Initiative (SI) 122
straw-man fallacy 28
Sunshine, Spencer 296–7
sustainability 4
Swede, climate friendly 129–30
Sweden: and climate change 7–8; climate change denial in 121–35; Sweden Democrats 121–35

Sweden Democrats (SD) 121–35; climate change denial and parliament politics 124–5; and environmental issues 123–4; global climate change as a national issue 129; history of 123–4; marketing of climate change doubts 126–7; portrait of Swedes as role models 129–30
Swedes, portrait of as role models 129–30
Swedish Meteorological and Hydrological Institute (SMHI) 124, 127
symbolic actions 243
symbolic dimension 8–9
Szyszko, Jan 244, 247

tabooed ideologies, normalization of 22–3
Tanton, John 277
Tea-Parties 23
terra redenta (redeemed soil) 96
thin ideology 25–6
Third Positionists 282
This England 58–60
threat scenarios 26
Three-Percenters 293
tipping points 32–4
topoi: topos of alienation 229–30; topos of analogue consequence 80; topos of autarky 228; topos of induction 81, 83; topos of (quasi-religious) irrationalism 225; topos of naturalness 227; topos of responsibility 80; topos of scientific untrustworthiness 225; topos of sustainability 226; topos of the consequential 80; topos of the FN'S vision of France 81, 83; topos of threat 80; topos of 'we' first 225
topos of alienation 229–30
topos of analogue consequence 80
topos of autarky 228
topos of induction 81, 83
topos of (quasi-religious) irrationalism 225
topos of naturalness 227
topos of responsibility 80
topos of scientific untrustworthiness 225
topos of sustainability 226
topos of the consequential 80
topos of the FN's vision of France 81, 83
topos of threat 80
topos of 'we' first 225
Toroczkai, László 186
Tory Party 24
traditional Christian conservative-reactionary agenda 22

traditional national culture and nature 175
transnational links 316–17
transportation 175–6
tribal anarchism 281
True Cascadia 279
Trump, Donald 2
Turner, Fredrick Jackson 298–9
Tyndall, John 67

uncertainty-framing 43
UN Climate Panel 127
United Kingdom Independence Party (UKIP) 23–4, 63–6; anti-immigration stance 65–6; environmental policies of 58–9, 64–5; overview of 64
US: alt-right ecology in 275–92; alt-right rhetorical landscapes 293–309; climate scepticism in 257–74; Donald Trump 2; far-right populism in 23; Greenback Party 23; populist parties in 23; Populist Party 23; Progressive Party 23; Share Our Wealth movement 23; Tea-Parties 23

Vandas, Tomáš 205, 207, 209
van de Cleen, Benjamin 25
Vanguard America 297
Vávra, Filip 206–7
VDU 23
veganism and animal rights 285–6
Vichy regime 74
visual environmental communication 46–7
Voice of Freedom 68
Volk 297–8
völkisch ideas 218
völkisch movement 60–1

Volkswagen emissions scandal of 2015 42
volonté générale 26
Vona, Gábor 186–8

Wallop, Gerard 61–3
waste management 176–7
Waszczykowski, Witold 244
watermelon metaphor 125
Welander, Per 128
welfare chauvinism 26
White Aryan Resistance 278
white habitat loss 277–8
White-Power Skinheads 202, 204, 206
white wilderness 299
Wilders, Geert 210
wild nature 112
wind power, campaign against 142–4
wind turbines, health hazards of 143–4
wolf-debate 107–8, 111–12
wolves: actual wolves in Denmark 111–12; classical myth of 112–13; cultural semiotics of 112–13; in fairy tales, mythology, and religion 112; legitimacy of fear of 114–15; lupophobia 113–17; rejection of 113–14; zoo-semiotics of 112
Wolves of Vinland (WoV) 283–4
(wo)man-nature relations 229–30
Workers' Party (Dělnická strana) 202, 207–10
Workers' Youth (Dělnická mládež) 202, 208–9

Yobbik 30

Zbela, Martin 209
Zegers, Peter 6
Zerzan, John 283–4
Zeta Zero Alfa 94

Printed in Great Britain
by Amazon